PROTECTIVE ARMOR ENGINEERING DESIGN

PROTECTIVE ARMOR ENGINEERING DESIGN

Magdi El Messiry

Apple Academic Press Inc.
3333 Mistwell Crescent
Oakville, ON L6L 0A2
Canada USA

Apple Academic Press Inc.
1265 Goldenrod Circle NE
Palm Bay, Florida 32905
USA

First issued in paperback 2021

Exclusive worldwide distribution by CRC Press, a member of Taylor & Francis Group

ISBN 13: 978-1-77463-464-6 (pbk)
ISBN 13: 978-1-77188-787-8 (hbk)

Library and Archives Canada Cataloguing in Publication

Title: Protective armor engineering design / Magdi El Messiry.

Names: El Messiry, Magdi, 1942- author.

Description: Includes bibliographical references and index.

Identifiers: Canadiana (print) 20190146710 | Canadiana (ebook) 20190146729 | ISBN 9781771887878 (hardcover) | ISBN 9780429057236 (ebook)

Subjects: LCSH: Body armor—Design and construction. | LCSH: Ballistic fabrics—Design and construction. | LCSH: Protective clothing—Design and construction.

Classification: LCC U825 .E6 2020 | DDC 623.4/41—dc23

CIP data on file with US Library of Congress

Apple Academic Press also publishes its books in a variety of electronic formats. Some content that appears in print may not be available in electronic format. For information about Apple Academic Press products, visit our website at **www.appleacademicpress.com** and the CRC Press website at **www.crcpress.com**

Dedication

This work is dedicated to my wife and sons for all their endless love and support.

About the Author

Magdi El Messiry, PhD
Professor of Textile and Composite Material Engineering,
Faculty of Engineering, Alexandria University, Alexandria, Egypt

Dr. El Messiry is currently Professor of Textile Composite Material Engineering in the Faculty of Engineering at Alexandria University, Egypt. He held the position of Vice Dean for Community Services and Environmental Affairs at Alexandria University from 1994 to 1998 and served as the Chair of the Textile Department (2000–2002 and 2005–2011). For his experience in the textile field, he was appointed as a member of the Directing Board of the Spinning and Weaving Holding Company, the main textile company in Egypt, and is also a technical adviser to the owner board. He has led many projects on energy conservation, textile technology, and nanotechnology, funded by National Science Foundation (USA), ARST of Egypt, and Alexandria University, Egypt. His list of publications exceeds 175 papers in the different fields of textile and materials science and nanotechnology. He participated in the establishment of a several textile departments in Egypt and in Arab countries and carried out several granted projects at the international level with the colleagues from the United Kingdom, France, Spain, the United States, Czech Republic, Algeria, Tunisia, and Morocco. He also participates on the scientific boards of several journals. In the last decades, he acted as an international expert in innovation and technology transfer as well as textile technology. He is the author of a several books in the field of braiding, textile technology, and industrial innovation. His last published book is *Natural Fiber Textile Composite Engineering*. Dr. El Messiry was awarded by the Alexandria University with scientific achievement and lifetime achievement awards.

Contents

Abbreviations

ABL	armor ballistic limit
ACP	automatic colt pistol
AI	angle of incidence
ANSI	American National Standards Institute
AP	armor piercing
ASPF	antistabbing protective fabric
ASTM	American standards for Testing of Material
BFS	back face signature
BL	ballistic limit
BR	bending rigidity
BSA	blade sharpness index
CFF	cross-over firmness factor
COM RR	compressional resilience rate
CPA	critical perforation analysis
CPP	cut protection performance
CTP	compliance testing program
CTR	compliance test report
CV	severity of consequences value
DOP	depth of penetration
EMR	electromagnetic radiation
EMT	extensibility
FMJ	full metal jacket
FPF	flexible protective fabrics
FYF	floating yarn factor
HC	heat transfer by convection to the environment
HITS	Hornady Index of Terminal Standards
HM	high modulus
HO	home office
HO1	protection level 1
HO2	protection level 2
HO3	protection level 3
HO4	protection level 4

HOSDB	Home Office Scientific Development Branch
HPF	high-performance fibers
HR	heat transferred by radiation to environment
HS	high strength
HT	high tenacity
HTPES	high tenacity polyester
HV	Vickers hardness
JHP	jacketed hollow point
JSP	jacketed soft point
KES	Kawabata evaluation system
KR	knife resistance (protection level)
LR	long rifle
LRN	lead round nose
MH	metabolic heat
MMF	manmade fibers
MMT	moisture management tester
MW	heat converted into mechanical power
MWR	maximum wetted radii
NIJ	National Institute of Justice
OMMC	overall moisture management capacity
OV	probability of occurrence
OWTC	accumulative one-way transport capacity
P-BFS	perforation and back face signature
PA	protective armor
PAN	polyacrylonitrile
PBI	polybenzimidazole
PBO	polybenzoxazole
PBT	poly(p-phenylene-2,6-benzobisthiazole)
PC	polycarbonate
PE	polyethylene
PEN	polyethylenenapthalate
PF	protective fabric
PIPD	poly[2,6-diimidazo(4,5-b-4′,5′-e)pyridinylene-1,4 (2,5-dihydroxy)phenylene
PMC	polymer matrix composite
PMD	protection materials-by-design
PMMA	polymethyl methacrylate
PPTA	polyparaphenyleneterephthalamide

PU	polyurethane
PVB	polyvinyl butyral
RBV	rigid ballistic vest
RN	round nose
RR	risk rating
S&W	Smith & Wesson
SG	shotgun (protection level)
SJHP	semijacketed hollow point
SJSP	semijacketed soft point
SK 1	protection level K1
SK 2	protection level K2
SK 3	protection level K3
SK 4	protection level K4
SK L	protection level KL
SP	spike resistance (protection level)
SPL	single penetration limit
SRBV	semirigid ballistic vest
SS	spreading speed
STF	shear thickening fluid
TCS	tactile comfort score
THV	total hand value
TPU	thermoplastic polyurethanes
UHMWPE	ultrahigh molecular weight polyethylene
UM	high modulus
UV	ultraviolet

Preface

Recently, there has been an increasing interest in the area of protective vests, either against bullets or protection from the most realistic threats within domestic frontline operations: edged weapons, knives, and medical needles. Also, protective armor has been extended to civilian and armored fighting vehicles and aircraft to withstand the impact of shrapnel, bullets, missiles, or shells.

Flexible stab-resistant fabrics have been widely used in military and civilian fields; however, numerous efforts have been taken to construct better lightweight soft body armors, for instance, using high-performance fibers.

The book discusses the following subjects:

1. **History of protective armor**
   The art of protective armor manufacturing has a long history, such as in ancient Egypt and China civilizations, and has continued as the development of weapons techniques advanced.

2. **Materials used for body armor**
   Many studies have been carried out on the designs and materials for soft armor to increase its cutting and perforation resistance. The stab resistance of soft armors attested to the need for improving not only the fiber's properties but also the armor fabric design. Multilayer high-density fabric with high-performance fibers is used for antiballistic armors. The fibers are high-modulus, high-tenacity (HM-HT) fibers, such as Kevlar, Zylon, Twaron, Spectra, or high-performance polyethylene (HPPE). Needle-punched non-woven antiballistic fabric structure allows greater compressibility than woven fabrics and therefore gives more impact protection. Also, the fiber orientation can be controlled and aligned to improve strength while still maintaining flexibility. The different fabric structures and fiber type properties are thoroughly outlined.

3. **Anti-stab and anti-bullet armor design**
   In the field of protective clothing, different materials are used to be bulletproof and tear- and puncture-resistant where high levels of stiffness and shear resistance are important for designed vests for protection; moreover, a soft vest should provide comfort for the user and be lightweight and cost-effective. Resistance to penetration is one of the most essential parameters considered for estimation of usage properties of the fabrics designed for protection against various types of mechanical impact. Nevertheless, current stab-resistant clothes have several weak points, such as heavy weight, bulky shape, and uncomfortable characteristics. In addition, without a good customized fit, body armor can become a distraction to the duties of the female wearer as she may feel uncomfortable or even pain on some parts, which will obstruct her mobility and leave her life in danger.

   The different parameters required for the design of flexible armor to stop high-velocity projectiles are discussed thoughout the book.

4. **The comfort of the body armor design**
   Protection and comfort are the main aspects for body armor. It is understandable that there is a need for protection of the body with the armor, but it should have enhanced comfort performance, taking into consideration the armor's flexibility, thermal resistivity, and evaporative moisture resistivity through the fabric. The parameters and measurements of the armors comfort are explored.

5. **Method of testing the flexible body amors**
   In this part of the book, the methods of testing of the fibers, yarns, and protective fabrics from different threats, that is, chemical, heat, fire, etc., while emphasizing on the flexible body anti-stabbing and bulletproof armor, are discussed according to the level of the protection required, such as NIJ 0115.00 and NIJ Standard—0101.06 and other international standards.

This book is a valuable resource for scientists, industry professionals, universities, laboratories, and others who want information on high-performance textile materials and the design of armor structure that provides the best protection against damage from impacts, explosions, and bullets.

Introduction

Protection is necessary for the safety of persons in certain professions where they might be exposed to chemical hazards, sharp tools, needles, knives, explosives, and bullets. Personal protective armor may be classified as industrial, agricultural, military, medical, sports, or space protective armor, according to their specific application. The basic purpose of the armor is to protect the body against the external hazards and, at the same time, maintain safety and comfort.

The design of body armor is a very old art and science; it uses the best materials and technology to provide protection and comfort for the wearer. It should protect against a wide range of ammunition of different calibers, including those used by handguns, rifles, automatics guns, as well as small fragments from explosives. The design of the bulletproof vest allows it to absorb the bullet energy through its penetration resistance and reduces the impact force that is delivered to the body, causing trauma.

Materials used for bulletproof vests started with leather, bronze, and iron and developed into a flexible armor with the use of multilayered textile fabrics from different fibers. Despite the progress of the inventions of the new types of fibers with very high strength and toughness, there is a continuous need for improvement to manage the trend of increasing weapons' capability by providing high bullet muzzle energy. Therefore, the bulletproof vest should have a low ratio of bulletproof vest weight to penetration resisting force.

Stab-resistant armor design is different from that of bulletproof vests. It is designed against the penetration by sharp tipped objects, such as knives, needles, and other sharp tools. Material used for stab-resistant flexible armor is made of multilayered textile fabrics from different fibers.

The historical review of protective armor, in Chapter 1 of the book, provides valuable information on both the development of the body armors along the timeline of the human history and the development of the ideas and inventions of the protective vest, and why we need more development in the science of materials for armor design. In Chapter 2, analysis of the different textile fibers used for the manufacturing of the body armors assigned for the protection against the various types of threats is provided, such as, for instance, Nylon, Kevlar, Kevlar® 29, Kevlar® 129, Kevlar® Protera, Spectra®, Spectra Shield®, GoldFlex®, Twaron®, Dyneema, and Zylon®.

The book has the following four parts:

Part I gives a brief view of the historical background of the protection armor and its principal designs over human history, indicating the necessity for developing protective armor for different end uses.

Part II discusses the physical, mechanical, and other different textile material properties to be utilized as protective fabrics with the emphasis on the material requiments and design principles for flexible protective armor.

Part III explains the design criteria of stab resistance of flexible protective armor and bulletproof flexible protective armor with the stress on the different construction ideas as well as the theoretical background of the mechanism of failure in each case. The new trends for building modern designs of flexible protective armor are considered.

Part IV explains thoroughly the protective armor testing methods and their different elements, fibers, yarns, fabrics, and gives a comprehensive analysis of the different testing procedures required for the performance of the armor according to the standards.

Several standards for testing the performance of the body armors were developed to ensure compliance with the minimum requirements for body armor protection energy levels. The most popular are NIJ Standard, HOSDB Body Armor Standards for UK Police, and German SK1 Standard. The testing procedures are discussed in Chapter 7.

The current book is concerned with providing the reader with a profound understanding of the basic engineering knowledge of the different types of body armor design and manufacturing and, finally, the method of testing and standards.

The author has done his best to enable the reader to have broad view about the science and technology of body armor design. However, the results of the research cannot be applied to build your own body armor without being tested, certified, and approved by the authorized organization, such as NIJ or other recognized international laboratories.

—Magdi El Messiry, PhD

Acknowledgments

The author wishes to thank all those who contributed so many excellent ideas and suggestions for this book. Special appreciation is extended to N. Smirnova for her thorough review and assistance in preparing the manuscript for publication.

PART I

Historical Background and Developments of Body Armor

CHAPTER 1

History of Protection Armor

ABSTRACT

The body armor has been invented to protect the wearer from the surrounding environmental hazards such as chemical, fire, UV, radiation, heat or to protect warriors from impact of bullet, sharp fragments or being stabbed by knife or spikes. Through the evolution of the mankind, thousands of types of the body protection were developed for a personal safeguard. In the modern era, there is an increase attention to develop a body armor that ensure the safety, comfort, and light weight by the application of modern high performance fibers. In this chapter, a historical background of the body armor is given with a special emphasize on the contemporary bullet proof and stab resistance armors classifications and their future development.

1.1 INTRODUCTION

In the last 7000 years of the history, mankind passed through several periods which may be classified as Classical Era (3600 BC–500 AD), Postclassical Era (500–1500), Early Modern Period (1500–1750), Mid-Modern Period (1750–1914), and Modern Period (1914–2018).[1] It can also be chronicled according to the main events; the historian deals generally with the big events which happen and also the decay of the empires in the different parts of the known world. History, as the action of human beings, is always the result of big events, such as wars, natural catastrophes, inventions, discoveries, or social and political changes.

The start of civilization in a certain area was due to the availability of the living resources that permit the gathering of communities and their unification to form a nation with enough numbers that was able to use the resources and also develop an army to defend them, set rules and regulations to arrange their daily activities, and finally, the capability of science and technology to develop and innovate new methods and tools to increase their competencies.

Figure 1.1 illustrates the period of the start and decay of each civilization. Usually, it depended on several factors either geographical, wars, natural disasters, or the internal troubles and division of their under-controlled states and finally most important, the rise of another civilization that starts to compete and try to conquer. The civilizations in North Africa, Mediterranean, and north Asia started to grow, flourish, and decade due to several reasons. The most important one was the collision between the existing civilizations and the growing new civilizations that conquered through the application of new weapons, war strategy, or allies.[2] Figure 1.2 illustrates different empires and their existing period. It is interesting to notice that in each period, several empires can exist and because of that, they fought with each other, leading to the disappearance of one and strengthening of the other. This happened to the Mesopotamian, Egyptian, Babylonian, Persian, and even the Roman empires. Generally, the development of an empire is about a central state, extending political control over territory and people as much as it can by military, economic, or cultural means.[3]

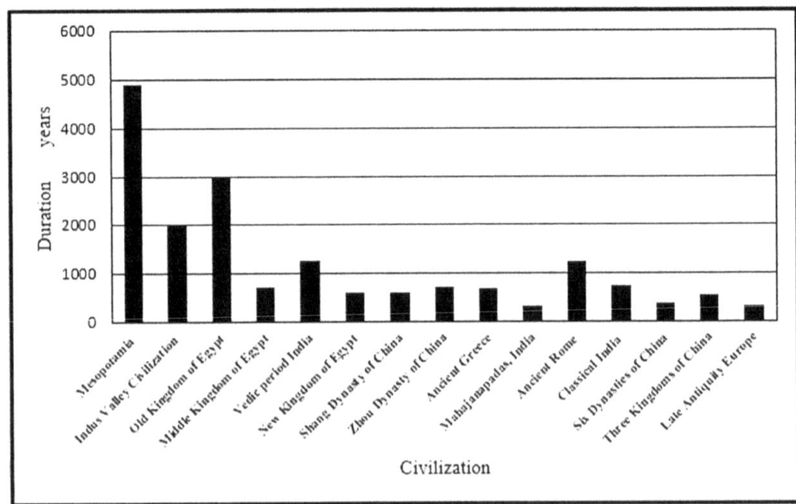

FIGURE 1.1 The duration of the different ancient civilizations (6000 BC–600 AD).

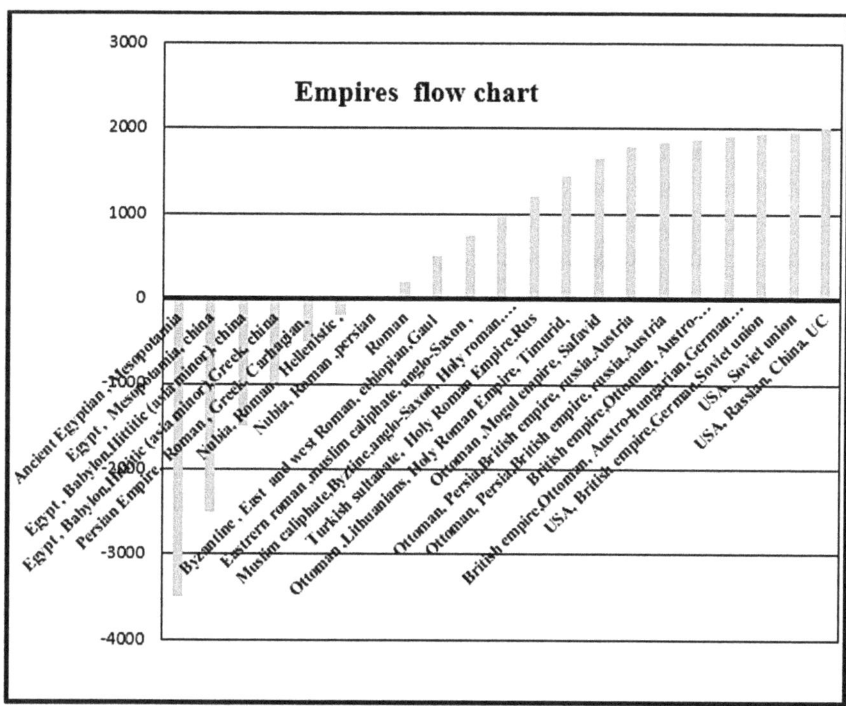

FIGURE 1.2 The main empires in the Middle East, Europe, and North America through different ages.

History tells that the humans are in a constant fight and all the time there is an increased demand for more advanced weapons and protective armors.

1.2 THE BODY ARMOR HISTORY AND EVOLUTION

Defense budget report found that the outlays on weapons and equipment rose in FY 2017 to a global total of $1.57 trillion. The United States is still the biggest global spender.[4,5] The military expenditure as a share of GDP of the different countries varies between 1% and 13.7%.[6]

The armor history and advancement are accomplished with the development of weapons that have been continuously progressive over the history of humanity. Over the centuries, different cultures developed body armor to use during battle.[7] Egyptian, Mesopotamian, Persians, and

Greeks used multilayer fabric or leather. Several hundred millions of dollars are yearly spent on the development of weapons. The progress in the types of weapons has been going in parallel with the achievements of science and technology. The shape and effectiveness of the weapons were resultant of the degree of threat, starting from the spears invented in 40,000 BC and developed through the different periods of the human civilizations.[7] It can be looked back at 25,000 BC, flexible dart was invented; at 23,000 BC, boomerang; at 20,000 BC, bows and arrows; at 5300 BC, projectile weapons; at 5000 BC, first metal daggers and swords; at 500 BC, siege weapon, the ballista (a kind of scaled-up crossbow); in between 800 and 1300 AD, gunpowder, fire arrow, and primitive bombs; in between 1200 and 1600, large swords, hand cannons, fire arms; in between 1368 and 1750, firearms technology moved forward; in between 1750 and 1800, rockets became a permanent fixture and submarines were used in battles; in between 1800 and 1850, shrapnel shells, revolving gun; in between 1850 and 1900, first machine guns, military submarine, fully automatic machine gun, and first bulletproof vest was invented; in between 1900 and 1950, gun silencer, first tanks, ballistic missile, first nuclear bomb, first successful test of a nuclear bomb, long-range missile; in between 1950 and 2000, the first hydrogen bomb, ray gun, laser weapons, supercavitating torpedo, antisatellite laser; at 2000 and 2017, missile-defense shield, high-energy lasers, kinetic energy projectile, tank that fires a NUKE (nuclear artillery shell), electronic warfare weapon, direct energy weapons, self-steering bullets, unmanned submarine hunter, laser cannons, unmanned aircraft, robot warriors, space war weapons and so on.

The progress rate in the invention of the new weapons in the last 50 years is tremendous and faster than ever in human history. Modern warfare can be divided into different elements, as illustrated in Figure 1.3, and can use all the available resources to destroy the enemy. More recently, the concept of battlespace, as the integrated information management of all significant factors that impact on combat operations by armed forces for the military theater of operations, including information, air, land, sea, and space, was developed.[8] Also, it is essential to develop protective armors to shield against the different possible threats.

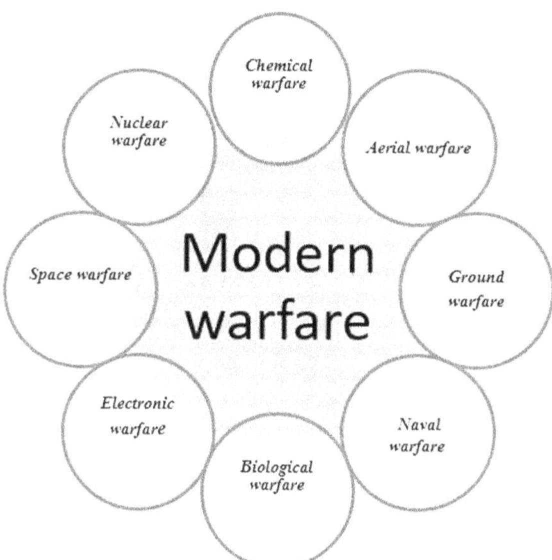

FIGURE 1.3 Modern warfare elements.

1.3 HISTORICAL REVIEW OF THE BODY ARMOR

There is a strong relationship between the development of the armor and the progress of the used weapons; the design of the armors and their materials were made to reduce the probabilities of the soldiers to be hurt during fights.[9] The first Egyptian soldiers carried a simple arming consisting of a spear with a copper spearhead and a large wooden shield covered by leather skin. The spear was provided for stabbing the enemy, allowing larger reach to the soldier.[10,11]

Figure 1.4 illustrates the basic shapes of spears, axes, short swords, daggers, and arrows developed in the Early Dynastic Period in Egypt over the years. Also, Figure 1.4 shows examples of the found weapons, exposed in the Cairo museum in Egypt. The main weapon used by the ancient Egyptians was the bow and arrow with arrowheads made early from hard stones, and later the bronze arrowhead had been introduced. The weight of arrow varied between 15 and 25 g, the arrow velocity could reach 50 m/s and the average impact force might reach 50 N.[10,11]

The spearmen were supported by archers carrying a composite bow and arrows with arrowheads made of hard stone, copper sharpened to a

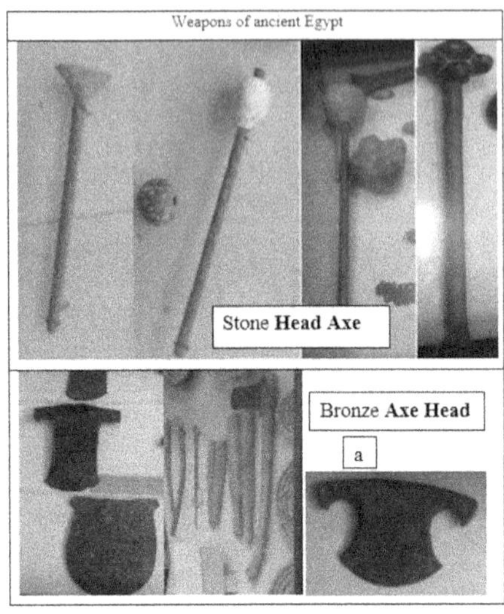

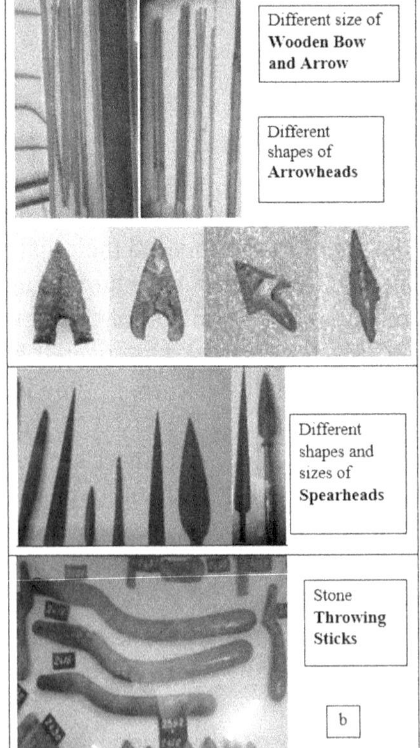

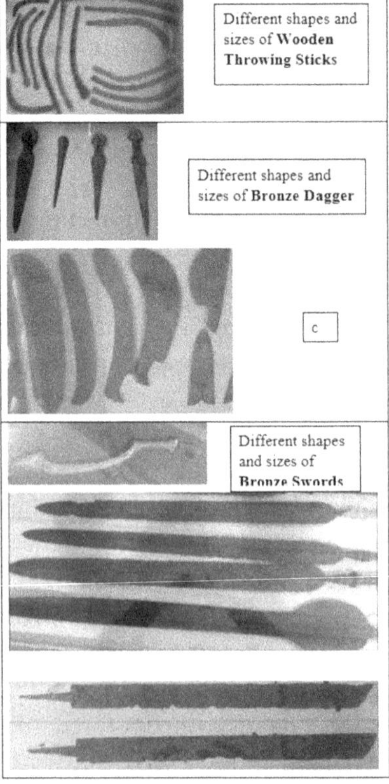

FIGURE 1.4 Some ancient Egyptian weapons (Egyptian Museum, Cairo, Egypt).

point, Figure 1.5, which was the most important weapon in the Egyptian army. The bronze swords were also used in the battles of the Egyptian army as one of the weapons of close-range fights.

Figure 1.6 illustrates Ramses II with his large bow fighting his enemy. The bow and arrows of different constructions and dimensions were able to hit the enemy from long distance. The bow shape was quite different in both dimension and type of material, depending on the range of the arrow. Composite bow structure was developed to allow more distance, power, and flexibility.[12–15]

The body armor of the Egyptian soldier was no more than leather stripes wounded around his chest and he mostly depended on a shield to protect himself from the flying enemy's arrows. Shields of different shapes were made of a wood frame covered with leather. The shield size varied to cover either all the body of the soldier or his chest only. The material was developed according to the state of the art of the material technology in the different periods: bronze, iron, or steel to afford better protection.[16]

FIGURE 1.5 Ancient Egyptian soldiers carrying their weapons Egyptian Museum, Cairo, Egypt.

FIGURE 1.6 Ramses II attacks his enemy.

Source: Adapted with permission from Ref. [15].

The development of the material science reflected on the technology of invention of the weapons, too. The bronze short sword was replaced by iron swords in 13th century B.C. The size and the shape were advanced strongly in the Roman Empire as well as the Persian Empire, although the Chinese used single-edged sword in late 12th century B.C. In the middle age, the sword becomes the main weapon in all the battles between different empires. By developing new steel technology, single-edged weapons became popular.

1.3.1 ANCIENT BODY ARMOR DESIGNS

The body armor is the over-body cover to protect it from being wounded during the engagement in the battle. The old civilizations invented several types of body armor, starting with the use of the leather cover and developed further to protect the soldiers by the use of hard armors from bronze and iron. The Romans and Chinese civilizations developed several types of flexible and hard body armors to cover all the parts of the body. The problem was to provide full protection of the body and at the same time allow the flexible and quick motion of the warriors. The heavy armor sometimes was the cause of the army defeat due to its slow motion against quick movement of the enemy warriors, even if they were less protected. Ancient China invented the interlocking chainrings to be used as body armor, while the Roman and Greek introduced plated armor.

The shield was used for the distant protection of the warrior's body from spears and crossbow bolts. It was also developed as the body armor using the hard leather, wood, bronze, iron to protect soldiers. Thousands of ideas were invented to develop a body armor to defend the warriors from the enemy weapons and be effectively impenetrable to sword slashes, spears, or pike thrusts and blunt trauma. Chinese and Japanese civilizations have a long history in the invention of the armors and weapons. Their soldiers attired flexible armors since 210 B.C. The Chinese first introduced the flexible armor by joining pieces of leather by cord yarns to form a flexible area that accords with the body shape (lamellar armor). Also, in the 14th century, a plated armor was used made of overlapped metal riveted inside a cloth or leather garment.[17,18]

Another design of the armor was the scale armor. The scales were made of leather, bronze, iron, or any other hard material overlapped on each other in successive rows fixed on fabric or leather material. Incas sandwiched cotton between layers of cloth and leather and stitched the whole thing together, creating quilted vests and bodysuits. This cotton armor was very dense and could be two fingers thick, and it repelled arrows and spears almost as well as a Spanish steel breastplate armor but was much lighter, cooler, and more flexible.[19]

Full-body armor appeared in combination with the appearance of mail armor, which secures the body with less weight and high flexibility. There are several types of mail with different designs and combination of metal plate, chain, leather, plated mail, plated chainmail, splinted mail/chainmail, such as[20–51]

- Lamellar armor formed from small pieces of leather, bronze, or steel fixed to each other by coarse plied yarns in a different pattern.
- Laminar armor structured by horizontal overlapped rows of solid plates.
- Scale armor: Armor consisting of many individual small armor scales (plates) of various shapes attached to each other and to a backing of cloth or leather in overlapping rows. The scales may be made of bronze, iron, or leather.

- Mail armor: The small metal rings are attached to each other in a manner to form area covering the body armor.
- Mail and plate armor: In this structure, the chainrings were splinted by a metal plate to increase the protection in a particular area of the armor. The type of joining the chainrings to the plate may or may not allow the overlap of the metal plates for breast reinforcement.
- Plate armor: It was designed from forged steel plates to cover the different parts of the body and at the same time allow their movement.
- Lame armor is constructed from metal plates shaped to take the form of the part to be covered by the body armor and they are connected to each other by coarse ply yarns.
- Brigandine Karuta (Japanese armor): Heavy fabric or leather armor formed from multilayers of fabric with an iron plate in between.
- Kikko (Japanese armor): Hexagonal-shaped iron or leather plates connected together with iron rings and fixed to the surface of heavy canvas fabric to imitate the turtle shell.
- Kusari (Japanese mail armor): It is a category of Japanese mail armor.
- Boiled leather: The boiled leather is softer can be molded to give the required shape of the body and may be reinforced by metal plate for more protections.
- Ring armor: Metallic rings or buttons of several shapes sewn to leather or heavy fabric.
- Splint armor: The splint armor constructed from stripes of metal sewn to heavy fabric.
- Mirror armor is a metal armor worn over the other types of armor as enforcement to increase the protection.

Figure 1.7 illustrates the idea of different elements of the armor designs.[20–31]

The modern technologies have been developing the new armors based on the above ancient designs of the body armor using new advanced fiber materials. The effectiveness of the armor design is to have maximum protection to weight ratio and guarantee the flexibility and the easiness of movement.

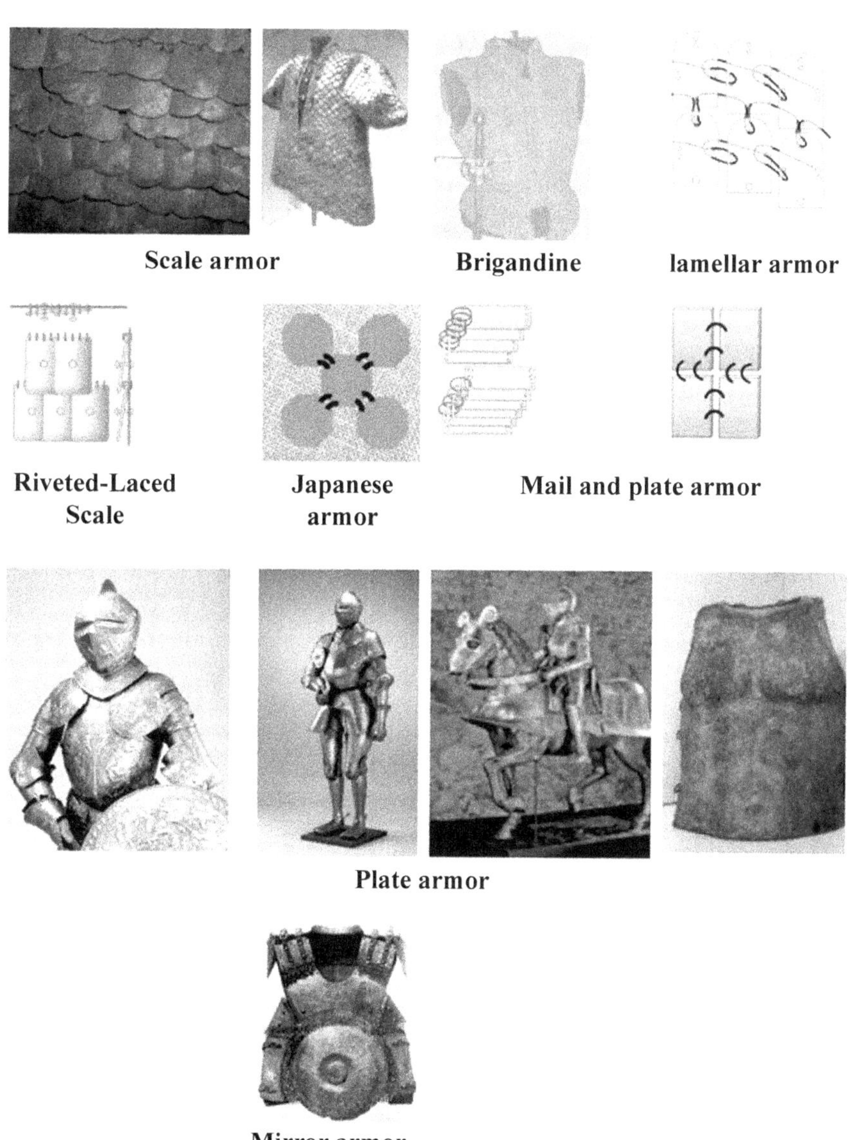

FIGURE 1.7 Elements of some ancient body armor designs.

Source: Adapted with permission from Refs. [20–31].

1.3.2 BODY ARMOR OF THE MODERN ERA

1.3.2.1 THE NEED FOR BODY ARMOR

In World War II, the military casualties increased in comparison with the World War I due to the enormous development of the weapons used and the lack of protection for the soldiers. As illustrated in Figure 1.8, the percentage increase of the military casualties in World War II reached 400% for some army than that in World War I.[52]

As the weapons become more sophisticated, the need for the protection of not only the soldiers but also the civilians turned out to be very vital. Figure 1.9 illustrates that the death percentage of civilians during World War II was more than the military personnel.

After World War I, the demand to have a new design of the body armor increased and elevated further after World War II. The percentage of the total military casualties surged up in some countries like Germany, the United States, and Russia almost tripled. During World War II, the number of the civilian's deaths in the allied countries increased, especially in Russia.[53]

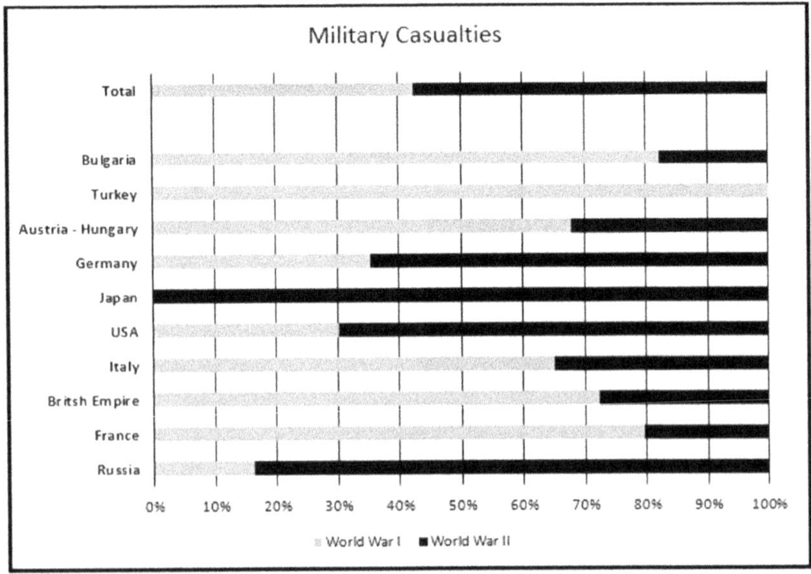

FIGURE 1.8 The percentage of military casualties during World Wars I and II (*note*: Japan participated in World War II, Turkey participated in World War I).

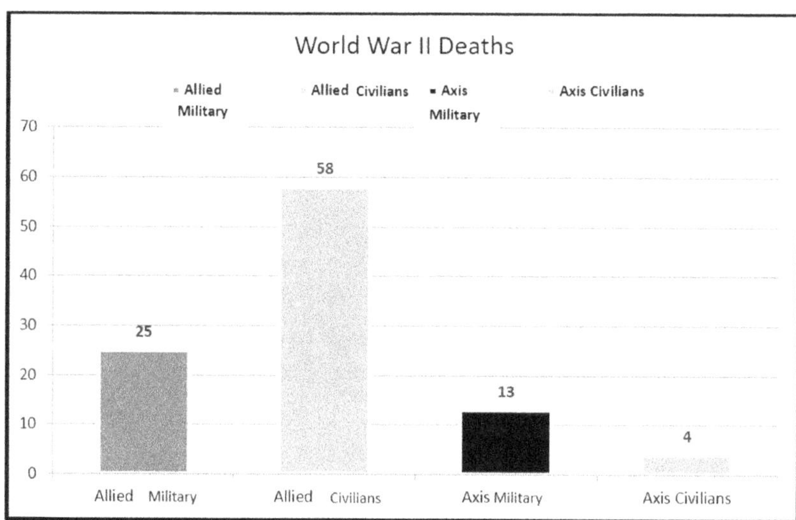

FIGURE 1.9 Distribution of the percentage of deaths for military and civilians during World War II.

Even during the last 70 years, several wars and conflicts occurred in the different areas of the new world. The number of wars and military conflicts, in average, reached 45 wars in the different zones of the continents as illustrated in Figure 1.10.

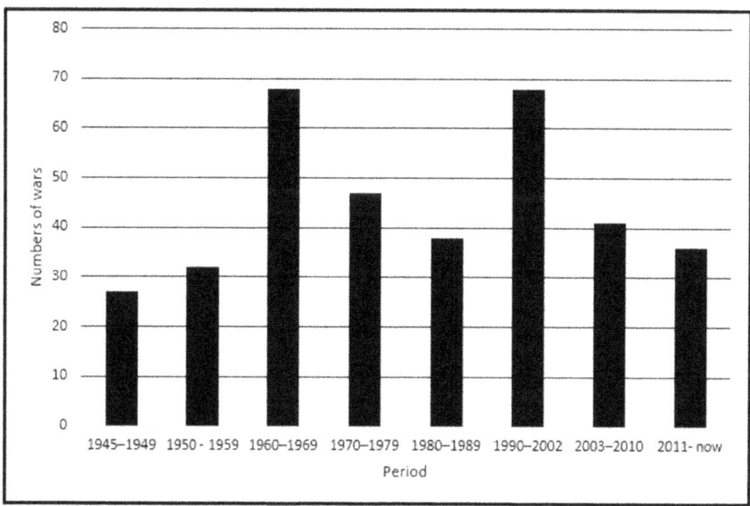

FIGURE 1.10 The frequency of war after World War II.

War in modern times has been the enclosure of civilians and civilian infrastructure as targets in destroying the enemy's ability to engage in war.[54]

1.3.2.2 MODERN BULLETPROOF ARMOR

After the invention of gunpowder and firearms, the body armor continuously developed with a new concept to be lighter and stronger to protect from the penetration of different calibers.[55] During World War I, the body armor was not used; however, experiments started on some designs later. An improved version was reported in 1916 to protect the wearer's chest, but it was heavy due to the use of steel plate. The US Patent and Trademark Office lists records dated back to 1919 for various designs of bulletproof vests and body-armor-type garments.

Figure 1.11 shows the tests of body armor at the beginning of the 20th century, which indicates the public awareness of the need for a bulletproof vest. In the middle of the 20th century, more efforts toward the development of a body armor were accelerated. At the early stage of the World War II, different models of the body armor of light and heavy weights were manufactured with the introduction of hard plates from fiberglass laminates composite (Doron Plate).[56]

FIGURE 1.11 Testing a bulletproof vest in Washington, DC, September 1923.
Source: Adapted with permission from Ref. [56].

The progress of the body armor in the last centuries may be summarized in the following landmarks[57–65]: 1800 AD, silk body armor, two steel plates inserted in ballistic armor; 1919–1923, silk body armor, thick textile vest of multiple layers of cotton padding and cloth; 1942, Nylon bulletproof vests, bulky flak jackets consisted of manganese steel plates sewn into a waistcoat made of ballistic nylon, fiberglass laminate composite plates; 1967, Nomex bulletproof vests; 1970, Kevlar 29 bulletproof vests; 1988, Kevlar 129 bulletproof vests; 1989, Spectra bulletproof vests; 2014, nonwoven Spectra rigid plates, shield panel incorporating Spectra as the reinforcing fibers in flexible resin.

There also are several innovations using recently developed materials to enhance armor products in various ways:

1. New fibers: Dyneema, Zylon, Twaron, Vectran, carbon, glass fibers
2. Ceramic disks, each individually mounted on a vest carrier, and all overlapping each other
3. Ceramic plates
4. Polyurethane–graphene laminar composite
5. Gold Shield GV-2017
6. Tegris™ technology
7. Innegra™ SPP-based ballistic materials
8. Long carbon nanotubes—in the millimeter length range—in yarn and nonwoven sheet form
9. Kevlar® KM2 Plus fiber, Kevlar®149
10. Flexible composite
11. Nanofibers structure composite
12. Liquid body armor

In the last decade, great attention was given to the development of a new protective armor system from the high-performance fibers composites.

1.3.3 CLASSIFICATION OF MODERN BODY ARMOR

The protection armor specifications depend on the expected threats. The different available types of body armor may be classified as:

- Ballistic protection armor
- Edged blade protection armor
- Spike protection armor.

1.3.3.1 BALLISTIC PROTECTION ARMOR

The main function of the body armor is to absorb the impact energy of bullet to reduce or stop the bullet from penetration into the body. The modern bulletproof vest is formed from several layers of fabric or laminated fibers or both capable to protect the warriors from small caliber bullet gun projectile and fragments of a grenade.

Soft vests can be classified into several levels according to their capability of protection that defines vest performance. The levels of protection of the armor standards are recommended by National Institute of Justice (NIJ),[66] UK Home Office Scientific Development Branch (HOSDB),[67] or German Standard VPAM.[68] The rating of vest depends on its resistance to penetration and blunt trauma protection.[69] The NIJ standard divided the level of protection for the bulletproof vest into five classes depending on the weight of the bullet and its velocity. The muzzle energy (the kinetic energy of a bullet as it is expelled from the muzzle of a firearm in Joules)

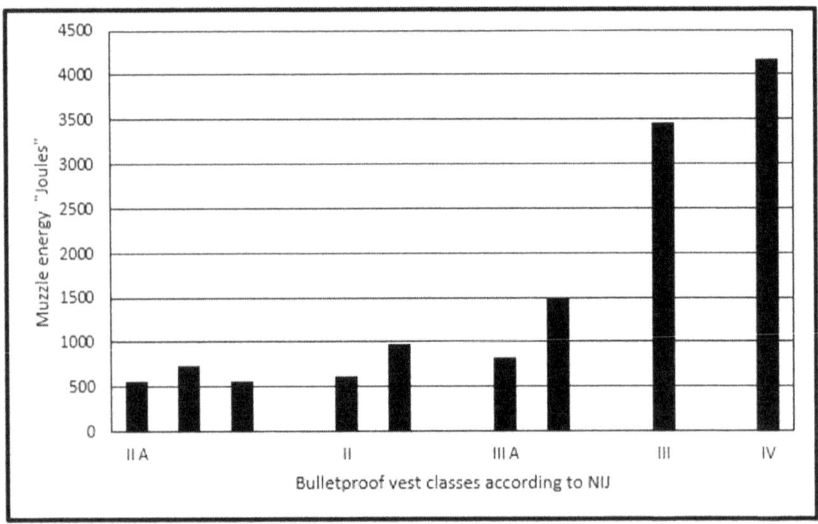

FIGURE 1.12 The levels of muzzle energy according to the National Institute of Justice.

that the bulletproof vest should absorb is illustrated in Figure 1.12. Class IV should withstand 4162.77 J. The muzzle energy depends on the weight of the bullet and its velocity when impacted the bulletproof jacket surface. For the different bullet calibers, the muzzle energy is illustrated in Figure 1.13.[70] The class level of bulletproof armor indicates the highest muzzle energy bullets it can absorb. The handgun bullet muzzle energy varied according to its weight and velocity of shooting. It may reach 2000 J for a shotgun bullet with weight 12 g, while for rifle, muzzle energy may reach 6500 J.

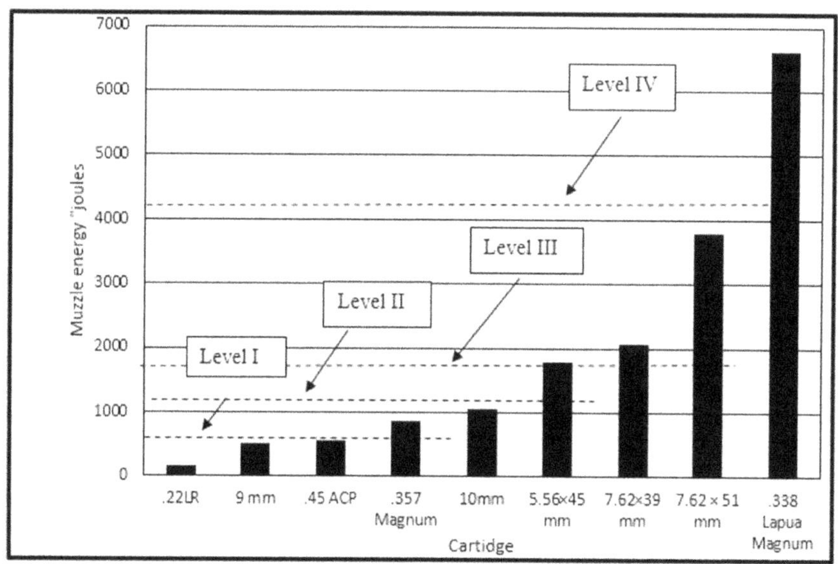

FIGURE 1.13 The bullet muzzle energy for the different types of bullet caliber.

The NIJ levels for bulletproof armor outline the various strength of bullet caliber it can protect from. Each level can stop lower level muzzle energy. The commonly used calibers (the approximate internal diameter of the gun barrel[70]) varied between 4 and 12.7 mm. The bulletproof vest can protect the body from the different caliber according to its level, for example[71]:

Level IIA can protect from bullet of caliber 9 mm (8 g) at velocity 373 ± 9.1 m/s, 557 J

Level II can protect from .357 Magnum (10.2 g) at velocity 436 ± 9.1 m/s, 965 J
Level IIIA can protect from .44 Magnum (15.6 g) at velocity 436 ± 9.1 m/s, 1483 J
Level III can protect from .308 caliber (9.6 g) at velocity 847 ± 9.1 m/s, 3444 J
Level IV can protect from .30-06 caliber (10.8 g) at velocity 878 ± 9.1, m/s, 4163 J
Level IV vest will protect against most ammunition[72] as illustrated in Figure 1.14.

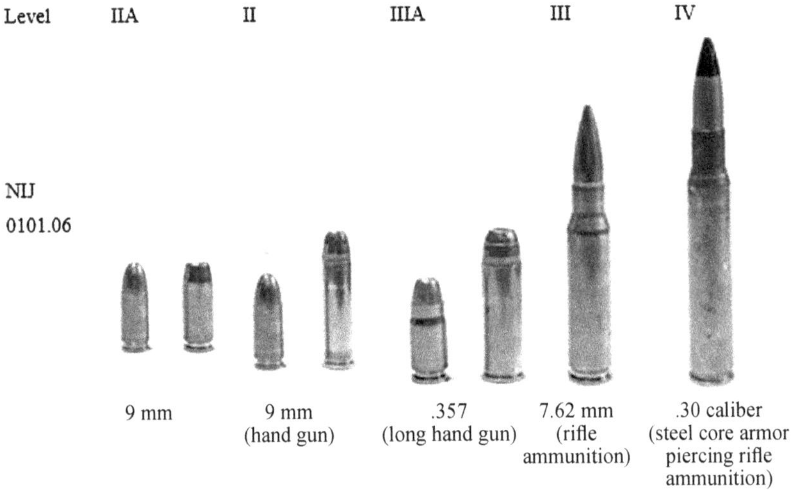

FIGURE 1.14 Ammunition that should be stopped by the different armor incompliant with NIJ 0101.06 standards.

To fulfill the requirements of the NIJ or UK standards, several layers of high-performance fibers fabric are used with different structures and areal density to reach fabric layers thickness 4–20 mm.[71] The areal density of the fabric varies between 100 and 160 g/m^2. Modern types of bullet-proof vest are provided with an armor plate (0.7-cm thick), which is a rigid structure, for protection against the high-energy bullet and placed in the front of the vest. The hard plate is usually a composite structure of high-performance fibers or ceramic layer that capable of reducing the areal density of the fabric and in the same time sustaining high level of protection.[73] The soft armor usually protects most handgun threats (Level IIIA) from bullets (.357 SIG FMJ FN, full-metal jacketed flat nose), with a nominal mass of 8.1 g at a reference velocity of 430 m/s and .44 Magnum SJHP (semijacketed hollow point) bullets, with nominal mass of 15.6 g at a reference velocity of 408. While hard armor plate protects against .30 caliber AP (armor piercing) bullets (US Military designation M2 AP) with nominal mass of 10.8 g at a reference velocity of 878 ± 9.1 m/s.[74] To maximize the effectiveness of soft armor, trauma plates may be added with a goal of developing lighter weight, higher performance vest that can improve mobility, performance, and comfort.

Material used for bulletproof armor started with iron vest and developed to a flexible armor through the use of multilayered textile fabrics from different fibers, such as cotton, silk, nylon, Kevlar, Kevlar® 29, Kevlar® 129, Kevlar® Protera, Spectra® fiber, GoldFlex®, Spectra Shield®, Twaron®, Dyneema, Zylon®. Numerous numbers of designs of the modern bulletproof vests exist on the market to comply with the standards and satisfy the comfort of the wearer. Figure 1.15 illustrates an example of such bulletproof vest.[55]

FIGURE 1.15 (See color insert.) Modern bulletproof vest.

1.3.3.2 EXPLOSIVE ORDNANCE DISPOSAL BODY ARMOR

The disposal military explosives, as well as other explosive devices, are interrelated functions in military fields (unexploded ordnance and bomb disposal), as well as public safety bomb disposal, such as bombs encountered in terror threats. To minimize this risk, a very strong armor to protect front and the back is usually applied, Figure 1.15. In this case, the mass of the armor is not so important as the full protection of the body of the wearer with less trauma. Explosive ordnance disposal technicians are required to wear protective clothing to protect themselves from the threat of overpressure, fragmentation, impact, and heat.[75] Cushioning the spine and head in case the wearer is knocked over by a blast, thermal heat protection should be granted. Freedom of motion to work efficiently required maximum weight restriction. Axillary protectors and the enhanced side ballistic inserts are needed for more protection[76,77,78] as illustrated in Figure 1.16.

FIGURE 1.16 Bomb suit.
Source: Adapted with permission from Ref. [78].

To effectively stop a blast wave, thick layers of Kevlar foam and plastic are needed to prevent serious body harm. The suit weight may reach 40 kg or more.

1.3.4 FUTURE DEVELOPMENT OF THE BODY ARMOR

Development of high-speed projectiles and explosive materials led to an evolution in advanced ballistic body armor with requirements for the new systems to be fabric-damage resistant, flexible and lightweight, and with a high and efficient energy-absorbing capacity.[79] The recent trends for the modern body armor include

- supercarbon nanotube vest
- dragon-skin type of ballistic vest
- imbricated overlapping configuration of high-tensile properties
- New fibers development
- Multilayer fabric improved designs (3D)
- New fabric–polymer composite
- Ceramic armor
- Biomimetic material system
- Integral laminated composite armor

1.4 STAB-RESISTANCE ARMOR

The number of terrorist attacks has been increased rapidly in the last decades. There were 500,000 cases reported during 2009–2015 against illegal carriage of the restrained knives.[79] The stab-resistance body armor should afford protection against injury from penetration of edge blades, such as knifes or sharp-point weapons, while spike protection armor should afford protection against injury from penetration of lower quality knife blades and spike-style weapons. Stab-resistant armor design is basically different from that of bulletproof vest. Consequently, in less degree, it can protect against bullets. Knives and needles have sharper points than bullets and require a different kind of material and fabric construction. Antiknife body armor is designed for stab protection from front, back, and side slash.

Stab-resistant armor is a flexible armor, using multilayered textile fabrics from different fibers, for instance, polyester, nylon, Cardura®,

Coolmax®, Gore-Tex ®, and Kevlar®. The vest will allow the knife blade to pass between the fabric threads without cutting it and applying enough pressure on the knife blade absorbs the stabbing energy of the knife as it passes through the vest layer. An attempt is made to develop flexible body armor by using different types of thermoplastic films to impregnate Kevlar fabric.[80,81]

Several standards for testing the performance of the body armors have been developed to ensure its compliance with the minimum requirements for body armor protection levels. The most common are NIJ standard, HOSDB body armor standards for UK police, German SK1 standards.[82,83] According to NIJ stab-resistant body armor classification, three levels of protection is presented in this standard. The standard for the level of protection depends on the maximum penetration of the stabbing object in the body armor (less than 7 mm and for high energy 20 mm). Performances have been assessed in terms of the amount of penetration occurring as a function of the impact energies; for low-protection armor, the striking energy is 36 J, for the second level is 50 J, and the high energy is 65 J. Typical terminal velocity of a knife attack is 6–10 m/s. Impact loading on the knife approaches 1000 N.

In the last UK standards 2017, there are two levels: E1—maximum penetration 8 mm or below and only one penetration 9 mm for energy 24–33 J, and E2—maximum penetration 20 mm or below and only one penetration 30 mm for energy level 36–50 J. Spike is only evaluated at energy level E1 and no penetrations are permitted.

Another type of body armor design was developed, concealable stab-protective body armor,[84] for protection against homemade weapons and handguns and crude stab/slash weapons as ice pick, blade, or slash.

KEYWORDS

- **body armor**
- **bulletproof vest**
- **stab resistant vest**
- **muzzle energy**
- **Explosive ordnance disposal (EOD) body armor**
- **NIJ standard**
- **HOSDB Body Armor Standards for UK Police**

REFERENCES

1. *History by Period.* https://en.wikipedia.org/wiki/History_by_period (accessed May 1, 2017).
2. *Ancient Near East.* https://en.wikipedia.org/wiki/Ancient_warfare#Ancient_Near_East (accessed May 1, 2017).
3. *Comparing the Rise and Fall of Empires.* https://www.khanacademy.org/humanities/world-history/ancient-medieval/classical-states-and-empires/a/rise-and-fall-of-empires (accessed May 1, 2017).
4. *The Telegraph.* http://www.telegraph.co.uk/business/2017/12/12/1570000000000-much-world-spent-arms-year/ (accessed May 1, 2017).
5. *Timeline: Weapons Technology.* https://www.newscientist.com/article/dn17423-timeline-weapons-technology/ (accessed May 1, 2017).
6. *Who Is Spending the Most on Weapons?* https://www.weforum.org/agenda/2017/04/the-world-is-spending-more-on-weapons/ (accessed May 1, 2017).
7. Marshall, M. *Timeline: Weapons Technology.* https://www.newscientist.com/article/dn17423-timeline-weapons-technology/ (accessed May 1, 2017).
8. *Modern Warfare.* https://en.wikipedia.org/wiki/Modern_warfare (accessed May 1, 2017).
9. *Military of Ancient Egypt.* https://en.wikipedia.org/wiki/Military_of_ancient_Egypt (accessed May 1, 2017).
10. Dunn, J. *The Equipment of Pharaoh's Military.* http://www.touregypt.net/featurestories/weapons.htm (accessed May 1, 2017).
11. Blyth, P. *Ballistic Properties in Ancient Egyptian Arrows.* http://margo.student.utwente.nl/sagi/artikel/egyptian/egyptian.html (accessed May 1, 2017).
12. *Bow and Arrow.* https://en.wikipedia.org/wiki/Bow_and_arrow (accessed May 1, 2017).
13. Ramsey, S. *Tools of War: History of Weapons in Ancient Times* (accessed May 1, 2017).
14. *History of Archery.* https://en.wikipedia.org/wiki/History_of_archery (accessed May 1, 2017).
15. *Military of Ancient Egypt.* https://en.wikipedia.org/wiki/Military_of_ancient_Egypt, https://commons.wikimedia.org/wiki/File:C%2BB-Egypt-Fig4-RamsesAttacksHittiteDapur.PNG (accessed May 1, 2017).
16. *Weapons.* http://www.reshafim.org.il/ad/egypt/weapons/missiles.htm (accessed May 1, 2017).
17. *Terracotta_Army.* https://en.wikipedia.org/wiki/Terracotta_Army (accessed May 1, 2017).
18. *Coat of Plates.* https://en.wikipedia.org/wiki/Coat_of_plates (accessed May 1, 2017).
19. *Pre-Columbian Cotton Armour: Better than Steel.* https://pintsofhistory.com/2011/08/10/mesoamerican-cotton-armour-better-than-steel/ (accessed May 1, 2017).
20. *Mail Armour.* https://en.wikipedia.org/wiki/Mail_(armour) (accessed May 1, 2017).
21. *Mail and Plate Armour.* https://en.wikipedia.org/wiki/Mail_and_plate_armour#Types (accessed May 1, 2017)
22. *Laminar Armour.* https://en.wikipedia.org/wiki/Laminar_armour (accessed May 1, 2017).
23. *Lamellar Armour.* https://en.wikipedia.org/wiki/Lamellar_armour (accessed May 1, 2017).

24. *Scale Armour*. https://en.wikipedia.org/wiki/Scale_armour (accessed Jan 10, 2017).
25. *Lame Armour*. https://en.wikipedia.org/wiki/Lame_(armour) (accessed Jan 10, 2017).
26. *Brigandine Armour*. https://en.wikipedia.org/wiki/Brigandine (accessed Jan 10, 2017).
27. *Karuta Armour*. https://en.wikipedia.org/wiki/Karuta_(armour) (accessed Jan 10, 2017).
28. *Kikko Armour*. https://en.wikipedia.org/wiki/Kikko_(Japanese_armour) (accessed Jan 10, 2017).
29. *Kusari Armour*. https://en.wikipedia.org/wiki/Kusari_(Japanese_mail_armour). (accessed May 1, 2017).
30. *Boiled Leather Armour*. https://en.wikipedia.org/wiki/Boiled_leather (accessed Jan 10, 2017).
31. *Ring Armour*. https://en.wikipedia.org/wiki/Ring_armour (accessed Jan 10, 2017).
32. *Splint Armour*. https://en.wikipedia.org/wiki/Splint_armour (accessed Jan 10, 2017).
33. *Mirror Armour*. https://en.wikipedia.org/wiki/Mirror_armour (accessed Jan 10, 2017).
34. *Plat Armour*. https://en.wikipedia.org/wiki/Plate_armour (accessed Jan 10, 2017).
35. *Armour*. https://en.wikipedia.org/wiki/Armour (accessed May 1, 2017)
36. *Elements of a Light-Cavalry Armour*. https://commons.wikimedia.org/wiki/File: Elements_of_a_Light-Cavalry_Armour_MET_DT780.jpg (accessed May 1, 2017).
37. https://commons.wikimedia.org/w/index.php?curid=3741921 (accessed May 1, 2017).
38. *Mail and Plate Armour*. https://en.wikipedia.org/wiki/Plated_mail#/media/File: Bechter_Diagramm.jpg (accessed May 1, 2017).
39. *Bronze Cuirass*. https://commons.wikimedia.org/wiki/File:Bronze_cuirass_BM_GR1873.8-20.223.jpg (accessed May 1, 2017).
40. *World War II Casualties*. http://fathersforlife.org/hist/wwiicas.htm (accessed Jan 10, 2017).
41. *World War II Death*. https://upload.wikimedia.org/wikipedia/commons/8/8d/World WarIICasualties.svg (accessed May 1, 2017).
42. *Modern Warfare*. https://en.wikipedia.org/wiki/Modern_warfare (accessed Feb 1, 2017).
43. *Bulletproof Vest*. https://en.wikipedia.org/wiki/Bulletproof_vest (accessed May 1, 2017).
44. *Doron Plate*. https://en.wikipedia.org/wiki/Doron_Plate (accessed May 1, 2017).
45. *A History of Bullet Proof Vests and Body Armour*. http://thetravelinsider.info/defense/historyofbodyarmour.htm (accessed May 1, 2017).
46. *New Options in Personal Ballistic Protection*. https://www.compositesworld.com/articles/new-options-in-personal-ballistic-protection (accessed May 1, 2017).
47. Burton, S. *Honeywell's Spectra Shield® Ballistic Material to Provide Ballistic Protection for Next-Generation Law Enforcement and Military Body Armour*. http://www.bodyarmournews.com/honeywells-spectra-shield-ballistic-material-provide-ballistic-protection-next-generation-law-enforcement-military-body-armour/ (accessed May 1, 2017).
48. Military Soft Body Armour—Protecting the Protectors. http://www.dupont.com/products-and-services/personal-protective-equipment/body-armour/uses-and-applications/military-body-armour.html (accessed Feb 1, 2017).
49. *Ballistic Materials for High-Performance Body Armour Protecting the Protectors*. https://www.honeywell-spectra.com/?document=body-armour-overview&download=1 (accessed May 1, 2017).
50. *Honeywell Advanced Fibers and Composites*. http://www.honeywell-advanced fibersandcomposites.com/product/gold-shield-2017/ (accessed May 1, 2017).

51. *Aramid Shield Material for Hard Armour Applications.* http://www51.honeywell.com/sm/afc/common/documents/PP_AFC_Honeywell_gold_shield_gv_2017_Product_information_sheet.pdf (accessed May 1, 2017).
52. *Understanding Armour Plates.* http://www.policemag.com/channel/patrol/articles/2013/01/understanding-armour-plates.aspx (accessed May 1, 2017).
53. Floccare. M.; Forman, C.; Hochuli, I.; Lopez, K.; Shyu, D. R.; Urban, D. *Polyurethane-Graphene Laminar Composite as Transparent Armour.* http://www.mse.umd.edu/sites/default/files/documents/undergrad/enma490/Graphene%20PU%20Bulletproof%20Shield-Report.pdf (accessed May 1, 2017).
54. *Ballistic Resistance of Body Armour NIJ Standard-0101.06. NIJ Standards.* United States Department of Justice, July 2008. https://www.ncjrs.gov/pdffiles1/nij/223054.pdf (accessed Feb 1, 2017).
55. *HOSDB Body Armour Standards for UK Police.* http://ped-cast.homeoffice.gov.uk/standards/39-07_A_HOSDB_BodyArmourStandards_%282007%29_Part1_GeneralRequirements.pdf (accessed May 1, 2017).
56. *German Schutzklasse Standard Edition 2008, Level SK1-SK2.* http://www.bodyarmournews.com/wp-content/uploads/2013/10/VPAM__Ballistische__Schutzwesten_2006_14_05_2009_1.pdf (accessed May 1, 2017).
57. National Institute of Justice. *Body Armour Performance Standards.* https://www.nij.gov/topics/technology/body-armour/Pages/standards.aspx (accessed Jan 2, 2017).
58. *Caliber.* https://en.wikipedia.org/wiki/Muzzle_energy (accessed Feb 1, 2017).
59. *NIJ and Cast Body Armour Levels and Standards.* https://www.safeguardarmour.com/uk/support/body-armour-protection-levels/ (accessed May 1, 2017).
60. *Understanding NIJ 0101.06 Armor Protection Levels.* https://justnet.org/pdf/Understanding-Armor-Protection.pdf (accessed May 1, 2017).
61. *Ceramic Armour for Ballistic Protection of Personnel, Vehicles and Assets.* https://www.ceramtec.com/ballistic-protection/ (accessed May 1, 2017).
62. *Knights Templar's Guide to Ballistic Armour/Body Armour, Sewing Including a Step-by-Step Guide to Construct "Loki's Armour".* https://sites.google.com/site/breivikmanifesto/2083/book-3/032 (accessed May 1, 2017).
63. Stewart, I. B.; Stewart, K. L.; Worringham, C. J.; Costello, J. T. Physiological Tolerance Times While Wearing Explosive Ordnance Disposal Protective Clothing in Simulated Environmental Extremes. *PLoS ONE* **2014,** *9* (2), e83740. http://journals.plos.org/plosone/article?id=10.1371/journal.pone.0083740 (accessed Jan 2, 2018).
64. Costello, T.; Stewart, K.; Stewart, I. Inside the 'Hurt Locker': The Combined Effects of Explosive Ordnance Disposal and Chemical Protective Clothing on Physiological Tolerance Time in Extreme Environments. *Ann. Occup. Hyg.* 2015, *59* (7), 922–931. doi:10.1093/annhyg/mev029. https://www.ncbi.nlm.nih.gov/pmc/articles/PMC4580838/ (accessed Jan 2, 2018).
65. Foust, C. Industry Analysis for Body Armour Procurement. Theses. Naval Postgraduate School: Monterey, CA, 2006.
66. *Bomb_suit.* https://en.wikipedia.org/wiki/Bomb_suit (accessed Jan 2, 2018).
67. Yadav, R.; Naebe, M.; Wanga, X.; Kandasubramanian, B. Body Armour Materials: From Steel to Contemporary Biomimetic Systems. *RRSC Adv.* **2017,** 116–125.
68. Yuan,Q.; Liu, Y.; Gong, Z.; Qian, M. The Application of PA/CF in Stab Resistance Body Armour. In 2017 Global Conference on Polymer and Composite Materials

(PCM 2017) IOP Publishing. *IOP Conf. Series: Mater. Sci. Eng.* **2017,** *213*, 012027. doi:10.1088/1757-899X/213/1/012027.

69. Hosur, V; Mayo J.; Wetzel, E.; Jeelani, S. Studies on the Fabrication and Stab Resistance Characterization of Novel Thermoplastic-Kevlar Composites. *Solid State Phenom.* **2008,** *136*, 83–92. www.scientific.net (accessed Dec 1, 2017).
70. Fangyu, C.; Li, Z.; Yihao, T.; Xiaoming, Z. An Analytical Modeling for High-velocity Impacts on Woven Kevlar Composite Laminates. *JME.* December 2017, Vol. 5, Issue 4, 249–256.
71. Payne, T.; Rourke, O.; Malbon, C.Body Armour Standard (2017). https://www.gov.uk/government/uploads/system/uploads/attachment_data/file/634517/Home_Office_Body_Armour_Standard.pdf (accessed Jan 2, 2018).
72. *Stab Vest.* https://en.wikipedia.org/wiki/Stab_vest (accessed Jan 2, 2018).
73. Pangolin. [online], https://commons.wikimedia.org/wiki/File:Pangolin_(Gir_Forest,_Gujarat,_India).jpg (accessed Mar 20, 2018).
74. Johnson, A. Establishing Design Characteristics for the Development of Stab Resistant Laser Sintered Body Armor. PhD Thesis, Loughborough University, UK, November, 2014.
75. Browning, A.; Ortiz, C.; Boyce, M. Mechanics of Composite Elasmoid Fish Scale Assemblies and Their Bioinspired Analogues. *J. Mech. Behav. Biomed.* **2013,** *19*, 75–86.
76. Nayak, R.; Crouch, I.; Kanesalingam, S.; Ding, J.; Tan, P.; Lee, B.; Miao, M.; Ganga, D.; Wang, L. Body Armor for Stab and Spike Protection, Part 1: Scientific Literature Review. *Text Res. J.* **2018,** *88* (7), 812–832.
77. Park. H.; Branson, D.; Petrova, A.; et al. Impact of Ballistic Body Armour and Load Carriage on Walking Patterns and Perceived Comfort. *Ergonomics* **2013,** *56*, 1167–1179.
78. Yoo, S.; Barker, R. Comfort Properties of Heat Resistant Protective Workwear in Varying Conditions of Physical Activity and Environment. Part II: Perceived Comfort Response to Garments and Its Relationship to Fabric Properties. *Text Res. J.* **2005,** *75*, 531–539.
79. Phillips, M.; Bazrgari, B.; Shapiro, R. The Effects of Military Body Armour on the Lower Back and Knee Mechanics During Toe-Touch and Two-Legged Squat Tasks. *Ergonomics* **2015,** *58*, 492–503.
80. Armstrong N.; Gay L. The Effect of Flexible Body Armour on Pulmonary Function. *Ergonomics* **2016,** *59* (5), 692–696.
81. Majumdar, D.; Srivastava, K.; Purkayastha, K.; Pichan, G.; Selvamurthy, W. Physiological Effects of Wearing Heavy Body Armour on Male Soldiers. *Int. J. Ind. Ergon.* **1997,** *20* (2), 155–161.
82. Carr, D.; Lankester, C.; Peare, A.; Fabri, N.; Gridley, N. Does Quilting Improve the Fragment Protective Performance of Body Armour? *Text Res. J.* **2012,** *82* (9) 883–888.
83. Bilisik, A.; Turhan, Y. Multidirectional Stitched Layered Aramid Woven Fabric Structures and Their Experimental Characterization of Ballistic Performance. *Text Res. J.* **2009,** *79* (14), 1331–1343.
84. El Messiry, M; El-Tarfawy, S. *Effect of Fabric Properties on Yarn Pulling Force for Stab Resistance Body Armour.* Sixth World Conference on 3D Fabrics and Their Applications. North Carolina State University (NCSU), Raleigh, NC, USA, May 26–28, 2015.

PART II

Textile Materials for Flexible Protective Armor

CHAPTER 2

Textile Materials for Flexible Armor

ABSTRACT

Since 15th century, the application of protective fabric as main material for the body armor used silk fabric. The increase in the demand for the advanced fibers for the protective body armor has been directly influenced by the modern armor designs, which in their turn, accomplished the increase of the bullet Muzzle energy. This chapter acquaints the reader with the analyses of the different types of the textile fibers and their physical, mechanical, chemical, thermal, tribological properties that define the armor performance and comfort. Also, it permits to determine the cost coupled with body armor protective capabilities and the risk rating. That will assist a designer to designate armor performance depending on the threat intensity and the implication of the harm on the wearer.

2.1 INTRODUCTION

In the modern era, the textile fibers found increasing implementations in the area of comfort clothing, technical, smart, and intelligent textiles. They are widely used as a grotextiles, geotextiles, sport textiles, medical textiles, transtextiles, protective textiles, and so forth, with annual growth rate up to 8%. The application of textile fiber-reinforced polymer was found as the innovative solutions in the designing of protective armor system, such as bulletproof plates, military wagon armors.

The body armor industry flow diagram, given in Figure 2.1, starts from the chemical mill that supplies continuous multifilament yarns or staple fibers to a weaving or nonwoven mill for fabric manufacturing. Both woven and nonwoven products are used in the manufacturing of the armors for bulletproof, stab resistance, chemical resistance, heat resistance, and so forth. The hard plates are produced from the fabrics as a multilayer fabric reinforced polymer composite using the same fabrics or ceramic plates or a combination of different materials.

The samples of the armor must comply with the standard tests and the regulations for each type of protection to be certified.

The armor performance will defiantly determine its cost and couple with its protective capabilities and the risk rating. Designated armor performance depends on the threat intensity and the implication of the harmon the wearer. A simple model in the decision on aprotective armor level of performance is shown in Figure 2.2.

The architecture of the protective armor with a combination of fibers and other materials properties results in the products that respond to the threats adequately and reaching the balance between comfort and safety. New fibers and architecture systems will lead to the new directions for armor of light weight and higher levels of safety.

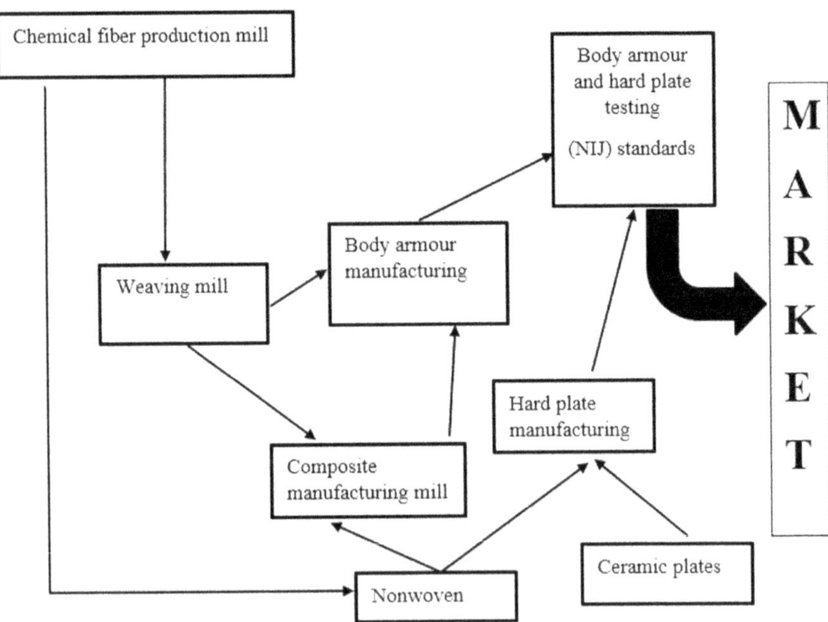

FIGURE 2.1 Bulletproof vest industry product flow diagram.

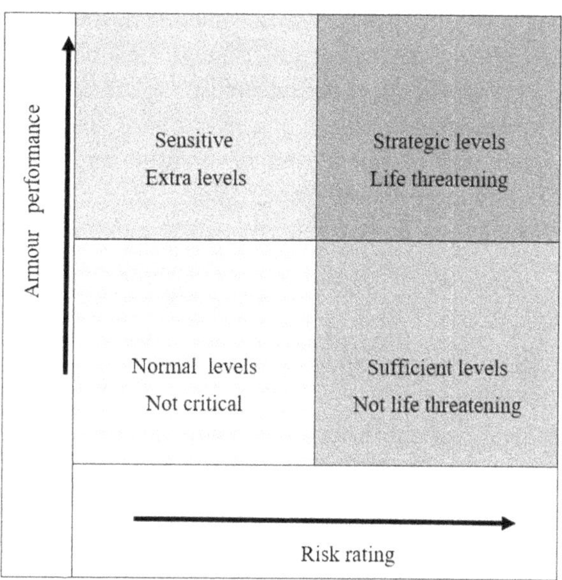

FIGURE 2.2 Model for the strategy of the protective armor choice.

2.2 TEXTILE FIBERS

2.2.1 INTRODUCTION TO TEXTILE FIBERS

Yarns are manufactured from fiber which may be:

- Short staple
- Long staple fibers
- Continuous filament
- Nanofibers

The arrangement of the fibers in the body of the yarn emphasizes the yarn's physical, mechanical, thermal, and otherproperties.

2.2.2 TEXTILE RAW MATERIALS FOR PROTECTIVE ARMOR

The protective fabric (PF), which may be woven, knitted, braided, and so forth, can be manufactured from the yarns architected in different ways to form a sheet of fabric. The mechanical, chemical, thermal, and physical properties of the PFs are a direct function of the fibers and yarns properties and fabric specification and structure. All mentioned above have the imperative influence on an armor performance. The strain velocity is found to be a function of fiber Young's modulus and density and has a significant effect on the protective armor performance (PAP).

Figure 2.3 illustrates the combination of the elements that define the components of the armor design and consequently affects its performance.

2.2.3 MATERIAL SELECTION FOR PF

To get a soft armor at the lower cost with the highest service ability and comfort, it is required to choose the suitable armor design, material, and selection of the manufacturing process.The optimum solution that satisfies the functional properties and economic problems should be found out. Fibers areexisted eitherin the form of microfibers or nanofibers.

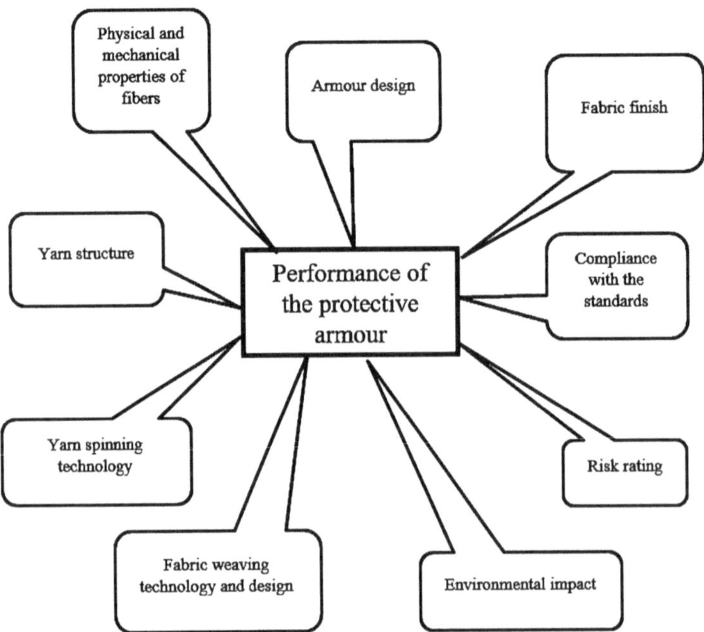

FIGURE 2.3 Factors impact PAP.

The final fabric for high PAP depends on the following textile fibers properties:

- *Physical properties*
1. Moisture absorption
2. Morphological properties
3. Degradability
4. Density

- *Mechanical properties*
1. Strength
2. Shear strength
3. Ductility
4. Young's Modulus
5. Creep resistance
6. Fatigue strength

- *Thermal properties*
1. Thermal expansion coefficient
2. Thermal conductivity
3. Fiberflammability
4. Glass transition temperature
6. Melting point

- *Chemical properties*
Chemical resistance

- *Electrical properties*
1. Capacitance
2. Dielectric constant
3. Dielectric strength
4. Electrical conductivity
5. Piezoelectric constants

- *Fabrication properties (tailorability)*
1. Ease of machining
2. Hardening ability
3. Formability
4. Joining techniques

- *Environmental impact*
- *Radiation impact*
- *Cost*

2.2.4 CLASSIFICATION OF FIBERS

Textile fibers can be categorizedas the following according to El Messiry[1]:

Natural fibers

- *Vegetable fibers*: Cotton, Jut, Flax, Hemp, Rami, Kenaf, Sisal, Coconut Fibers, etc.
- *Animal fibers*: Wool, Silk, Alpaca, Camel, Lama, Mohair, Cashmere, Angora, etc.
- *Mineral fibers*: Asbestos
- *Inorganic fibers*: Glassand Carbon.

MMF

- *Regenerated cellulose fibers*: Viscose rayon, cellulose acetates, modal, bamboo, rubber, lyocell

- *Regenerated protein fibers*: Casein, azlon
- *Synthetic fibers:* Polyester, nylon, acrylic, vectran, kevlar, synthetic rubber, polyvinyl chloride, elastin, etc.

Figure 2.4 illustrates the production of the natural and chemical fiberin the last two decades, which indicates the tendency in decline of the first, 26.5 million metric tons,[2–7] and the continuous rise in the demand for manmade fibers (MMF). The fiber production in 2016 reached 99 million tons as illustrated in Figure 2.5.

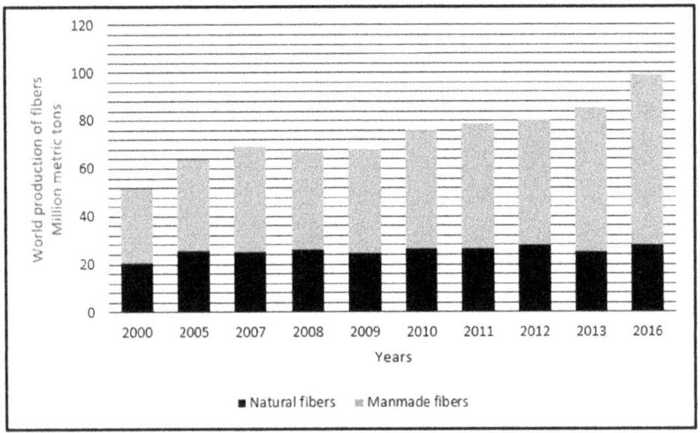

FIGURE 2.4 World production of fibers.

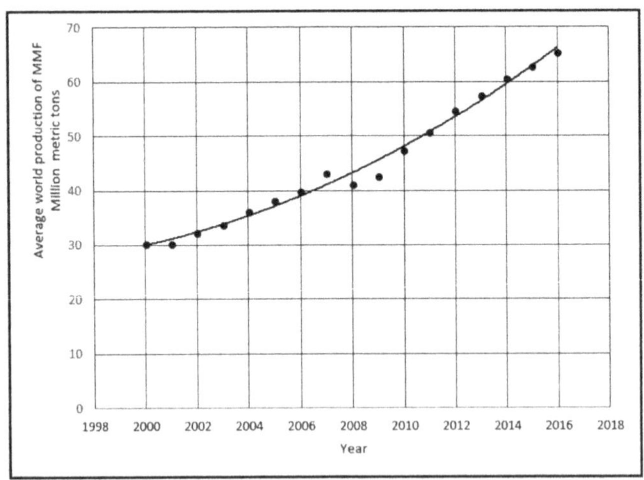

FIGURE 2.5 World productions of manmade fibers.

The MMF can be classified as organic and inorganic. Figure 2.6 shows the generic MMF classification.

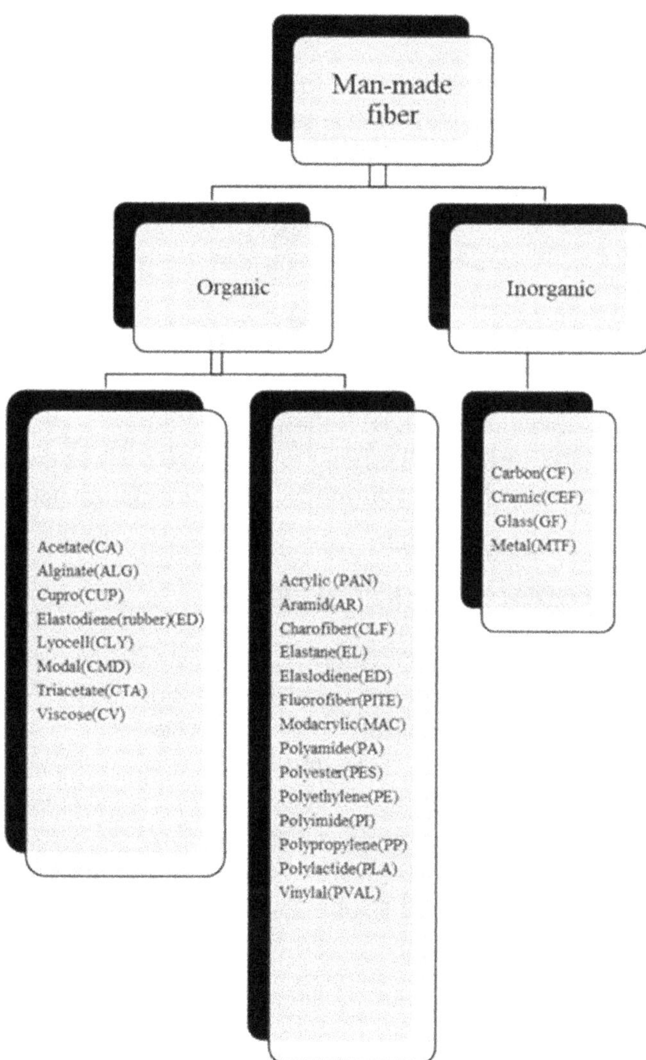

FIGURE 2.6 Generic manmade fiber classification.

As it is mentioned above, each PF requires a fiber or blend with the particular specifications to fulfill its end use.

2.3 FIBER PROPERTIES

An increased emphasis has been placed on the development of improved lightweight body-armor and lightweight vehicle-armor systems, as well as on the development of new high-performance armor materials.[8] Studies of the ballistic PFs have been concerned with micro size instead of macro size, from the fabric and yarn length scales to the sub-fiber length scale which enabled establishment of the relationships between the fabric penetration resistance and various fiber-level phenomena, such as fiber–fiber friction, transverse properties of the fibers, fiber morphology, rheological properties of textile fibers, the stochastic nature of fiber strength, and fiber packing density.

2.3.1 FIBERS TECHNICAL PROFILE

The primary problem to be solved is the choice of the appropriate material to get the final designed PF satisfying PAP requirements at the minimum cost and environmental impact during the lifetime of the protective armor.

2.3.2 FIBER GEOMETRICAL PROPERTIES

Generally, the fibers can be classified as a staple with defined length or continuous filaments with undefined length. Most of the natural fibers have a defined length, except silk, while the synthetic fibers are continuous filaments and can be stapled when needed. So, the first property which describes the fibers is the fiber length. However, there are the other parameters that define the fiber geometry, for instance, fiber diameter, fiber cross-section shape, fiber crimp, fiber aspect ratio (l/d_f), fiber length, and fiber fineness. Each of these parameters has a significant consequence on the fabric properties.

2.3.2.1 FIBER DIAMETER

The fiber diameter (d_f) range varies from micrometer to nanometer. It determines the fiber's, yarn's and fabric's following properties: Mechanical, frictional, bending stiffness, creep, resilience, bulkiness, porosity, thermal, water vapor transition rate, moisture absorption, electrical conductivity, and so forth. The multifilament yarn diameter (d_y), consisting of n_f filaments of a density (ρ_f), filament count (tex_f), and fiber volume fraction (V_f), is:

$$d_y = 0.00357\ [n_f tex_f / \rho_f V_f]^{0.5}. \tag{2.1}$$

Theoretical value for yarns with square packing is $V_f = 0.785$, while for hexagonally packed fibers is 0.906.[9,10]

The multifilament yarn has anumber of the continuous filaments depending on its count. Kevlar 29 multifilament yarn of 44 *tex* may have from 120 to 267 filaments with a filament diameter between 10 and 18 μ. For instance, the diameter of Kevlar 49 d_f is 10–11.9, of PAN d_f is 6–8.4, of Pitch d_f is 10, of E-glass d_f is 10, and of Nextel d_f is 11 μ. For Kevlar, Nomex, carbon, high-performance polyethylene, and polybenzoxazole (PBO) fibers, their tensile properties depend on the filament fiber diameter.

2.3.2.2 FIBER CROSS-SECTION SHAPE

In the MMF, the diameter is constant along the filament length, but it can have several shapes depending on the fiber end use. MMF cross-section shape can be produced using various spinneret shapes to get the cross section that suits a certain application. Figure 2.7 gives the cross-section shapes of some MMF. Shapes of the fiber cross-section also affect the following fiber properties:

- Fiber surface area
- Fiber bending stiffness
- Fiber tenacity
- Fiber–water absorption
- Fiber frictional properties
- Fiber abrasion resistance.

The cross-section profile plays a pronounced role on the mechanical, thermal, frictional, abrasion, and several other properties of the yarns.

The fabric tenacity, tear strength, shear strength, punching resistance, surface friction of the fabric, and thermal properties are the function of fiber cross-section shape. Fiber cross-sectional shape has a more apparent effect on tearing of fabrics than on breaking strength.[11,12]

The thermal comfort properties of a protective armor, that is, thermal conductivity, thermal absorption, and thermal resistance, water vapor, and air permeability of woven fabrics are influenced by the different fiber cross-sectional shapes. The thermal conductivity increases in the fabrics woven with hollow fibers compared to those woven with solid fibers. As a result, the thermal insulation property of the fabrics produced from hollow fibers is lower.[13]

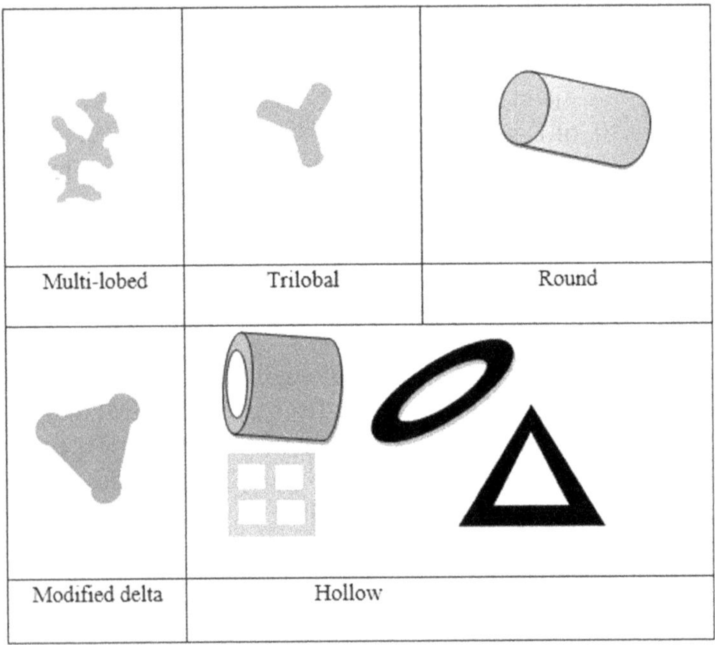

FIGURE 2.7 Filament cross-section shapes.

2.3.2.3 FIBER LENGTH

MMF, either filament ($L = \infty$) or staple, are usually manufactured byone of the following spinning methods: wet, dry, melt, gel spinning, and electrospinning. The fibers that havelimited or finite length are spun into continuous yarns than proceeded to weave, knitting or braiding into a fabric or be directly converted to nonwoven fabric.

2.3.3 MECHANICAL PROPERTIES OF TEXTILE FIBERS

The manmade textile fibers are viscoelastic materials, which can affect the mode of deformation of the PF under the applied loads, constant or dynamic.

The tensile strength of the MMF is the function of their polymer,the degree of polymerization, and the polymer arrangement of molecular structure.

The armor materials are subjected to various environmental parameters during its lifetime, and they may cause deterioration in their mechanical properties. Thus, the choice of textile material for PFs should take into considerations all the expected situations.The fabrics used for the protection against heat, their mechanical properties and strength retention are unaffected by the working temperature and the time of the heat exposure. The main fiber mechanical properties are:

1. Breaking load "N" or tenacity (g/*den*, g/*tex*, N/*tex*, or MPa),
2. Breaking elongation percentage,
3. Work of rupture (Joule, J/m^2),
4. Initial modulus, modulus of elasticity (g/*den*, g/*tex*, N/*tex*or GPa),
5. Work factor,
6. Elastic recovery,
7. Creep,
8. Fatigue limit (MPa),
9. Cutting force (N, N/*tex*, or MPa),
10. Coefficient of friction,
11. Strength heat retention, and
12. Flammability.

2.3.3.1 FIBER STRENGTH

PFs are subjected to various types of forces during their service time. These forces differ depending on armor design and its end use. Accordingly, fiber mechanical properties are highly imperative. For organic MMF, their tensile properties are a function of the exposure temperature, increasing temperatures reduce the fibers modulus, tenacity, and increase break elongation. For example, Kevlar 29 has a reduction of strength from 275.79 to 262.0 MPa after exposure to 160°C for the duration of 500 h, and the elevation of temperature to 250°C reduces its strength to 106.87 MPa when exposed for 80 h.[14] The modulus is reduced by 25% when the temperature increases from 30°C to 200°C. The stress–strain curve of Kevlar fibers illustrates the changeability of the strength properties according to the fiber molecular structure, Figure 2.8, but fiber's stress–strain curve is almost a straight line.[15]

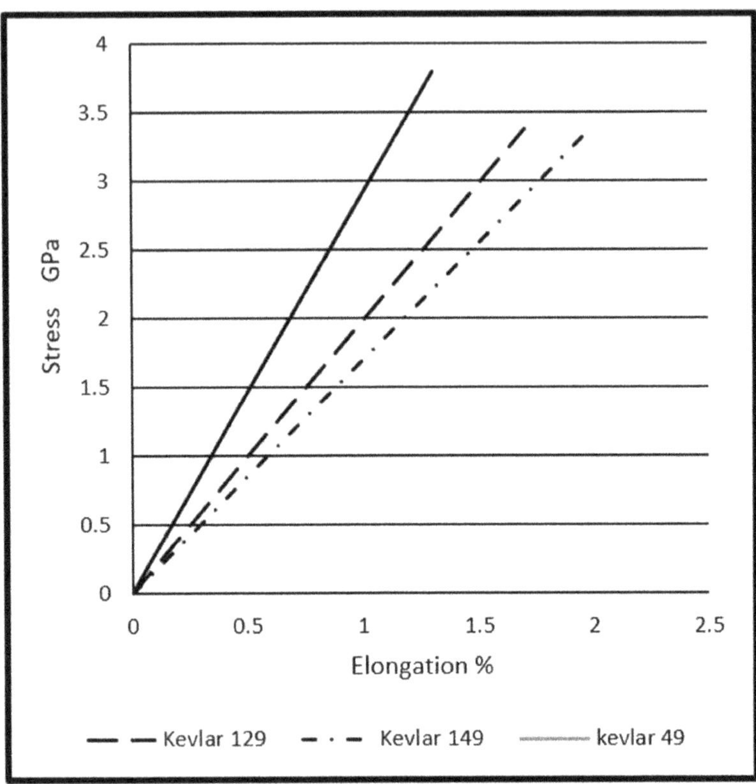

FIGURE 2.8 Load elongation curve of Kevlar fibers.

The yarn mechanical properties are influenced by several other parameters, such asmechanical properties, frictional properties and yarns specifications, and structural parameters of fibers.

The fabric is 2D or 3D sheet formed by the interlacement of two or more sets of yarns (warp and weft yarns). The yarn interlacements will give the fabric its integrity. The woven fabric mechanical properties are a function of the fiber properties, yarn count, yarn/yarn frictional properties, fabric's areal density, and fabric's tightness and weave pattern. For instance, MMFs' tensile properties are affected by the value of load extension rate, therefore, it will reflect on the fabric made from them. It was revealed that Young's modulus, tensile strength, maximum breaking strain, and toughness of Kevlar 49 fabric are increasing with the increase in theload-extension rate.[16,17]Woven fabrics of high strength and toughness are ideal materials for use in the structural and body protection systems where high energy absorption is required.

2.4 HIGH-PERFORMANCE FIBERS (HPF)

The mechanical properties of MMF fibers pivot on their molecular structure, which can be arranged in crystalline or amorphous. The molecules chains arearranged mostly along the fiber axis and are bonded with intermolecular bonds.[15,18] The HPF are the technical fibers that possess high strength and modulus with low density and capable of the high temperatures resistance, chemical resistance, abrasion resistance, fatigue, thermal resistance, and cut resistance.The HPFs are characterized by their high strength (3–7 GPa) and Young's modulus (50–400 GPa).Their properties make them ideal for manufacturing PFs for ballistic and stabarmor applications, chemical protective clothing, EMI shielding, etc.

These properties are gained through the strong and continuous axial bonding and ultra-high molecular weight with a linear string of covalently linked carbon atoms and hydrogen bonds.

HPFs are polymer fibers (organic chemical material; Kevlar, Twaron, Technora, Zylon, M5, Nomex, UHMWPE, Spectra, Dyneema), and Carbon. Inorganic HPFs are glass and alumina borosilicate ceramic fibers (Nextel). The methods for spinning of the MMF fibers[5,15] are given in Table 2.1, the Polymer fibers can be spun on one or more systems, each results in a fiber with different mechanical properties.The fundamental sequence of the MMF spinning has the following process starting with the monomer and proceeds with its polymerization, extrusion, stretching, heat setting, and finish application. For electrospinning, polymers are completely dissolved in the appropriate solvents, and then a high voltage is used to create an electrically charged jet of a polymer solution.The electrostatic field overcomes the droplet fluid surface tension and leads to evaporation of the solvent, a fine fiber will be formed and can be collected on the grounded collector.For polymers that arehard to dissolve at the room temperature, gel-electrospinningwas introduced.

HPF are developed fabrics used for special technical functions, such as heat resistance armors, chemical resistance protective cloth, cutting resistance gloves, impact energy absorption hard plates, electromagnetic shielding, flame retardant vest, radiation shield vest, stab-resistant vest, bulletproof vest, etc. The HPF mechanical properties are given in Table 2.2.

TABLE 2.1 MMF Fiber Spinning Systems.

Fiber	Polymer	Method of spinning	Short process description
Nylon	Polyamide	Melt	*Melt spinning*: either molten polymer or polymer pellets are melted down through spinneret. As filaments are extruded, they are cooled by air or gas
Vectran	Polyester	Melt	
Glass		Melt	
PET	Polyethylene terephthalate	Melt	
PBI	Polybenzimidazole	Dry	*Dry spinning*: the polymer is dissolved in a volatile solvent and extruded through the spinneret. The solution goes through a drying process in the tower, where the solvent is evaporatedin a stream of hot air
PBO	Polybenzobisoxazole	Gel	*Gel spinning*: the polymer is in a "gel" state. It goes through polymer dissolution,the spinning of a solution,gelation and crystallization, solvent extraction. The fibers are first air dried, then cooled further in a liquid bath. The polymer solution is then extruded and quenched to form gel-like fibers, followed by super drawing and removal of the remaining solvent
Dyneema	Polyethylene	Gel	
Spectra	Polyethylene	Gel	
Kevlar	Aramid	Gel	
PAN		Gel	
Nomex	Aramid	Wet	*Wetspinning:* a non-volatile solvent is used to convert the raw material into a solution which is extruded through the spinneret, and the solvent is removed in a liquid coagulation medium
PAN		Dry-jet-wet	*Dry-jet-wet:* the polymer is dissolved in an appropriate solvent to make the fiber solution. This solution is then extruded under heat and pressure into a small air gap before it enters a coagulation bath

TABLE 2.1 *(Continued)*

Fiber	Polymer	Method of spinning	Short process description
PBI PEO PANPLA, etc.	Conducting polymers Polyethylene terephthalate Poly(2-hydroxyethyl methacrylate	Electrospinning	*Electrospinning:* it has been mostly applied to polymer solutions. Nevertheless, processing from melt has been also performed.The high positive voltage is applied to a polymer fluid, usually contained in a syringe, with respect to a grounded collector
PE PP PET	Polymersnot properly solvent at room temperature: polyethylene, polypropylene, and polyethylene terephthalate	GelElectrospinning	*Gelelectrospinning:* polymer filament is formed by electrospinning at elevated temperature. During this process, the filament undergoes electrostatically driven drawing and whipping processes at high temperature.(US 2017/0101726 A1)

PAN, polyacrylonitrile; PBI, polybenzimidazole; PBN, polybutylene naphthalate; PBO, polybenzoxazole; PE, polyethylene; PEO, polyethylene oxide; PET, polyethylene terephthalate; PLA, polylactic acid; PP, polypropylene.

TABLE 2.2 Mechanical Properties of Selected Fibers.

Fiber	Density (g/cm^3)	Tensile strength (Gpa)	Young's modulus (Gpa)	Breaking strain (%)	Specific strength [$Gpa/(g/cm^3)$]	Specific modulus [$GPa/(g/cm^3)$]
Asbestos	2.5	1.36	172	0.9	0.5	68.8
Boron	2.61	3.5	380	0.7	1.3	145.6
Carbon HM	1.8	3.7	387	0.95	2.1	215.0
Carbon HS	1.8	3.1	275	1.35	1.7	152.8
Carbon IM	1.75	4.45	296	1.5	2.5	169.1
Carbon UHM	1.9	2.65	633.5	0.45	1.4	333.4
Pan	1.8	5.6	300		3.1	166.7
E-Glass S. Filament	2.61	3.79	70.5	4.8	1.5	27.0
Kevlar 129	1.45	3.38	99	1.7	2.3	68.3
Kevlar 149	1.47	3.4	186	2	2.3	126.5
Kevlar 29	1.44	3.6	83	2.2	2.5	57.6
Kevlar 49	1.45	3.8	131	1.3	2.6	90.3
Kevlar 68	1.45	2.9	107	3.3	2.0	73.8
Kavlar MK2	1.45	4.52	84.62	4.52	3.12	58.35
Nextel 312	2.7	1.63	135	–	0.6	50.0
Nextel 440	3.05	2.1	189	–	0.7	62.0
Nextel 480	3.05	2.25	224	–	0.7	73.4
Nextel 610	3.96	2.27	328	–	0.6	82.8
Nomex	1.38	0.6	17	21–30	0.4	12.3
Nylon 6	1.14	0.0862	2.8	23–42	0.1	2.5

TABLE 2.2 *(Continued)*

Fiber	Density (g/cm^3)	Tensile strength (Gpa)	Young's modulus (Gpa)	Breaking strain (%)	Specific strength [Gpa/(g/cm^3)]	Specific modulus [GPa/(g/cm^3)]
Nylon 66	1.13	0.092	3.3	26–32	0.1	2.9
Pbo "Zylon"	1.56	5.8	270	2.5	3.7	173.1
Polycarbonate	1.2	0.07	3	-125	0.1	2.5
Polyester	1.38	0.683	9.8	40	0.5	7.1
Polyester H. modulus	1.38	0.83	12.7	25	0.6	9.2
Polyethylene H. density	0.95	0.586	1.2	up to 500	0.6	1.3
Polyethylene L density	0.92	0.241	0.12	up to 1000	0.3	0.1
Polypropylene	0.9	0.7	6.8	up to 45	0.8	7.6
S-Glass single filament	2.485	4.826	85.6	5.4	1.9	34.4
Silk	1.36	0.65	4.3	25	0.5	3.2
Spectra 2000	0.97	3.25	116	2.9	3.4	119.6

Source: Adapted with permission from Refs. [15–36].

Fiber technologies have revolutionized the performance of fiber by developing new material to meet a specific criterion for protective armor, its durabilityand wearer comfort. Figure 2.9 illustrates the comparison of fiber density for various types of fibers which indicates that all the fibers have a density much lower than steel.

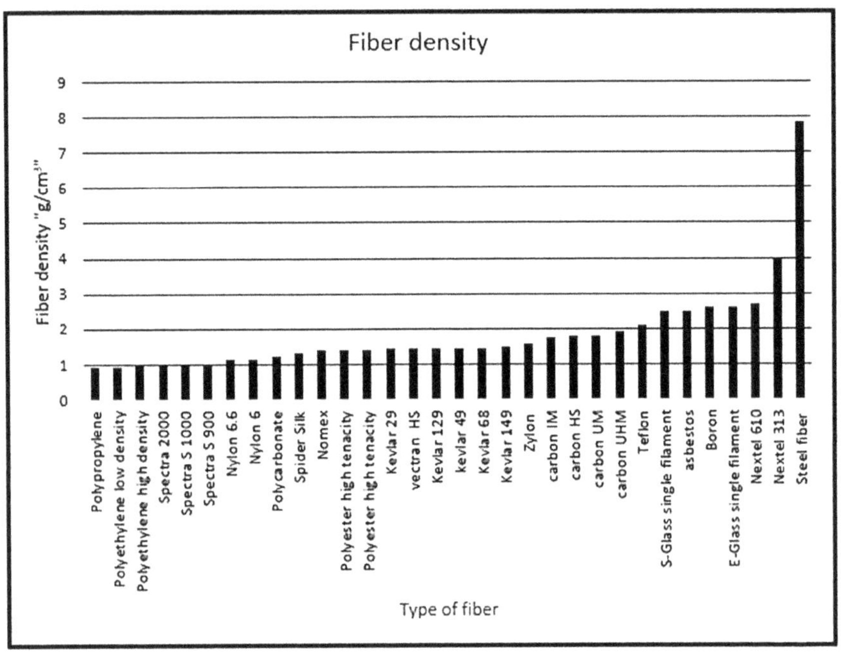

FIGURE 2.9 Fiber density.

The specific strength and modulus of some selected fibers are illustrated in Figure 2.10a,b, demonstrating the specific strength of the MMF fibers and their specific modulus.

Figure 2.11a,b indicates the existence of a relationship between the fiber strength and modulus and fiber density.[15]

For high protective armor performance fibers has high specific strength and modulus are preferable and to have lighter armor weight. As illustrated in Figure 2.12, a linear regression is noticed.

In fibers science, one property of a fibers may be enhanced but it concomitant with a decline in another characteristic, for instance the increase inthe polymer degree of polymerization increase the fiber tenacity but losing its toughness.[37] The theoretical maximum values for tenacity and modulus shown that for many polymer fibers the substantial improvements in properties are still possible.[38]

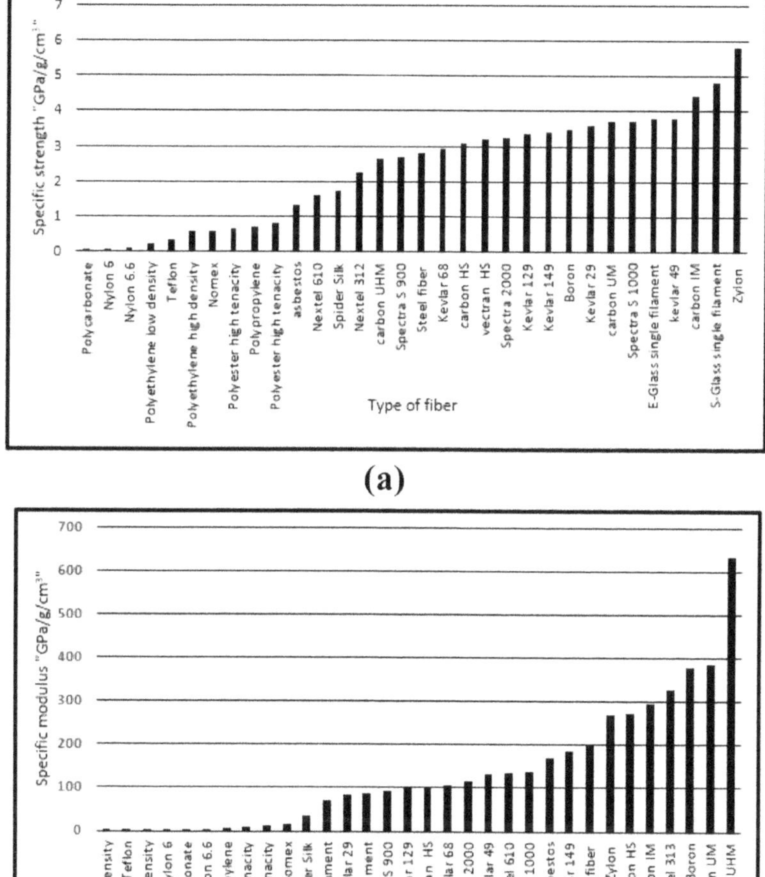

FIGURE 2.10 Fibers properties (a) fibers specific strength and (b) fibers specific modulus.

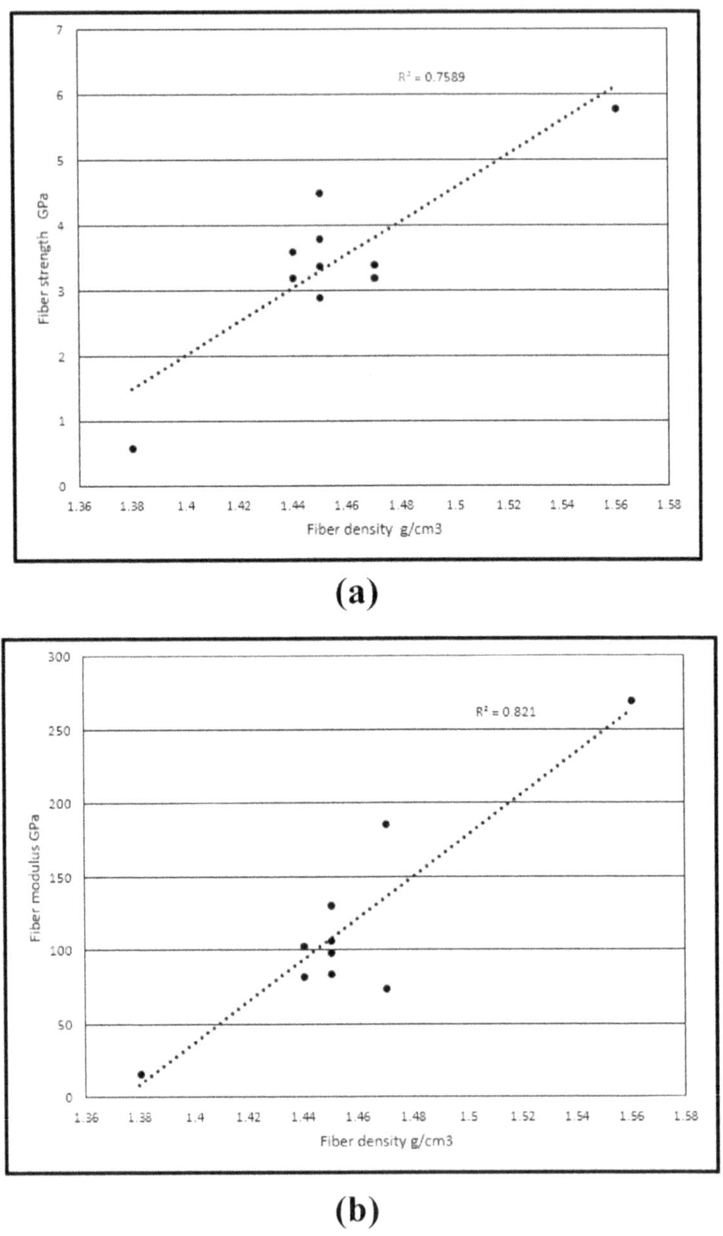

FIGURE 2.11 Effect of fiber density (a) fiber strength versus fiber density and (b) fiber modulus versus fiber density.

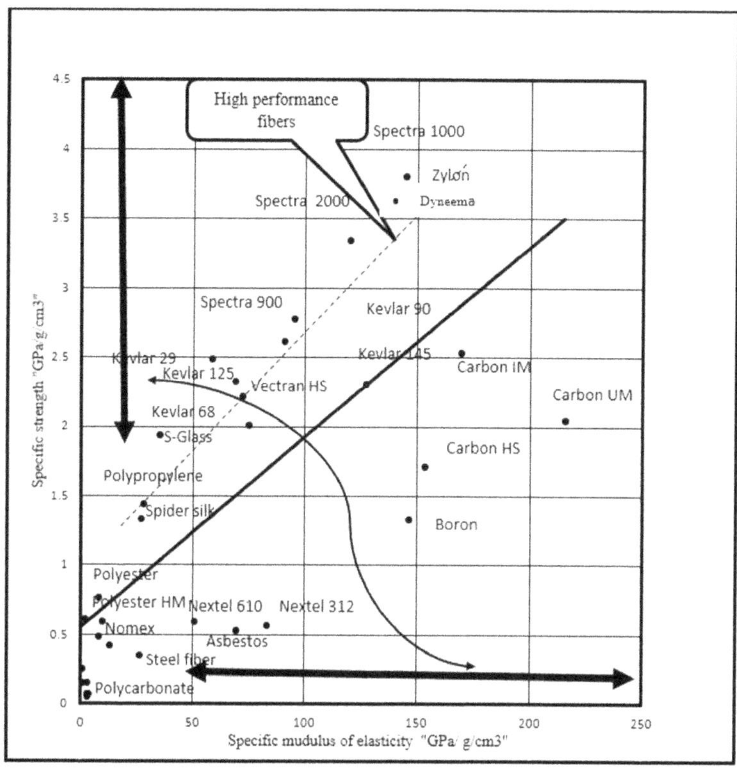

FIGURE 2.12 Specific strength versus modulus of HPF elasticity.

2.5 FIBER WORK OF RUPTURE

The fiber work of rupture is an indicator of how the fiber is strong and extensible to withstand sudden impact force determines energy absorption of woven fabrics when subjected to ballistic impact.The impact properties of the PF depend on the ability of fabric weave to absorb the high-speed impact. The fabric absorption energy is contingent on strain longitudinal wave speed propagation under impact force.[39] The speed of strain propagation in the fibers relies on their specific elastic modulus. The higher projectile impact velocity on the fibers with lower strain propagation speed results in stress concentration at the area of the impact. It was verified that yarns and fabrics strain propagation speed is lower than that of the fibers, owing to the complex structure of the yarns and fabrics, so the strain wave will propagate differently.[40]

2.6 FIBER CREEP PROPERTY

When the textile material is subjected to stress, the material is instantaneously deformed and will continue to deform under the same value of stress.The strain during creep will take the shape as illustrated in Figure 2.13. Creep strain starts with the instantaneous deformation—to primer deformation—to secondary deformation and ends in a zone characterized by a high rate of deformation completed by rupture. When the stress removed, the material will recover its deformation in the same order, ending by the permanent deformation. The creep phenomena start with the instantaneous deformation, followed by secondary deformation, and as the load removed there is the instantaneous recovery followed by permanent deformation, which differ for fibers type, yarn and fabric specifications, temperature, and the applied stress. The creep property is of MMFwhen used in fabric is working at elevated temperature. It has been found that strain is expressed as a function of time, for MMF, at up to 70% of their breaking load a primary creep mechanism dominates and at higher loads a secondary creep becomes increasingly important.[41]

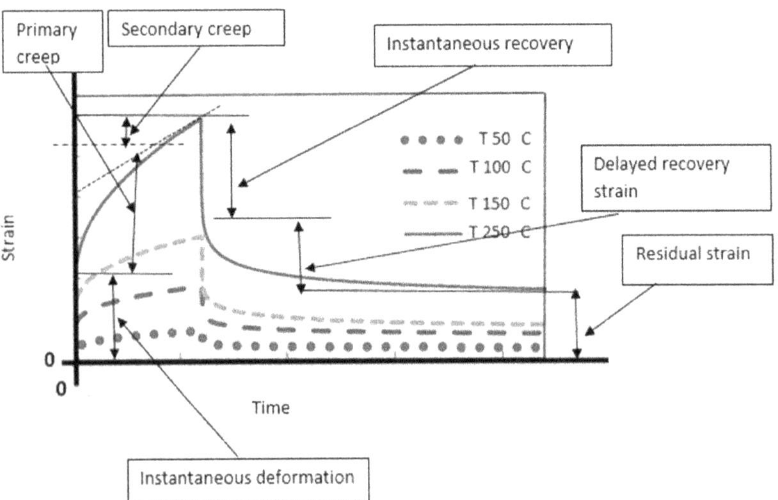

FIGURE 2.13 Creep strain—time curve.

The elevating of the surrounding temperature, primary creep and residual strain increases as illustrated in Figure 2.13. The creep properties of p-aramid fibers differ expressively from the other highly oriented fibers; Vectran HS (liquid crystalline polymer) creeps at a rate four times slower than Kevlar 29.[42] There is the increasing interest in the fibers creep properties for the thermal shield PF.

2.7 FIBER TRIBOLOGICAL PROPERTIES

Frictional property is one of the properties which affect all the mechanical properties of the yarns and fabrics as well as the processing parameters. The frictional property of the fibers differs from the metal friction hencethe fibers in most cases are not smooth or straight consequently, under no normal load, the friction forces areactivated (cohesion forces).

Coefficient of friction dependson fiber count, the morphology of the fiber surface, fiber crimp, fiber cross-sectional shape, and fiber finish. Figure 2.14 gives the coefficient of friction for some fibers.[43] Coefficient of friction (μ) of Vectran is 0.12–0.15[44] like for most of HPF, except PBO fiber, which attained a higher coefficient of friction (μ) of 0.18.

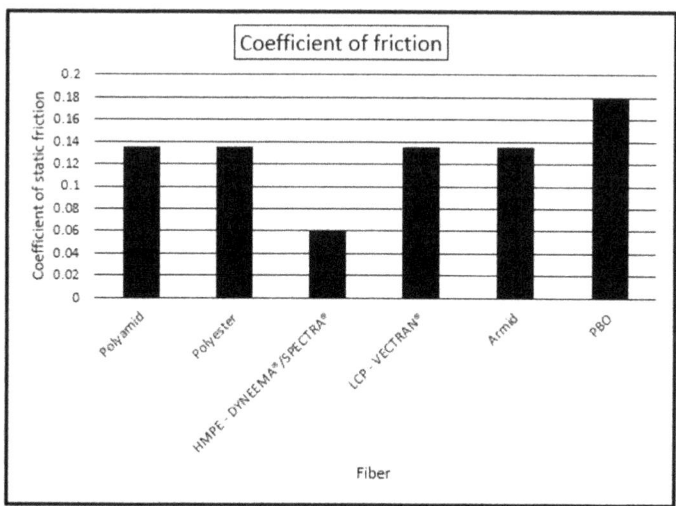

FIGURE 2.14 HPF coefficient of friction.

The coefficient of friction of the fibers, yarns, and fabrics is quite different. The yarn coefficient of friction depends on the fiber type, yarn specification, yarn structural parameters, fiber packing density, and twist factor, while the fabric coefficient of friction depends on the fiber type, yarn specification, weave design, packing density, fabric cover factor, fabric tightness, and crimp percentage. Generally, the fabric coefficient of friction is higher than the fiber coefficient of friction due to the structural parameters of the yarn and surface roughness of the fabric. This relates to the variability in the morphology of the fabric surface during the application of a friction force.

2.8 FIBER THERMAL PROPERTIES

Thermal resistance is the critical property of PF used for firefighter and heat shields to keep heat out but also retain the wearer's heat in. It was found that the thermal conductivities (*k*) of most polymer materials are lower than those of metals, depending on the fiber molecular structure; amorphous polymers are smaller semiconductors.[45] But, for example, highly crystallized polymer materials, including high-strength polyethylene (PE) fiber and high-strength polypara-phenylene-benzo-bis-oxazole (PBO) fiber, are known to possess high (*k*) value like those of metals.

Polymer fibers (*k*) value in fiber direction depends on the crystallinity, orientation, crystal size, length of molecular chains, chemical bridge points, and morphologies composed of crystal and amorphous.[15,45,46] Thermal conductivities(*k*) value toward the direction of the molecular chain of PE, polyethyleneterephtalate and polypropylene increase by the increasing crystallinity and orientation of the crystal.

Thus, for HPF their thermal conductivity will increase linearly in proportion to tensile modulus[45] as illustrated in Figure 2.15. The thermal conductivity increases for the thicker multifilament yarns. It should be mentioned that the (*k*) value of the fibers is a function of the surrounding temperature. High tenacity fibers have higher thermal conductivity longitudinally and lower transverse thermal conductivity.[47] The (*k*) value of the yarns and fabric (fiber-air assembly) depends on their specifications and bulk density as well as the fiber packing density, fiber arrangement, and fiber thermal conductivity. However, the thermal resistance of the yarns and fabrics is higher than that of their constituent fibers; consequently, the (k) value of the fibers will be of less importance than their location in the space of bulk of yarn or fabric.[48] Table 2.3 gives the thermal conductivity (k) value ofsome selected fibers.

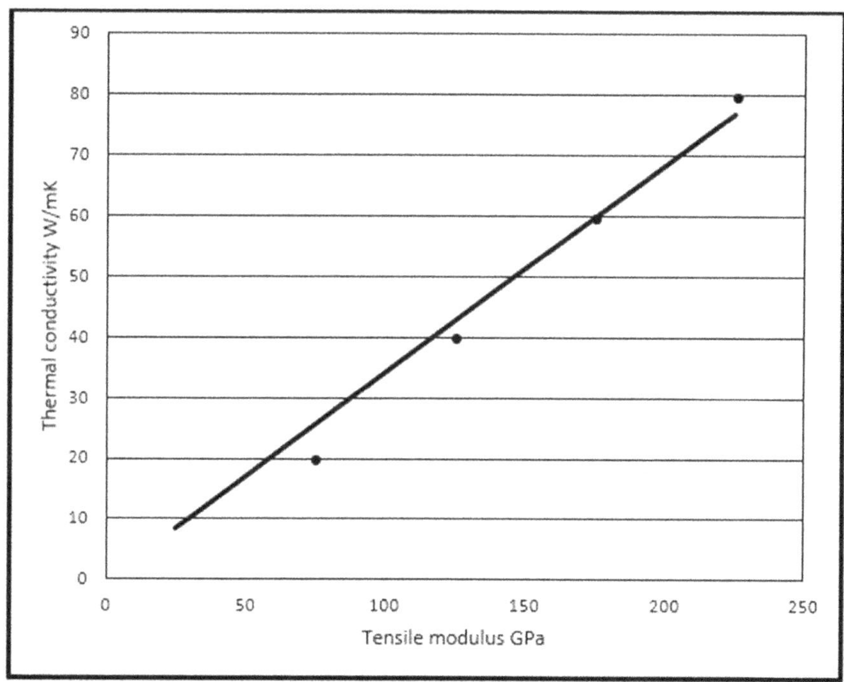

FIGURE 2.15 Thermal conductivities versus elastic modulus of high-strength PE fibers.

Heat resistance of fibers is important for their end use as in fireproof vest or heat shield. The fibers should not deteriorate in their properties under an elevated temperature environment. Generally, the fiber strength reduces when the fiber is exposed to a high temperature for a long service time. Figure 2.16 illustrates that the properties retention percentages decrease with the temperature; Kevlar fiber tenacity retention reaches 55% at 300°C, while PBO fiber still have tenacity retention 80%,[55] Vectran fiber strength will lose 50% of its strength at temperature 135°C and Vectran HT 145°C, while Vectran UM at 150°C, Aramid standard fibers will lose 50% of their strength at 430°C and HM Aramid fibers at 230°C.[56]The Zylon strength retains 40% at 500°C, its modulus at 400°C retains 75% of its modulus at room temperature (melting temperature of Zylon is 600°C).[57] After exposing NOMEX fiber to dry air at 260°C for 1000 h and then returning them to room temperature, the breaking strength and toughness are approximately 65% of that exhibited before exposure.[31,58]

TABLE 2.3 Thermal Conductivity (k) of Fibers.

Fiber	Thermal conductivity (W/mK)	Fiber	Thermal conductivity (W/mK)
PAN (according to its modulus)	5–90	PBI	0.41
Pitch (according to its modulus)	120–800	Dyneema (HDPE)	20
Nylon	0.25–0.7	Kevlar	0.0400
High strength PE	28	NOMEX	0.139–0.25
Polystyrene	2	Glass fiber	1.28–1.32
PBO (Zylon)	14–20	Polypropylene	–
Vectran	1.5	Alumina-silica ceramic fibers	0.09[a]–0.22[b]
Aramid	2.5	E-Glass	1.04

[a]Temperature—316°C.
[b]Temperature—982°C.
HDPE, high-density polyethylene; PAN, polyacrylonitrile; PE, polyethelene.

Source: Adapted with permission from Refs. [15,45–54].

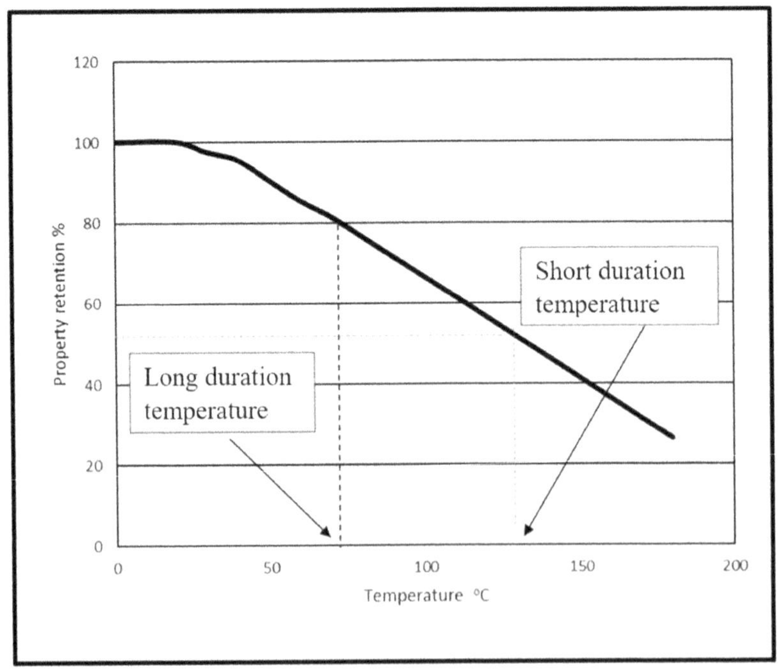

FIGURE 2.16 Fiber strength retention versus temperature.

Heat deuteration of fibers at the different rates contingent on the fiber chemical structure and time of exposure, thereby ultimately destroying their integrity.[19] Heat resistance (the temperature at which the fiber is melted or recomposited)

The m-Aramid fiber is melted or recomposited at 400°C, while for PBO and UHMWPE fibers it is 650°C and 150°C, respectively,[59] for p-Aramidand PBI fibersis 550°C, and for Polyester it is 260°C. Carbon fiber heat resistance is elevated up to 2000°C. The maximum recommended continuous operation temperature (COT) should be less than these limits. For Spectra fibers, advisable long duration temperature limit is 70°C and for short duration 130°C.[59]

Besides the fiber thermal properties, the thermal behavior of fabrics is also influenced by fiber arrangement in yarns, the fabric design, its thickness, fluffiness, and porosity.

The recommended values of the operational temperature for selected fibers are demonstrated in Table 2.4.

TABLE 2.4 Recommended Continuous Operation Temperature (COT)°C.

Fiber	COT (°C)	Fiber	COT (°C)
HPPE	70–82	PBI	260–400
PAN (according to its modulus)	300–537	Lastan	200
Pitch (according to its modulus)	300–537	Dyneema (HDPE)	120
Nylon	260	Kevlar	149–177
High strength PE	130	NOMEX	204
Polystyrene	–	Glass fiber	260
PBO (Zylon)	287–316	Polypropylene	107
Vectran	145	Dyneema	70
Aramid; standard	430	Aramid; high modulus	230
Spectra	70		

HPPE, high-performance polyethylene; PBI, polybenzimidazole; PBO, polybenzoxazole; PE, polyethylene.

Source: Adapted with permission from Refs. [15,18,31,49–64].

2.9 FIBER CHEMICAL RESISTANCE

Chemical resistance of the fibers is their ability to protect against chemical attack or solvent reaction. Several factors affect the fiber chemical resistance, such as chemical concentration, time of exposure, and temperature.

The chemical PFs are designed to prevent the chemical liquid or gases from penetration through using fibers that resist the degradation by the chemical liquids and proactive fabric not allow the chemicals or gas to penetrate through its fabric pores.

Most of the HPFs have a good to excellent resistance to acids, alkalis, and organic solvents. The chemical resistance rating of:

- polytetrafluoroethylene (PTFE) is excellent;
- high-densitypolyethylene (HDPE) is high;
- NOMEX is very good (degraded by strong alkalis at high temperatures);
- PBI is good;
- Zylonis good; and
- ultra high molecular weight polyethlyne UHMWPE is low.

Kevlar is acceptable, certain strong aqueous acids bases and sodium hypochlorite can cause degradation, particularly at elevated temperatures.

For the PF, the maximum value of permeation rate, μg/(m^2/min) (ASTM F903, ASTM F739), the chemical protection armor is not one-piece cloth, the presence of seams might be of higher permeation rate for chemicals or gases compared to PF.[65]

KEYWORDS

- **protective fabric**
- **stab resistance**
- **chemical resistance**
- **bulletproof vest**
- **stab-proof vest**
- **heat resistance**
- **fireproof fabric**

REFERENCES

1. El Messiry, M. *Natural Fiber Textile Composite Engineering*; Apple Academic Press Inc.: USA, 2017.
2. Wisniowski, T.Natural versus Synthetic Fibers and Dyes Opinion Piece. Marie. http://artquill.blogspot.com.eg/2014/12/natural-versus-synthetic-fibers-and.html (accessed May 6, 2016).
3. European Man-Made Fiber Association. http://www.cirfs.org/KeyStatistics/World ManMadeFibresProduction.aspx (accessed May 6, 2016).
4. Carmichael, A. Man-made fibers continue to grow. *Text World* January/February 2015. http://www.textileworld.com/textile-world/fiber-world/2015/02/man-made-fibers-continue-to-grow/ (accessed May 6, 2016).
5. Spinning (Polymers). https://en.wikipedia.org/wiki/Spinning_(polymers) (accessed May 6, 2015).
6. Natural and Man-Made Fiber Overview. https://ihsmarkit.com/products/fibers-chemical-economics-handbook.html (accessed May 6, 2016).
7. Wisniowski, T. Natural versus Synthetic Fibers and Dyes Opinion Piece. Marie. http://artquill.blogspot.com.eg/2014/12/natural-versus-synthetic-fibers-and.html (accessed May 6, 2017).
8. Grujicic, M.; Hariharan, A.; Pandurangan, B.; Yen,C.; Cheeseman, B.; Y. Wang, Y.; Y. Miao, Y.; Zheng, J. Fiber-Level Modeling of Dynamic Strength of Kevlar KM2 Ballistic Fabric. *J. Mater Eng. Perform.* **2012,** *21*, 1107–1119.
9. Ning, P. Theoretical Determination of the Optimal Fiber Volume Fraction and Fiber–Matrix Property Compatibility of Short Fiber Composites. *Polym. Compos.* **1993,** *14* (2), 85–93.
10. El Messiry, M. Theoretical Analysis of Natural Fiber Volume Fraction of Reinforced Composites. *AEJ.* **2013,** *52*, (3), 301–306.
11. Matsudaira, M.; Tan, Y.; Kondo, Y. The Effect of Fiber Cross-sectional Shape on Fabric Mechanical Properties and Handle. *J. Text. Inst.* **1993,** *84*, (3), 376–386.
12. Karaca, E.; Omeroglu, S.; Becerir, B. Effects of Fiber Cross-Sectional Shapes on Tensile and Tearing Properties of Polyester Woven Fabrics. *Text. Confect.* **2015,** *25*(4), 313–318.
13. Karaca, E.; Kahraman, N.; Omeroglu, S.; Becerir, B. Effects of Fiber Cross Sectional Shape and Weave Pattern on Thermal Comfort Properties of Polyester Woven Fabrics. *Fibres Text. East. Eur.* **2012,** *20*, *3*(92), 67–72.
14. Technical Guide for Kevlar®. Aramid Fiber DuPont™ Kevlar® Aramid Fiber, http://www.dupont.com/content/dam/dupont/products-and-services/fabrics-fibers-and-nonwovens/fibers/documents/Kevlar_Technical_Guide.pdf (accessed Jan 7, 2018).
15. Hearle, J. (Ed) High-performance fibers. Woodhead Publishing Limited in association with The Textile Institute, UK, 2001.
16. Zhu, D.; Mobasher, B.; Rajan, S. Dynamic Tensile Testing of Kevlar 49 Fabrics. *J. Mater Civil Eng.* **2011,** 1–10.
17. Zupin, Z.; Dimitrovski, K. Mechanical Properties of Fabrics Made from Cotton and Biodegradable Yarns Bamboo, SPF, PLA in Weft. Woven Fabric Engineering, Polona Dobnik Dubrovski (Ed.), InTech..http://www.intechopen.com/books/

woven-fabricengineering/mechanical-properties-of-fabrics-made-from-cotton-and-biodegradable-yarns-bamboo-spf-pla-inweft. (accessed Jan 7, 2018).
18. Committee on Opportunities in Protection Materials Science and Technology for Future Army Applications; National Research Council. Opportunities in Protection Materials Science and Technology for Future Army Applications. National Academies Press; 2011. High-Performance Structural Fibers for Advanced Polymer Matrix Composites.
19. High Performance and High Temperature Resistant Fibers—Emphasis on Protective Clothing. http://www.intexa.com/downloads/hightemp.pdf (accessed Jan 2, 2018).
20. Vectran™ Fiber, Kuraray America, INC. (online) http://imattec.com/linked/vectran%20-%20technical%20data.pdf. (accessed June 6, 2017).
21. Dupen, B. *Applied Strength of Materials for Engineering Technology*; 6th ed.; 2014. http://opus.ipfw.edu/mcetid_facpubs/35 (accessed June 6, 2017).
22. Mayo, Jr.; Wetzel, E. Cut Resistance and Failure of High-Performance Single Fibers. *Text. Res. J.* **2014,** *84* (12), 1233–1246.
23. Boron. *The International Handbook of FRP Composites in Civil Engineering*, Zoghi, M., Ed. https://books.google.com.eg/books?id=jOHKBQAAQBAJ&pg=PA46&lpg=PA46&dq=kevlar+fibers+129+properties&source=bl&ots=J8kBLrNTXY&sig=N_ZwU9M-wTfVQ7gelrF23p3nUDk&hl=en&sa=X&ved=0ahUKEwiwiPrGppTSAhXFcBoKHZWtANM4ChDoAQgnMAI#v=onepage&q=kevlar%20fibers%20129%20properties&f=false (accessed Jun 6, 2017).
24. PYROFIL™—Typical Properties of Carbon Fiber. http://www.mitsubishi-rayon.de http://mrcfac.com/wp-content/uploads/2013/09/Pyrofil-Typical_Properties_of_Carbon_Fiber_03_2010.pdf (accessed Jun 6, 2017).
25. Huang, X. Fabrication and Properties of Carbon Fibers. *Mater* **2009,** *2*, 2369–2403; DOI: 10.3390/ma2042369. 2369-2403 (accessed Jun 6, 2017).
26. Walsh, P. *Carbon Fibers. ASM Handbook,* Composite; Vol. 21. http://ansatte.hin.no/ra/MatLinks/carbonfiber_overview.pdf (accessed Jun 6, 2017).
27. Kevlar, Teijin-Aramid Ballistics Material Handbook, TwaronBallistic Yarns. http://www.teijinaramid.com (accessed Jun 6, 2017).
28. Nextel Ceramic Fibers. Nextel, Ceramic Textile Technical Notebook. http://solutions.3mdeutschland.de/3MContentRetrievalAPI/BlobServlet?locale=de_DE&lmd=1270717655000&assetId=1258565166920&assetType=MMM_Image&blobAttribute=ImageFile (accessed Jun 6, 2017).
29. 3M™, Nextel™ Ceramic Structural Fabrics 610 and 720. http://multimedia.3m.com/mws/media/1242134O/3m-nextel-ceramic-structural-fabrics-610-720.pdf (accessed Jun 6, 2017).
30. Farced, A.; Schiroky, G. *Microstructure and Properties of Nextel™ 610 Fiber Reinforced Ceramic and Metal Matrix Composites*. Proceedings of the 18th Annual Conference on Composites and Advanced Ceramic Materials—A: Ceramic Engineering and Science Proceedings, Volume 15, Editor(s): John B. Wachtman Jr. 1994, 344–352. https://ceramics.onlinelibrary.wiley.com/doi/book/10.1002/9780470314500, (accessed June 3, 2017)
31. Nomex. Technical Guide for NOMEX® Brand Fiber. Technical Information. http://www.nakedwhiz.com/gasketsafety/nomextechnicalguide.pdf (accessed June 6, 2017).
32. Polyethylene Low Density. http://www.plasticmoulding.ca/polymers/polypropylene.htm (accessed Jun 6, 2017).

33. Ko, F.; Kawabata, S.; Inoue, M.; Niwa, M.; Fossey, S.; Song, J. Engineering Properties of Spider Silk. http://web.mit.edu/course/3/3.064/www/slides/Ko_spider_silk.pdf (accessed June 6, 2017).
34. Vectran™ Fiber, Kuraray America, INC. Kuraray America, Inc. Is a Subsidiary of Kuraray Co., Ltd. Kuraray America's Product Lines Include Vectran™ Liquid Crystal Polymer (LCP) Fiber. http://www.kuraray.us.com/products/fibers/vectran/ (accessed Jun 6, 2017).
35. Ceramic Fibers and Coating. Advanced Materials for the Twenty-First Century. Committee on Advanced Fibers for High-Temperature Ceramic Composites, Commission on Engineering and Technical Systems, National Research Council, Publication NMAB-494. http://www.nap.edu/catalog/6042.html (accessed Jun 6, 2017).
36. 3M™ Nextel™ Ceramic Fibers and Textiles: Technical Reference Guide. http://multimedia.3m.com/mws/media/1327055O/3m-nextel-technical-reference-guide.pdf (accessed Jan 7, 2018).
37. Kalebek, N.; Babaarslan, O. Fiber Selection for the Production of Nonwovens, Chapter 1, "Non-woven Fabrics,"Han-Yong Jeon (Ed.), IntechOpen, DOI: 10.5772/61977. Available from: https://www.intechopen.com/books/non-woven-fabrics/fiber-selection-for-the-production-of-nonwovens(accessed Feb 7, 2018).
38. Yasuda, H.; Ban, K.; Ohta, Y. *Advance Fiber Spinning Technology*; Woodhead, N. (Ed.), Cambridge: UK, 1994.
39. El Messiry, M.; Eltahan, E. Stab Resistance of Triaxial Woven Fabrics for Soft Body Armor. *J. Ind. Text.* **2016,** *45* (5), 1062–1082.
40. El Messiry, M.; Ibrahim, S. Investigation of Sonic Pulse Velocity in Evaluation of Knitted Fabrics. *J. Ind. Text.* **2016,** *46* (2), 455–472.
41. Lafitte, M.; Bunsell, A. The Creep of Kevlar 29 Fibers. *Polym. Eng. Sci.* **1985,** *25* (3), 182–187.
42. Fette,R.;MarjorieF.;Sovinsk,M.VectranFiberTime-DependentBehaviorandAdditional. https://code541.gsfc.nasa.gov/Uploads_recent_publications/04-12_Fette.pdf (accessed Jun 6, 2017).
43. Different Fibers, Different Characteristics, Different Result. http://www.alpha-ropes.com/uploads/files/alpharopes_technical_info.pdf (accessed Jun 6, 2017).
44. Vectran® 12 Strand and 12x12 tech sheet. https://www.cortlandcompany.com/sites/default/files/downloads/media/product-data-sheets-vectran-12-strand.pdf (accessed June 6, 2017).
45. Yamanaka, A.; Takao, T. Thermal Conductivity of High-Strength Polyethylene Fiber and Applications for Cryogenic Use. *ISRN Mater Sci.* **2011,** *2011*, Article ID 718761, 1–10.
46. Wang, X.; Ho, V.; Segalman, R.; Cahill, D. Thermal Conductivity of High-Modulus Polymer Fibers. *Macromolecules* **2013,** *46*, 4937–4943.
47. Kawabata, S.; Rengasmy, R. Thermal Conductivity of Unidirectional Fiber Composites Made from Yarns and Computation of Thermal Conductivity of Yarns. *IJFTR* **2002,** *27*, 217–223.
48. Bogaty, H.; Norman. S.; Hollies, N.; Harris, M. Some Thermal Properties of Fabrics Part I: The Effect of Fiber Arrangement. *Text. Res. J.* **1957,** *27*, (6), 445–449.
49. MatWeb. http://www.matweb.com/search/datasheet.aspx?matguid=22515cfc3a9a4f2f88a1a3f8086e94e2&ckck=1 (accessed Jun 6, 2017).

50. EmcoIndustrial Plastics INC. http://www.emcoplastics.com/uhmw-faqs/ (accessed Jun 6, 2017).
51. Vectra Liquid Crystal Polymer (LCP) http://www.nakedwhiz.com/gasketsafety/nomextechnicalguide.pdf (accessed Jun 6, 2017).
52. DuPont™ Kevlar® 29 Aramid Fiberhttp://www.dupont.com/products-and-services/fabrics-fibers-nonwovens/fibers/brands/kevlar.html (accessed Jun 6, 2017).
53. Toyobo Dyneema® SK60 High Strength Polyethylene Fiber. http://www.matweb.com/search/datasheettext.aspx?matguid=4481722d60e54cc3b3c112eb3d4b9d02 (accessed Jun 6, 2017).
54. Excellent Properties of Nomex® 410 Enable Long-term Reliability. http://www.dupont.com/products-and-services/electronic-electrical-materials/electrical-insulation/brands/nomex-electrical-insulation/articles/nomex-410.html (accessed Jun 6, 2017).
55. Bourbigot, S.; Flambard, X. Heat Resistance and Flammability of High Performance Fibres: A Review. *Fire Mater* **2002,** *26*, 155–168.
56. Vectran data sheet. http://imattec.com/linked/vectran%20-%20technical%20data.pdf (accessed Jan 14, 2018)
57. PBO fiber Zylon. http://www.toyobo-global.com/seihin/kc/pbo/zylon-p/bussei-p/technical.pdf (accessed Jul 10, 2017).
58. Ozgen, B. Physical Properties of Kevlar and Nomex Plied and Covered Yarns. *Text. Res. J.* **2012,** *83* (7), 752–760.
59. Dyneema® Heat Resistant of Fibers. Ultra High Molecular Weight Polyethylene Fiber from DSM Dyneema UHMWPE Fiber Combines Excellent Mechanical Properties with Low Density, Resulting in High Performance-on-Weight Basis. www.send-cockpit.com/appl/ce/software/code/ext/_fd.php?tdat_nl (accessed July 10, 2017).
60. Honeywell Spectra® Fiber: Lightweight Strength. https://www.honeywell-spectra.com/?document=fiber-capability-guide&download=1 (accessed Jan 14, 2018).
61. Liu, X; Yu, W. Degradation of PBO Fiber by Heat and Light. *RJTA* **2006,** *10* (1), 26–32.
62. Vectran. http://imattec.com/linked/vectran%20-%20technical%20data.pdf (accessed June 6, 2017).
63. Liu, X.; Yu, W. Evaluation of the Tensile Properties and Thermal Stability of Ultrahigh-Molecular-Weight Polyethylene Fibers. *J. Appl.Polym. Sci.* **2005,** *97*, 310–315.
64. Polybenzimidazole fiber. https://en.wikipedia.org/wiki/Polybenzimidazole_fiber#Thermal_Properties (accessed Jan 14, 2018).
65. *KleenGuard.* Chemical Resistance Information Guide. http://www.kcproductselector.com/~/media/RelatedMedia/PDFs/Apparel/ChemK0001_08_01.ashx (accessed Jun 6, 2017).

CHAPTER 3

Material Design Principles of Flexible Protective Armor

ABSTRACT

The modern body armor is made from the textile fibers for the different types of threat protection in several applications: military, medical, sport, industrial, etc. This chapter targets to give the basic concept for the choice of textile materials for a specific body armor design with the emphasize on threats and Risk Assessment calculations. A complete analysis of the protective fabric requirements for armor design are given for heat and flame resistant clothing, chemical protective garments, ballistic body armor, electrical resistance armor, radiation protective garments, and stab resistance armor. Some aspects for fiber selection for the protective materials are investigated.

3.1 INTRODUCTION

Generally, the body armors in the modern era are made from textile fabrics principally for their protection performance or functional characteristics that can be used in the design of an armor for different fields, such as military protective, medical protective, sports protective, space protective, agricultural protective, and industrial protective.[1] For each type of protective fabric (PF) materials, there are the necessary parameters that will satisfy a certain application. That represents a complex scientific area due to the diversity and contradiction of the parameters. The degree of protection and the comfort of the wearer are two contradicting parameters in all body protection armor. Numerous publications are dealing with the application and design of the new manmade fibers in the various protective armors from surrounding threats. The number of articles concerned with the PF (Google Scholar, 426,000 publications) is classified according to the subject as follows:

- Design of bullet-protective fabric
- Design of stab-protective fabric
- Design of electrostatic-protective fabric
- Design of electromagnetic-protective fabric
- Design of radiation-protective fabric
- Design of rain-protective fabric
- Design of heat-protective fabric
- Design of chemical-protective fabric
- Design of fire-protective fabric.

In this review, Figure 3.1signposts the growing interests in the application of protective fabrics (PFs) for the different types of armors.

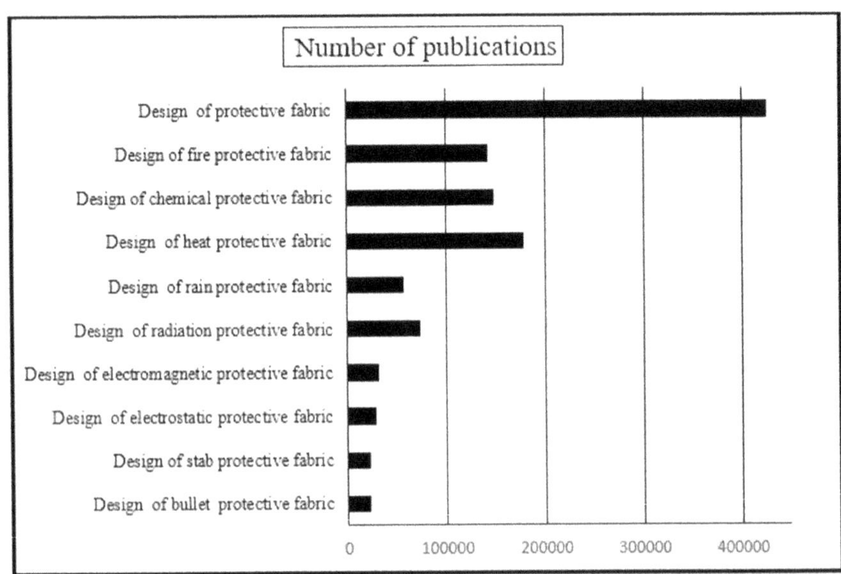

FIGURE 3.1 Number of articles for different protective fabric applications.

Although the number of publications for stab PFs and bullet PFs is the smaller class, because of the surged interest in such field of applications in the last decades, the design of the PFs shows the highest frequency. The textile material for the PFs should generally satisfy the different parameters as given in Figure 3.2. The importance of each parameter depends on the type of the armor and the nature of the threat and its intensity. If the armor does not satisfactorily protect from the threats, it might increase the injury risk. The increasing probability of various threats, terrorist activities and warfare situations around the globe place a demand for advanced protective armors.[2] The global ballistic protection market was valued at $7.91 billion in 2013 and is expected to reach $9–11.03 billion by the end of 2020, growing at an average rate of 5.11%.[3] The US market reached $4 billion in 2010 with an annual growth of 6.1%.[4–5]

The market of the PFs in the present time covers the needs of defense, homeland security, industrial, commercial, personal protective equipment, and vehicle armor such as:

- Heat- and flame-resistant clothing,
- Chemical protective garments,
- Ballistic body and vehicle armor,
- Biological armor,
- Electrical resistance armor,
- Gloves for industrial applications, and
- Ballistic shields.

The basic requirements for the PF are its protection performance, durability, and lightweight. The properties of the fabric and the design of the protective armor are closely related to the end use and the target. A key challenge for this market lies in comfort and the light weight of the armor.[6]

3.2 TEXTILE MATERIAL FOR ARMOR DESIGN

The armor design may be static or dynamic. The static design of the armor is a design to prospect a certain level of protection despite the change of the environmental conditions (speed, force, temperature, electrical current, moisture, etc.). The dynamic design can react to the threat and consequently can act. The armor is designed taking into consideration the maximum value of the threat it may be subjected to. The main requirements of the material for the PFs are given in Figure 3.2.

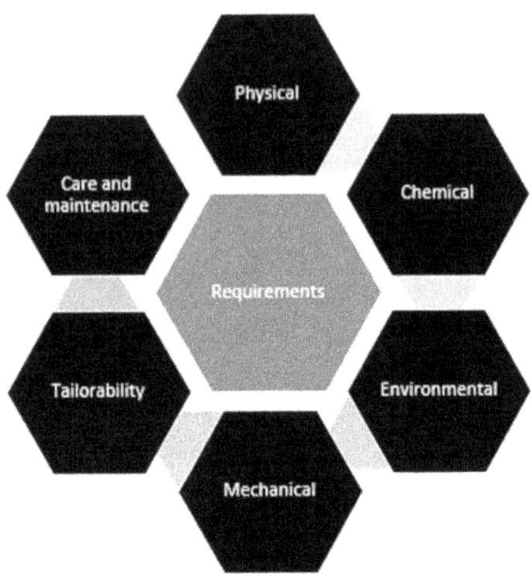

FIGURE 3.2 Requirements of the textile material for armor design.

Textile material for the design of PF should have the following properties:

- High strength and modulus,
- High-impact absorption energy,
- High fatigue resistance,
- High cut resistance,
- High tear resistance,
- High shear modulus,
- High pilling resistance,
- High abrasion resistance,
- High chemical resistance,
- Hydrophobic,
- High melting point,
- High tailorability,
- Low density,
- Low elongation,
- Low creep,
- Low coefficient of surface friction,
- Easy care, and
- Recyclable.

For each group of application, part of the above properties is more anticipated. A major challenge for engineers is to analyze armor structures that involve the dynamic properties of fibers and structural armor systems dynamics when subjected to dynamic loading.

3.3 THREAT AND RISK ASSESSMENT

It is very essential to evaluate the threat and the risk facing the armor wearer while designing the PF. Many methodologies how to perform a risk and threat analysis have been developed for the different applications.[7] The designer of the PF must analyze the following: The probability of the damage, the nature and the intensity of the threats, and the effect of the damage and how to minimize it. There is no completely secure armor; the acceptable risk is when the design of the armor can minimize harm, injury, and trauma. There are several standards the armor should comply to maximize the degree of protection and reduce the damage.

3.3.1 RISK ASSESSMENT

Risk assessment involves risk identification, risk analysis, and evaluation.[8] Risk assessment is designed to identify the sources of potential danger and the probability of exposure to these sources of danger.[9] The goal of the assessment process is to achieve the level of protection sought through implementation of justifying measures in the vest design to select mitigation measures for reducing potential risk. This analysis forms part of an essential approach to introduce the methods for controlling risks, including personal protective equipment.[10]

The risk may be defined as the level of danger multiplied by the probability of exposure:

$$\text{Risk} = \text{Level of danger} \times \text{Probability of exposure.} \tag{3.1}$$

Risk analysis must consider the characteristics of the task to be performed. This includes: The identification of the risks and their potential effects on the health and safety of the wearer and the probable impact of protective armor, such as a bulletproof vest, stab resistance vest, protective gloves, on the performance of the wearer itself. It is important to avoid either underestimating or overestimating the level of protection required.[11,12] Consequently, there is a level of risk (low, medium, and high) associated with physical harm expected. The National Institute of Justice (NIJ) notes, that there is no such thing as absolute bulletproof armor. Ballistic-resistant body armor provides protection against penetrating bullets and the blunt trauma associated with bullet impacts, but it will not stop all bullets. Some armor vests only protect against handgun bullets, while some others protect against rifle bullets.[12] The level of danger includes two elements: the probability of occurrence of a harm and how severe the harm could be.[13] Probability of exposure to the threat may be defined as the percentage of the time the wearer is subjected to the threat to the time of wearing the armor and the probability of occurrence of the threat.

Several methods were used for risk assessments: qualitative, quantitative, and semiquantitative.

The risk rating, according to the American Chemical Society, is given by:

$$\text{Risk rating (RR)} = \text{Probability of occurrence (OV)} \times \text{Severity of consequences value (CV)}. \tag{3.2}$$

Probability of occurrence varies with standard linear scaling from 1 to 4 (4 being the highest probability),[14] this corresponding to the OV, 0, up to 0.1, from 0.1 to 0.5, from 0.50 to 0.9, and from 0.90 to 1. The severity of CV rating is classified as No risk, minor risk, moderate risk, and high risk, and have the values of CV 1, 5, 10, and 20, respectively.[14] Another method was used for risk assessments[15] to evaluate the possibility of a risk occurrence—technically assessing a conditional probability. A risk event that is certain not to occur has, by definition, probability equal to zero. A risk event that is certain to occur has, by definition, probability equal to one.[16] Risk event probability is a subjective probability reflecting a subject expert's degree of belief that the risk event is, or is not, likely to occur. An alternative classification of the probability of the risk event provides a guide for assessing risk event probabilities and their rating.[17] The interpretation of the assessment of risk event probabilities was divided into eleven categories, each elevates probability by 0.05. The rating is devised into three classes: Low, medium, and high risk with probabilities of occurrence from 0 to $\leq$0.35, from >0.35 to >0.65, and from >0.65 to 1, respectively. When designing a protective armor, assign a categorical rating (High, Medium, or Low) to a risk event that is confident to occur. The problem in the calculation of the risk rating in the case of bullet and stab proof vests is the high value of uncertainty, for that reason it should be designed on the high-risk rating, that is, the risk event probability varies between 0.65 and <1. Table 3.1 provides a guideline for assessing risk by the expected impact of the damage on the armor wearer when subjected to the threat of stabbing or gunshot.

The risk rating can be evaluated using eq 3.1. For instance, if the probability of occurrence is 0.5, then the occurrence value will be 3. If the severity of consequences is medium, then the RR will be 30.

3.3.2 CYCLE OF RISK MANAGEMENT

Risk identification is the critical first step of risk management. Risk is assessment according to two components: The probability of failing to achieve specified objectives and the impact of failing to achieve those objectives. The protection level of body armor is selected through the analysis of harm, threat, threatening situations, expected usages, and residual risk. Risk management is a closed loop of systematic applying of the management policy, procedures, and practices to the tasks of: risk analysis, risk assessment, risk control, estimation of risk, risk controlling, and monitoring.[14–17] In the case of protective armor, the change of the armor structure and material is the only way to control the risk.

TABLE 3.1 Severity of Consequences, Weighted Value Scale for Body Armor.

Wound definition	Rating value		Equivalent numerical value
No injuries		1	0 < 0.35
Will cause minor injuries. (abrasions wound, surface puncture wound, light bruise, etc.)	Low	5	0.35 ≤ 0.65
Will cause noticeable injuries. (noticeable bruise, cut and stab wound, abrasions wound, puncture wound, etc.)	Medium	10	0.35 ≤ 0.65
Will cause significant injuries. (life-threatening wound, deep incised wounds, penetrating stab wound, gunshot wound, blunt trauma, etc.)	High	20	>0.65 – <1

3.4 ANALYSIS OF PROTECTIVE FABRICS

The PF properties can be evaluated by a defined criterion to be suitable for the tailoring the protective armor. The problem that complicated the choice of the suitable fabric properties to satisfy the required specification of the protective armor is that the specifications requirements are overlap but sometimes contradict. For example, the lightweight armor requirements are contradicting with its stab resistance, improving in one property may deteriorate the other.[18] There are several types of PFs to deal with the category of the protection designed for, as illustrated in Figure 3.3.

Optimization is a quantitative approach to produce overall best results by choosing the proper combinations of variables subjected to a set of constraints. The optimization of the fabric properties to satisfy the armor specifications can be achieved by applying one of the known models, such as regression analysis, linear programming, multiobjective optimization goal programming, and artificial neural network model.[19] Table 3.2 illustrates the main properties of the PF and the factors affecting it. Fabric structure may be 2D or 3D, manufactured on different weaving systems: Woven, knitted, braided, meltblow, spun bonded, needle bonded, chemical bonded, or thermal bonded nonwoven. The fiber properties are the common factor that defines all the PF properties.

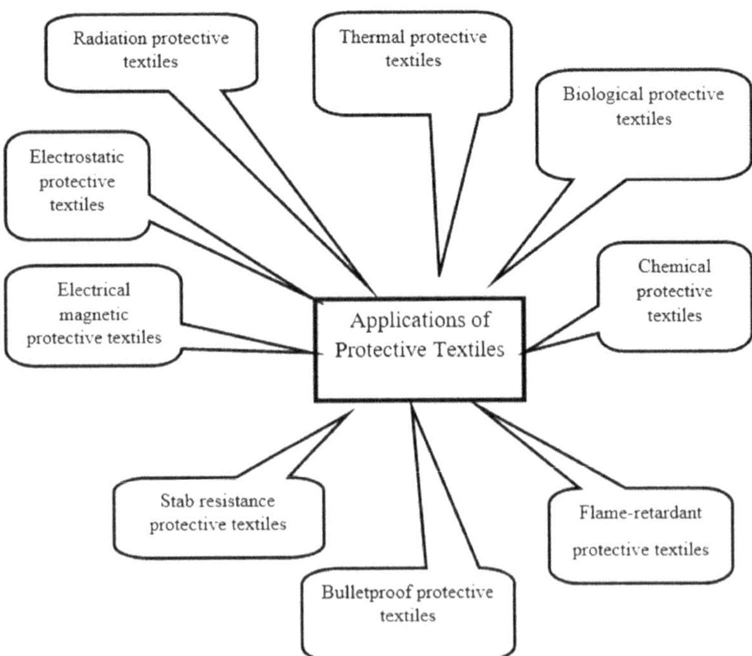

FIGURE 3.3 Classifications of different applications of protective textiles.

TABLE 3.2 Factors Affecting Fabric Properties.

Fabric property	Parameters
Tensile	Fabric structure and specifications, yarn tenacity
Extension	Fabric structure, yarn extension, fabric tightness
Bending	Fabric modulus of rigidity, fabric structure, and fabric tightness
Shear	Fabric shear modulus, fabric structure, fabric tightness, and yarn-to-yarn friction coefficient
Compression	Fabric structure, fiber volume fraction, and fabric tightness
Creep	Fabric structure, fiber volume fraction, fabric tightness, yarn specifications, and fiber creep property
Surface properties	Fabric coefficient of friction, surface roughness amplitude, and yarn specifications
Contraction	Fabric structure, fabric tightness, yarn specifications, and fiber properties
Fabric weight	Fabric structure, fabric density, and yarn count
Thermal property	Thermal conductivity, fabric tightness, yarn specifications (twist, count, yarn density), and yarn thermal properties
Wrinkle recovery	Fabric elastic recovery, stiffness, flexural rigidity, fabric tightness, yarn specifications, and yarn elastic recovery properties
Fabric porosity	Fabric structure and specifications, fabric density, fabric tightness, yarn specifications, fiber packing density, fiber diameter, method of fabric, and yarn formation
Fabric thickness	Fabric structure and specifications, fabric density, yarn specifications (twist, count, yarn density), and yarn structure
Drapability	Fabric initial tensile modulus, stiffness, fabric tightness, yarn specifications
Volume properties	Fabric areal weight, compressibility, thickness, fabric tightness, fabric design, yarn specifications, method of fabric, and yarn manufacturing
Water vapor transmission	Fabric porosity, fabric thickness, air permeability, fabric tightness, fabric structure, yarn structure, and specifications
Stab resistance	Fabric areal weight, fabric compressibility, fabric thickness, fabric structure, fabric shear modulus, fabric tightness, yarn specifications (twist, count, yarn density), and yarn cutting strength
Ballistic resistance	Fabric areal weight, fabric compressibility, fabric thickness, fabric structure, fabric shear modulus, fabric tightness, fabric tensile properties, and fabric specifications

TABLE 3.2 *(Continued)*

Fabric property	Parameters
Electrical conductivity	Fabric areal weight, thickness, fabric structure, yarn, and fiber resistance
Thermal conductivity	Fabric areal weight, thickness, fabric structure, air permeability, and yarn thermal conductivity
Flame retardant	Fabric area weight, thickness, fabric structure, fabric porosity, yarn, and fiber ignition temperature

3.5 PFs REQUIREMENTS

The essential parameters of protective fabric depend mainly on its end use and the different constraints of the surrounding environment. At the same time, the comfort and mobility of the wearer limit how long he/she can use it. Figure 3.4 gives the knowledge needed to satisfy the functionalities of PF to fulfill the design. For example, electromagnetic shielding PF is to limit the penetration of electromagnetic fields into space by blocking them with a barrier made of conductive material[20–22] Reflection and absorption are the main shielding mechanisms. Reflection results from the impedance mismatch between free space and shield. This phenomenon depends on the fiber material's conductivity, dielectric permittivity, and magnetic permeability.[23] In the case of bulletproof armor, the function of the PF is quite different. During the interval of impacting time, the value of punching resisting force of the bulletproof vest will increase till full penetration, and then it will gradually decline and vanish after a bullet passed or prevented from passing through the fabric. Most of the designs tend to use high tenacity yarns to stop bullet penetration.[24] Several synthetic-based PFs are commercially available for application in multi-layered panels designed to defeat high-velocity ballistic impacts of projectiles from handguns and high-powered rifles.[25] The yarn-to-yarn surface frictional force and that between the penetrating bullet and the yarns, the number of yarn intersections that contribute to the total frictional force resisting the penetration of the bullet, the structural parameters of a fabric, as well as the speed of propagation of strain in the fabric significantly affect its protective performance.[26, 27]

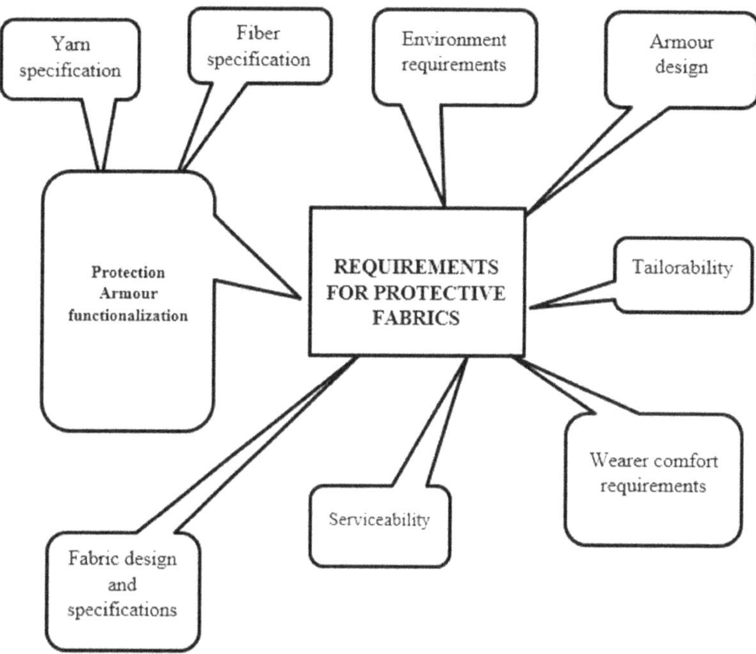

FIGURE 3.4 Requirements for protective fabrics.

The mechanism of the stab resistance of PF is different from that of the bulletproof PF. The depth of penetration of the stab resistance fabric by the sharp edge or spikes is one of the most essential parameters considered for estimation of the usage properties of fabrics designed for protection against various types of mechanical impact. Moreover, a soft vest should provide comfort for the user, as well as be of lightweight, and cost-effective. The puncture resistance is often calculated by taking an average value between the cutting and tearing or tensile force of the fabric.[28, 29] A fabric gripped at its edges and slashed with a knife may be cut in tension and shear. The fabric weave pattern, fabric tightness, fabric thickness, fabric specification, friction between the yarns and that between the penetrating tool and the yarns are the parameters that affect the stab resistance.[30, 31]

One of the oldest PFs used for the protection in the chemical industry and laboratories dealing with chemical materials and liquids is chemical PF. Liquid penetration is a process determined by both material and liquid variables, so including both parameters and quantifying their individual contributions to liquid penetration is essential in understanding how each of these factors contributes to the penetration process.[32] Interaction between liquid and fabric is of great importance in the liquid penetration process, which depends on repellency mechanism, which is driven mainly by the liquid medium surface interaction, air permeability, chemical liquid permeability. Nonwoven fabrics spunbonded, meltbown could be used as a chemical PF. Multilayer of different materials presents modern designs of the chemical protection armor.

Thermal PFs are used for clothing under the exposure of hot surface contact. In the last few decades, various thermal protective polymer-based fibers have been developed, like Nomex, Kevlar, Kermel, Lenzing, and polybenzimidazole (PBI), to assemble thermal protective clothing. Generally, a multilayered fabric system comprising thicker thermal liners has high thermal protective performance. Low air permeability, high areal density, high thickness, and high thermal resistance can possess a significant positive impact on the performance of thermal protective clothing.[33] Fabric thickness, bulk heat capacity, bulk conductivity, fiber to air ratio, and air void distribution become very characteristic parameters.[34] There are compelling indications that moisture is a critically important factor determining thermal protective insulation.[35, 36] These woven and/or nonwoven thermal PFs are used as raw materials in the garment manufacturing process to assemble firefighters' clothing.

Radiation shielding fabric, designed to facilitate attenuation of indirectly ionizing radiation, which includes neutrons, gamma rays, and X rays, will work as a barrier. It should have superior strength and modulus and low effect from the radiation on its properties. Generally, shielding reduces the intensity of radiation depending on the fabric thickness, for instance, textile base with a bilayer, nonlead coating opts the objective of achieving a 90% attenuation percentage.[37] A fabric, that provides X ray and gamma radiation protection intended for medical personnel, was developed with high-strength para-aramid fibers. Gamma radiation shielding effectiveness depends on the fabric fullness and its indenting and protruding structures.[38]

The best PF should have the optimized values of the above factors, which depend on the material and armor design. The fabric material properties should satisfy chemical stability, drapeability, water absorption, water repellency, crease recovery, resistance to microorganisms and fungus, cleaning ability, thermal, electrical, service life, and recyclability. The dilemma of fabric design for protective body vest arises when one adds to these general constructional concepts the variable details of yarn and fiber properties.[39] Today's marketplace contains both synthetic and natural fibers to perform a protective clothing function. Costs for protective clothing fabrics are related to all the above-mentioned factors. In all applications, the fabric should have excellent flexibility and workability.[40] Figures 3.5–3.11 illustrate the main requirements for some PFs.

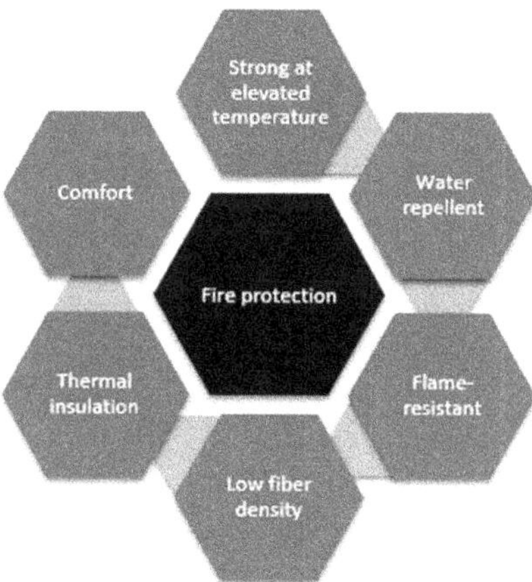

FIGURE 3.5 Fire-protective fabric.

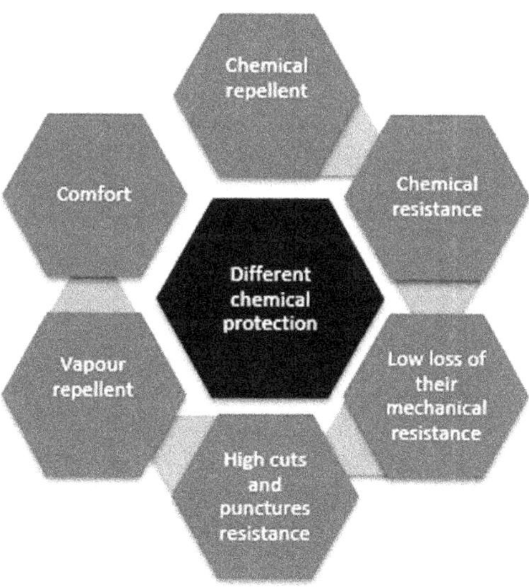

FIGURE 3.6 Chemical-protective fabric.

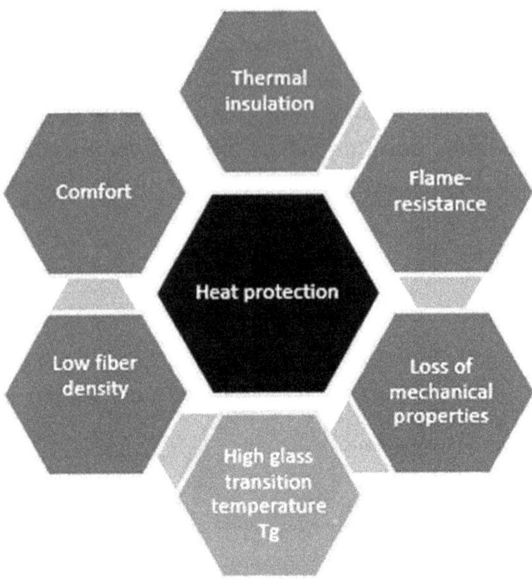

FIGURE 3.7 Thermal-protective fabric.

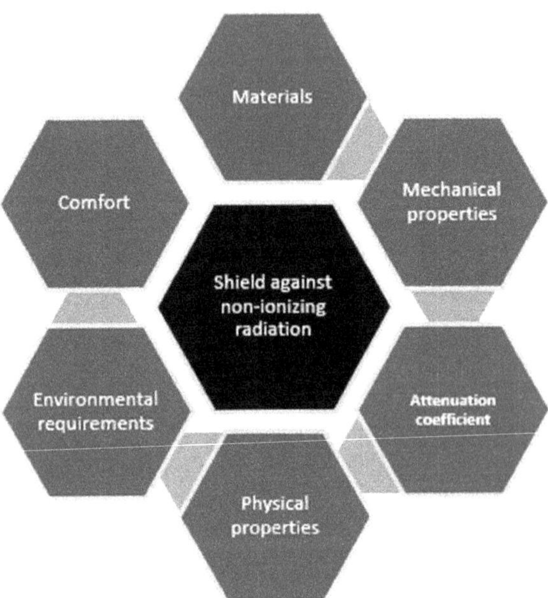

FIGURE 3.8 Radiation-protective fabric.

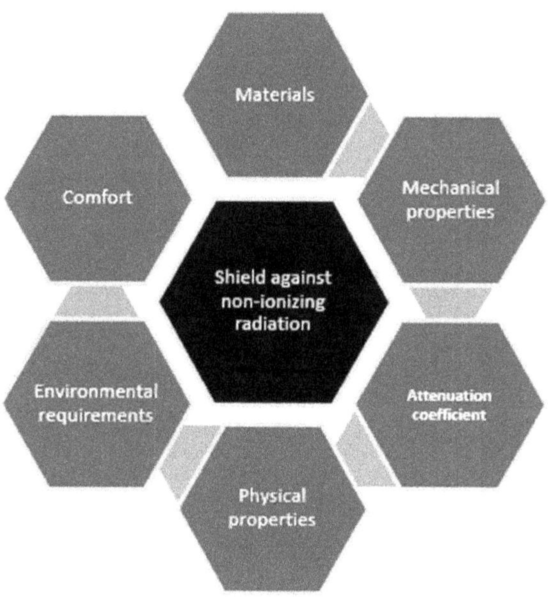

FIGURE 3.9 Electromagnetic-protective fabric.

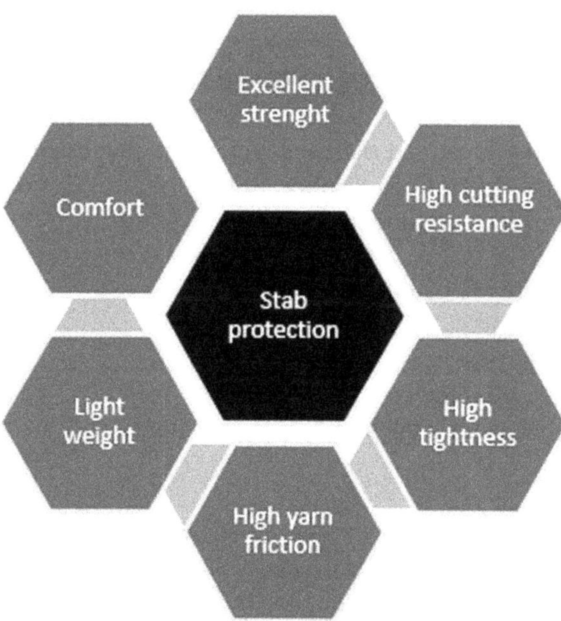

FIGURE 3.10 Stab-resistance fabric.

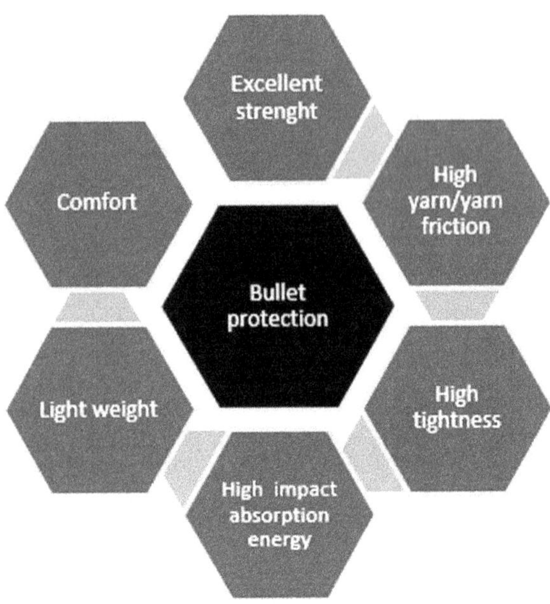

FIGURE 3.11 Bullet-protective fabric.

3.5.1 EXAMPLES OF TEXTILE ARMOR FOR DIFFERENT END USE

According to the threat and risk assessment, the requested parameters for each application of protective armor will be diverse. Besides that the PF itself should fulfill the following:

- Tailorability constraints,
- Weight limits,
- Degree of protection,
- Comfort parameters,
- Compliance with the standard requirements, and
- Cost

For instance, the fire fighter's armor should protect against such items as chemical splashes, hot metal splash, flame, thermal radiation, water penetration, heat, and at the same time be of lightweight and comfortable.

TABLE 3.3 Analysis of the Essential Parameters of Protective Armors.

Application	Police armor	Fire-fighters armor	Racing car drivers armor	Medical uniforms	Chemical armor	Radiation armor
Comfort	+++++	+++++	+++++	+++++	+++++	+++++
Functionality	+++++	+++++	+++++	+++++	+++++	+++++
Interaction with the working environment	+++++	+++++		+++++	+++++	+++++
Fashionable looks	+++	+++	+++	+++++	+++	+++
Fitted with built-in communication systems	+++++	+++++	++	++	+++	+++++
Chemicals splashes	+++++	+++++	+	0	+++++	0
Microorganisms	+	+	+	+++++	+++++	+
Hygiene	+	+	+	+++++	+	
Protection against hot metal splash	+++++	+++++	0	0	+++++	0
Electrostatic protection	+++++	+++++	0	0		
Stab resistance	+++++	+++++	++	+++	+++	++
Fire resistance	+++++	+++++	+++++	0	+++++	++
Flame resistance	+++	+++++	+++++	0	+++++	
Care and maintenance	+++++	+++++	+++++	+++++	+++++	+++++
Cost	++	++	++	+++	+++	++
Life cycle cost	++	++	++	+++	++++	++
Explosion resistance	+++++	+++++	+++++	0	+++++	+++
Electromagnetic	+	+	+	+++	++++	++++
Moisture management	+++++	+++++	+++++	+++++	+++++	++++
Tailorability	+++++	+++++	+++++	+++++	+++++	+++++
Mold resistance	+++++	+++++	+++++	+++++	+++++	+++++

0 Not at all important

+Slightly important, ++moderately important, +++very important, ++++extremely important.

Table 3.3 gives an evaluation of the importance of some parameters that define the overall effectiveness and serviceability of the textile armor.

3.6 SOME ASPECTS FOR FIBER SELECTION

The PFs provided for the variety of protective armors need different textile fibers to reflect this diversity. Apart from the direct performance of the PF, its impacts on the environment should be considered the life cycle analysis.[11] PF performance is a function of the following parameters:

Fiber parameters: Fiber material, physical properties (length, density, cross-section shape, fiber morphology) mechanical properties, strength, strain, elastic recovery, modulus, resilience, abrasion resistance, liquid absorption properties, thermal conductivity, heat resistant, chemical resistance, electrical resistance, etc.

Yarn parameters: Yarn structure, count, twist factor, number of fibers or filament in the cross-section, packing density, twist direction yarn bending stiffness, yarn strength, strain, and yarn cross-section shape.

Fabric parameters: woven fabric Weave design (plain, twill, satin, etc.), number of warp and weft /cm, cover factor, crimp %, thickness, 2D or 3D design, fabric porosity, surface texture, and frictional properties, areal weight, etc.

Nonwoven fabric: Method of manufacturing (needle punched, chemically bonded, thermally bonded, meltblown), thickness, porosity, fibrous structure of the nonwoven fabric, areal weight, etc.

Knitted fabric: Weave design, number of wales and courses per cm, loop shape factor, thickness, porosity, areal weight, etc.

The main criteria that the designer should take into consideration are the choice of the material to satisfy the product performance goals at minimum cost. The first step is to choose the suitable fiber through analyzing the required properties that are essential for the end users, followed by the construction of the fiber—PF properties matrix, as given in Table 3.4. The second step is to choose the fabric design and specifications which have a critical role in the final PF performance. The fiber cost plays an important role in the choice of the PF.

Applying the above matrix on the choice of materials for bulletproof PF, we must choose fibers with small diameter, low density, high strength, high modulus of elasticity, high energy dissipation, high coefficient of friction,

TABLE 3.4 Fiber-Protective Fabric Properties Matrix.

PF properties / Fiber property	Mechanical properties	Tailorability	Thermal properties	Weight	Sound absorption	Comfort
Diameter	+1[a]	+1	+1	0	+1	+1
Bending stiffness	−1[b]	−1	0[c]	0	+1	+1
Strength	+1	0	0	0	0	0
Modulus of Elasticity	+1	−1	0	0	+1	−1
Density	+1	0	0	+1	−1	−1
Cross section shape	+1	0	+1	0	−1	−1
Coefficient of friction	+1	0	0	0	+1	−1
Filament count (*den*)	−1	0	−1	0	−1	−1
Moisture absorption	+1	0	+1	0	0	+1
Thermal properties	+1	0	+1	0	0	+1
Cutting strength	+1	+1	0	0	0	0
Chemical resistance	+1	0	0	0	0	0
Sound absorption	0	0	0	0	+1	+1

[a]+1 positive correlation.

[b]−1 negative correlation.

[c]0 insignificant effect.

TABLE 3.5 Fiber Profile and Its Applications.

Fiber	Description and application	Molecular structure
Carbon	About 90% of the carbon fibers produced are made from polyacrylonitrile (PAN). The remaining 10% is made from rayon or petroleum pitch The properties of carbon fibers are high tenacity, modulus of elasticity, brittleness and low creeping tendency, heat expansion, and good electrical conductivity and chemical-resistant behavior. Used for carbon fiber-reinforced plastic composites. In the last decade, rapidly expanded into new uses for several industrial applications (aviation, automobile)	
Kevlar 129 *Kevlar 149* *Kevlar 29* *Kevlar 49* *Kevlar 68* *KM2*	*Para-aramid Kevlar*: It has high tenacity, tensile modulus, heat resistance, chemical resistance, toughness, cut resistance, flame resistance and low electrical conductivity, thermal shrinkage, thermal conductivity, and excellent dimensional stability. Several types are existing, each of different properties for a certain application. KM2 is designed as a material for bulletproof vests, helmets, ballistic vests. Kevlar is used in gloves, thermal insulation, heat shields, military vesicle armor, and in firefighter protective apparel	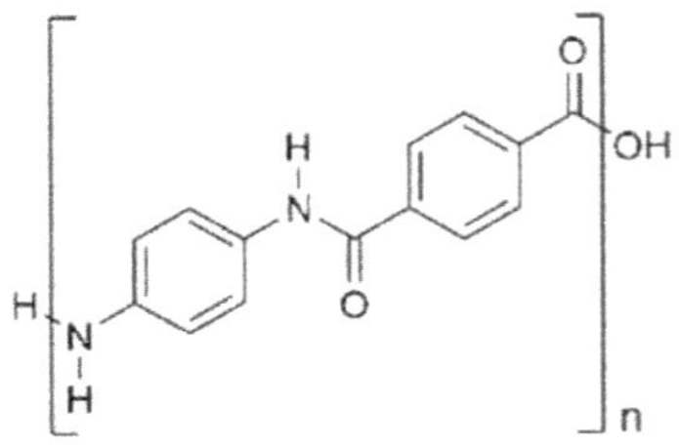 H, N, H, H, N, O, O, OH, n
Nextel 720 *Nextel 610* *Nextel 440* *Nextel 480* *Nextel 312*	Oxide ceramic fibers are mainly based on alumina Al_2O_3 and Silica SiO_2. The fibers retain their strength and modulus at higher temperatures up to 700°C, which meets demanding performance requirements in high temperature operating environments. Developed primarily for heat resistance applications and are recommended for the applications requiring thermal insulation. Very high-performance reinforcement materials for the composites, being used in aerospace structures	

TABLE 3.5 *(Continued)*

Fiber	Description and application	Molecular structure
Nomex	*Meta-aramid (Nomex)*: Poly m-phenyleneisophthalamide. Good heat resistance and strength, not ignite, melt or drip thus, better longstanding retention of mechanical properties at elevated temperatures. Good resistance to abrasion and cutting, heat and flame resistance, high ultraviolet resistance, chemical resistance, but low thermal shrinkage, breaking elongation, electrical conductivity. Used for Safety & Protective clothing, firefighter protective apparel, heat shields, flame-resistant clothing for protection where a flash fire or electric arc hazard	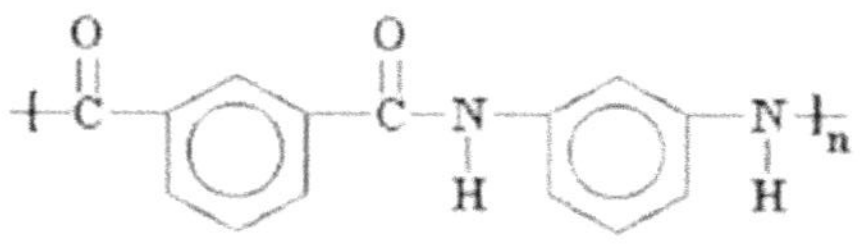
-Spectra 2000 *-Spectra S 1000* *-Spectra S 900*	Spectra fibers are ultra-high-molecular-weight polyethylene with high tenacity and modules and low density. Spectra fiber has a low coefficient of friction, low dielectric constant, exhibits high resistance to chemicals and UV. Used in military and police armor applications, flexible and ultra-lightweight panels, vehicle barriers, safety and protective fabric, and aerospace applications	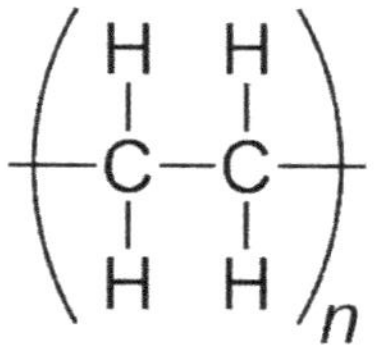
Vectran	*Vectran*: is a polyester-polyarylate fiber, spun from liquid crystal polymer. Vectran fiber is thermotropic, it melts at a high temperature (330°C). High strength and modulus, cut resistance, impact resistance, excellent chemical resistance and good chemical stability, and low creep. High properties retention over a wide range of temperature Used for airbags, for ropes, electrical cables, and high-performance composite material reinforcement for aerospace and military applications	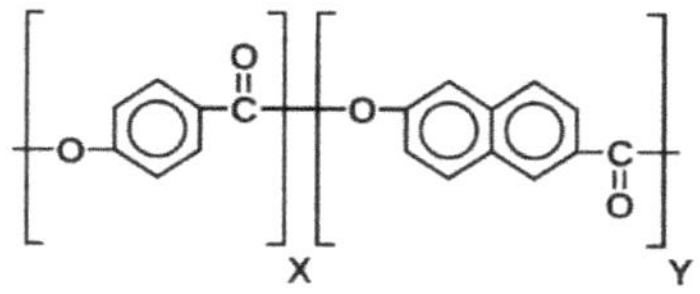
PBI	*PBI: Polybenzimidazole*: $(C_{20}H_{12}N_4)_n$ Excellent thermal resistant, flame retardant. Decomposition temperature 1300°C. Good chemical resistance, hig h moisture regain and low thermal conductivity. Used for safety and protective clothing and flame-retardant fabrics, thermal insulation. Firefighter thermal liner, astronaut space suits, gloves, hoods, and aircraft wall fabrics	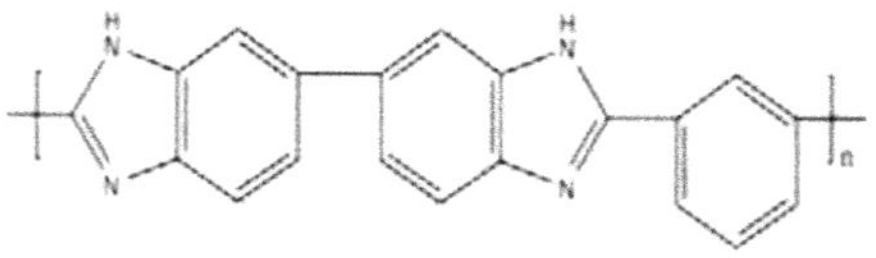

TABLE 3.5 *(Continued)*

Fiber	Description and application	Molecular structure
PBO (Zylon)	*Zylon*: poly(p-phenylene-2,6-benzobisoxazole. Zylon has the highest tensile strength and tensile modulus among high-performance fibers, outstanding thermal properties, good flame resistance, and good chemical resistance. Zylon is the material of choice due to its low weight, high tensile strength, and thermal properties. Used in safety and protective clothing, heat shields, flame retardant, ballistic applications	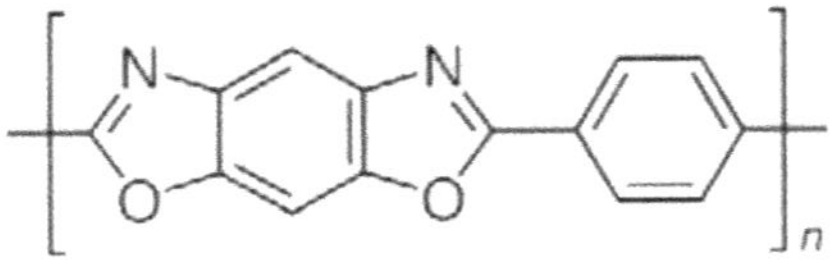
Dyneema	*Dyneema*: Ultra-high Molecular Weight Poly Ethylene (UHMWPE) fibers. Dyneema has the highest impact strength. Dyneema fiber is lower density than aramid fiber with improvement in cut resistance, puncture resistance and very resistant against chemicals. The strength and modulus increase at sub-ambient temperatures and decrease at higher temperatures, reaches 55% at 100˚C Used for ballistic helmets, vests inserts to protect against a wide range of ballistic threats, shields, vehicle armor, gloves, high-performance apparel and excellent materials for radomes	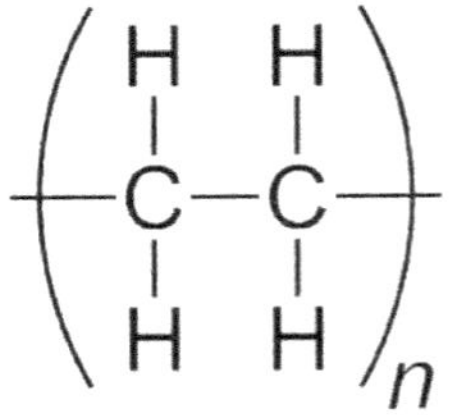
PAN	*Polyacrylonitrile PAN: (C3H3N)n*: The properties of the carbon fibers depend on the temperature of carbonization; the higher the temperature (2000˚C), the higher is the modulus, while at (1500˚C) is the highest strength, and low values of both properties are at (1000˚C). These fibers are strong, have high tensile strength, stiffness, chemical resistance, temperature resistance, fatigue strength and low weight, thermal expansion, and can be mixed with other materials. PAN fibers found various application, especially for high temperatures environment, and as reinforced composites material for aircraft, car, aerospace industries, and wind blade	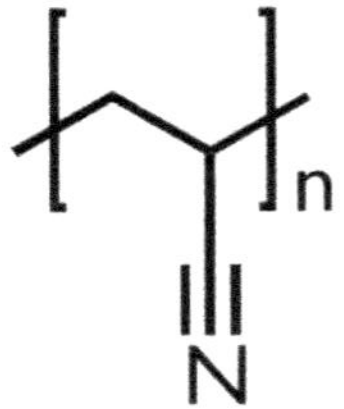

TABLE 3.5 *(Continued)*

Fiber	Description and application	Molecular structure
M5 (PIPD)[a]	*M5*: Fiber (polyhydroquinone-diimidazopyridine. Very high strength and modulus, higher compressive strength and excellent chemical and heat resistance. High energy absorption capacity. Low flammability. Suitable for bulletproof and stab-resistant light vest	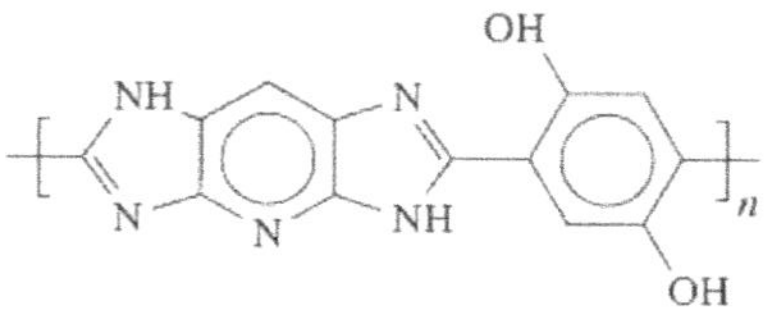 NH N N N NH OH OH n
Boron	*Boron*: Chemical Vapor Deposition (CVD) on a fine tungsten wire substrate is an inherently-brittle material. High values of density, strength, modulus and compression strength. The melting point 2000°C with low expansion coefficient. Used for composites for aircraft and space shuttle	
Nextel	*Nextel*: Ceramic Fibers. High strength and modulus, temperature resistance, fire, abrasion and chemical resistance. Used for protection against heat, flame and impact shields	

[a]Adapted with permission from Ref. [66].

Source: Adapted with permission from Refs. [41–67]. Armor must provide sufficient coverage to protect the five major organs in the torso, while still meeting the ergonomic needs of the wearer. This statement is true to protect the wearer from stab or to shoot from the gun or explosive fragments. While in the case of the bomb explosion it is not sufficient and full body area should be protected.

low filament count, high shear strength, good thermal properties, abrasion resistance, high energy absorption, and high moisture absorption. While for gloves material, the cutting strength, bending stiffness, heat and chemical resistance of the material will be the key attributes. Table 3.5 summarizes some properties and end uses of some of the high-performance fibers.

The weight of the armor governs its effectiveness at the same time restrict the movement of the wearer. Stab-resistant vest weight is 1.36–3.175 kg, for ballistic vest it is 3.175, besides the ballistic insert plate's weight of 1.18–4 kg. For explosive ordnance disposal (EOD) body armor, the weight is much higher to ensure complete protection from blast wave, fragmentation, fragmentation pulse, heat, and flames.[68] Figure 3.12 shows an example of a protective armor.[69–71]

The new generation protective armors for the soldiers should provide them with full spectrum protection system from the expected threats and environments; it should be designed to be chemical resistant, flame resistant, antimicrobial, antiexplosion, antistabbing, and bulletproof. Combination of stab and ballistic proof vests for the police force use is a new trend.

Bulletproof vest parts

FIGURE 3.12 (See color insert.) Protective armor.

Sunlight is another important factor affecting the material function when the materials are applied in the actual situation.

The above discussions indicate that the key attributes of the PF may be contradicting, and the designer needs an optimization technique to reach the best solution. Today, existing information and technology allows the interpreting of complex interactions between various properties of the fabric and final armor system performance as it occurs during the real service, in all cases, the armor should comply with the standards such as NIJ or Home Office Scientific Development Branch.[72–75]

KEYWORDS

- **protective fabric**
- **stab resistance**
- **bulletproof vest**
- **stab-proof vest**
- **radiation-protective fabric**
- **electrical-protective fabric**
- **fireproof fabric**

REFERENCES

1. Textiles for Protection at the Workplace: Developments in Textiles for a Safer Working Environment, saxion.nl/safety. https://www.saxion.nl/wps/wcm/connect/a3a7547e-19bb-4fa1-a2d3-a682b3e63f5d/Textiles+for+protection+at+the+workspace.pdf?MOD=AJPERES (accessed Dec 2, 2017).
2. Advanced Protective Gear and Armor Market Analysis, Market Size, Application, Analysis, Regional Outlook, Competitive Strategies and Forecasts, 2015–2022. GVR2338. http://www.grandviewresearch.com/industry-analysis/advanced-protective-gear-and-armor-market (accessed Dec 2, 2017).
3. Ballistic Protection Market by Type (Personal Protective Equipment, Vehicle Armor), Sub-Type (Soft Armor, Hard Armor, Personal Protective Head Gear), & by Application (Defense, Homeland Security, Commercial)–Global Forecasts & Analysis to 2014–2020. Report Code: AS 2798. http://www.marketsandmarkets.com/Market-Reports/ballistic-protection-market-30112278.html (accessed Dec 2, 2017).
4. AVM021G. http://www.bccresearch.com/market-research/advanced-materials/advanced-protective-gear-armor-avm021g.html (accessed Dec 2, 2017).

5. https://www.bizjournals.com/prnewswire/press_releases/2017/05/22/DE96550 (accessed Dec 2, 2017).
6. Ballistic protection material market. https://www.transparencymarketresearch.com/ballistic-protection-materials-market.html (accessed Dec 2, 2017).
7. SANS Institute InfoSec Reading Room. An Overview of Threat and Risk Assessment. https://www.sans.org/reading-room/whitepapers/auditing/overview-threat-risk-assessment-76 (accessed Dec 2, 2017).
8. Bell, M. Med-A Clinic Risk Assessment. Theses University of Maryland University College; 2016. https://megbell.io/wp-content/uploads/2017/05/MBell-Example-Case-Study-Risk-Assessment-for-Health-Care.pdf?189db0 (accessed Dec 2, 2017).
9. Mansdorf, Z. Occupational Hygiene Assessments for the Use of Protective Gloves. In *Protective Gloves for Occupational Use,* 2nd ed; Boman, A., Estlander T., Wahlberg, J. E., Maibach, H. I. Eds.; CRC Press, USA, 2005, pp 5–14.
10. Dolez, P.; Soulati, K.; Gauvin, C.; Lara, J.; Vu-Khanh, T. Information Document for Selecting Gloves for Protection against Mechanical Hazards. Studies and Research Projects, Technical Guide RG-738. https://www.irsst.qc.ca/gants/en/InfoDocu.pdf (accessed Jan 3, 2018).
11. El Messiry, M. *Natural Fiber Textile Composite Engineering*; Apple Academic Press Inc: USA, 2017.
12. Nathan, J. Body Armor for Law Enforcement Officers: In Brief. Congressional Research Service 7-5700 www.crs.gov R43544. https://fas.org/sgp/crs/misc/R43544.pdf (accessed Jan 3, 2018).
13. Struszczyk, H. Risk Analysis in Designing of Body Armor. *Tech Text Prod* (Techniczne Wyroby Włókiennicze) 2009, 66–73. file:///C:/Users/magdy/Downloads/httpwww_moratex_euplikitww200923tww20092312-riskanalysisindesigning.pdf (accessed Jan 3, 2018).
14. American Chemical Society. Risk Rating and Assessment. https://www.acs.org/content/acs/en/about/governance/committees/chemicalsafety/hazard-assessment/fundamentals/risk-assessment.html (accessed Jan 3, 2018).
15. Risk Management Toolkit. http://www2.mitre.org/work/sepo/toolkits/risk/StandardProcess/definitions/occurence.html (accessed Jan 4, 2018).
16. Pamela, E.; Lansdowne, Z. Risk Matrix User's Guide, Version 2.2, Document MP 99B0000029, Nov 1999.
17. Garvey, R. Implementing a Risk Management Process for a Large-Scale Information System Upgrade: A Case Study. *INCOSE Insight* **2001,** *1*, 1–12.
18. Adanur, S.; Tewari, A. An Overview of Military Textiles. *IJFTR* **1997,** *22*, 348–352.
19. Majumdar, A.; Singh, S. Modelling, Optimization and Decision-Making Techniques in Designing of Functional Clothing. *IJFTR* **2011,** *36*, 398–409.
20. Maity, S.; Singha, K.; Debnath, P.; Singha, M. Textiles in Electromagnetic Radiation Protection. *J. Safety Eng.* **2013,** *2* (2), 11–19.
21. Koprowska, J.; Pietranik, M; Stawski, W. New Type of Textiles with Shielding Properties. *Fibres Text. East. Eur.* **2004,** *12*, 3, 47, 39–42.
22. Chen, H.; Lee, K.; Lin, J.; Koch, M. Fabrication of Conductive Woven Fabric and Analysis of Electromagnetic Shielding Via Measurement and Empirical Equation. *J. Mater Process. Technol.* **2007,** *184*, 124–130.

23. Hoghoghifard, S.; Mokhtari. H.; Dehghani, S. Improving EMI Shielding Effectiveness and Dielectric Properties of Polyaniline Coated Polyester Fabric by Effective Doping and Redoping Procedures. *J. Ind. Text.* **2018,** *47* (5), 587–601.
24. El Messiry, M.; El-Tarfawy, S. Performance of Weave Structure Multi-Layer Bulletproof Flexible Armor. The 3rd Conference of the National Campaign for Textile Industries NRC, Cairo, March 9–10, 2015.
25. Sinnppoo, K.; Arnold, L.; Padhye, R. Application of Wool in High-velocity Ballistic Protective Fabrics. *Text. Res. J.* **2010,** *80* (11), 1083–1092.
26. El Messiry, M.; Ibrahim, S. Investigation of Sonic Pulse Velocity in Evaluation of Knitted Fabrics. *J. Ind. Text.* **2016,** *46* (2), 455–472.
27. Liu, Y.; Hu, H.; Man Au, W. Protective Properties of Warp-Knitted Spacer Fabrics Under Impact in Hemispherical Form. Part II: Effects of Structural Parameters and Lamination. *Text. Res. J.* **2014,** *84* (3), 312–322.
28. Choi, N.; Hong, T.; Lee, E.; Paik, J; Yoon, B.; Lee, S. Stab Resistance of Aramid Fabrics Reinforced with Silica STF. The 18th International Conference on Composite Materials (ICCM18), Aug 21–26, 2011, Jeju Island, South Korea.
29. Militky, J. Cut Resistance of Textile Fabrics. Selected Topics of Textile and Material Science. TUL FT Liberec, 2011.
30. Kothari, V. Cut Resistance of Textile Fibres: A Theoretical and Experimental Approach. *Ind. J. Fibre Text. Res.* **2007,** *32*, 306–311.
31. El Messiry M. Investigation of Puncture Behavior of Flexible Silk Fabric Composites for Soft Body Armor. *Fibres Text. East. Eur.* **2014,** *22, 5* (107), 71–76.
32. Lee, S.; Obendorf, K. A Statistical Model to Predict Pesticide Penetration through Nonwoven Chemical Protective Fabrics. *Text. Res. J.* **2001,** *71* (11), 1000–1009.
33. Mandal, S.; Song. G. Characterizing Thermal Protective Fabrics of Firefighters' Clothing in Hot Surface Contact. *J. Ind. Text.* **2018,** *47* (5), 622–639.
34. Shalev, I.; Barker, R. Protective Fabrics: A Comparison of Laboratory Methods for Evaluating Thermal Protective Performance in Convective/Radiant Exposures. *Text. Res. Inst.* **1984,** 648–654.
35. Lee, Y.; Barker, L. Effect of Moisture on the Thermal Protective Performance of Heat-Resistant Fabrics. *J. Fire Sci.,* **1986,** 4, 315–331.
36. Mandal.S.; Song, G. Characterizing Thermal Protective Fabrics of Firefighters' Clothing in Hot Surface Contact. *J. Ind. Text.* **2018,** *47* (5), 622–639.
37. Pulford, S.; Fergusson, M. A Textile Platform for Non-Lead Radiation Shielding Apparel. *J. Text. Inst.* **2016,** 107, (12), 1610–1616.
38. Zdemir, H.; Camgo, B. Gamma Radiation Shielding Effectiveness of Cellular Woven Fabrics. *J. Ind. Text.* **2018,** *47* (5), 712–726.
39. Holmberg, R. Composite Fabrics for Protective Clothing. *JCTR* **1988,** *18*, 64–70.
40. Adanur S. *Military and Defense Textiles. Wellington Sears*; CRC Press: Pennsylvania, USA, 1995.
41. Hearle, J. (Ed.) *High-Performance Fibers;* Woodhead Publishing Limited in Association with the Textile Institute: UK, 2001.
42. Committee on Opportunities in Protection Materials Science and Technology for Future Army Applications; National Research Council. Opportunities in Protection Materials Science and Technology for Future Army Applications. National Academies Press, 2011. High-Performance Structural Fibers for Advanced Polymer Matrix Composites.

43. High Performance and High Temperature Resistant Fibers: Emphasis on Protective Clothing. http://www.intexa.com/downloads/hightemp.pdf (accessed Jan 2, 2018).
44. Huang, X. Fabrication and Properties of Carbon Fibers. *Mater* **2009,** *2*, 2369–2403.
45. Nextel ceramic fibers. Nextel, Ceramic Textile Technical Notebook. http://solutions.3mdeutschland.de/3MContentRetrievalAPI/BlobServlet?locale=de_DE&lmd=1270717655000&assetId=1258565166920&assetType=MMM_Image&blobAttribute=ImageFile (accessed Jun 6, 2017).
46. 3M™, Nextel™ Ceramic Structural Fabrics 610 and 720. http://multimedia.3m.com/mws/media/1242134O/3m-nextel-ceramic-structural-fabrics-610-720.pdf (accessed Jun 6, 2017).
47. Fareed, A. S. Microstructure and Properties of Nextel 610 Fiber Reinforced Ceramic and Metal Matrix Composite. 18th Annual Conference on Composites and Advanced Ceramic Materials-A, Wachtman, J. (Ed.).
48. Nomex. Technical Guide for NOMEX® Brand Fiber. Technical Information. http://www.nakedwhiz.com/gasketsafety/nomextechnicalguide.pdf (accessed Jun 6, 2017).
49. Polyethylene low density. http://www.plasticmoulding.ca/polymers/polypropylene.htm (accessed Jun 6, 2017).
50. Aniołczyk, H.; Koprowska, J.; Mamrot, P.; Lichawska, J. Application of Electrically Conductive Textiles as Electromagnetic Shields in Physiotherapy. *Fibres Text. East. Eur.* **2004,** *4* (48), 47–50.
51. 3M™ Nextel™ Ceramic Fibers and Textiles for Aerospace. https://www.3m.com/3M/en_US/company-us/all-3m-products/~/All-3M-Products/Advanced-Materials/Advanced-Ceramics/Ceramic-Fibers-and-Textiles/?N=5002385+5581327+8710684+8711017+8735201+8745513+3294857497&rt=r3 (accessed Jan 7, 2018).
52. Honeywell Spectra. https://www.honeywell-spectra.com/products/fibers/ (accessed Jan 7, 2018).
53. Marissen, R.; Nelis, M.; Janssens, M.; Meeks, M.; Maessen, J. A Comparison Between the Mechanical Behavior of Steel Wires and Ultra High Molecular Weight Poly Ethylene Cables for Sternum Closure. *Mater Sci. Appl.* **2011,** *2*, 1367–1374.
54. ZYLON®(PBO fiber) Technical Information (2005). http://www.toyobo-global.com/seihin/kc/pbo/zylon-p/bussei-p/technical.pdf (accessed Jan 8, 2018)
55. Marissen. Designwith Ultra Strong Polyethylene Fibers. *Mater Sci. Appl.* **2011,** *2* (5), 319–330.
56. Ultra-High Molecular Weight Polyethylene Fiber from DSM Dyneema. file:///C:/Users/magdy/Downloads/Dyneema_Folder_201611_Gruschwitz_19723.pdf (accessed Jan 8, 2018).
57. Honeywell Spectra® Fiber. https://www.honeywell-spectra.com/?document=fiber-capability-guide&download=1 (accessed Jan 8, 2018)
58. Synthetic Fibers and Fabrics Information. http://www.globalspec.com/learnmore/materials_chemicals_adhesives/composites_textiles_reinforcements/synthetic_fibers_fabrics_polymer_textiles (accessed Jan 2, 2018).
59. Aramid Fibers. http://www.chem.uwec.edu/Chem405_s01/malenirf/project.html (accessed Jan 2, 2018).
60. What Are Aramid Fibers? https://www.ipitaka.com/blogs/news/what-are-aramid-fibers-i (accessed Jan 9, 2018).

61. Vectran Engineering Data. http://www.swicofil.com/vectranengineering.pdf (accessed Jan 7, 2018).
62. Global Carbon Fiber Composites Supply Chain Competitiveness Analysis. https://www.nrel.gov/docs/fy16osti/66071.pdf (accessed Jan 9, 2018).
63. PAN-based Carbon Fiber & Composite Materials. https://www.m-chemical.co.jp/en/products/departments/mcc/cfcm/product/1201230_7502.html (accessed Jan 9, 2018).
64. Chemical Formula of Fibers. https://en.wikipedia.org/w/index.php?search=chemical+formulas++of+fibers&title=Special:Search&profile=default&fulltext=1&searchToken=6v08986ehkvdlhzrvm2hgel66 (accessed Jan 9, 2018).
65. Kumbasar, E. Protective Functional Textiles. MDT 'Protective textiles'. http://www.2bfuntex.eu/sites/default/files/materials/PROTECTIVE%20FUNCTIONAL%20TEXTILES_%20PERRIN.pdf (accessed Jan 12, 2018).
66. Chemical Structure of Polyhydroquinone Diimidazopyridine (M5). https://commons.wikimedia.org/w/index.php?curid=19313119 (accessed Jan 12, 2018).
67. Nomex. https://en.wikipedia.org/wiki/Nomex (accessed Jan 14, 2018).
68. Body Armor and Accessories for Military Forces. http://www.bulletproofvests.com/vests.html (accessed Jan 12, 2018).
69. Bomb Suit. https://en.wikipedia.org/wiki/Bomb_suit (accessed Jan 12, 2018).
70. Bulletproof Vest. https://en.wikipedia.org/wiki/Bulletproof_vest#Stab_and_stab-ballistic_armor (accessed Jan 12, 2018).
71. Body Armor Performance Standards. https://nij.gov/topics/technology/body-armor/Pages/standards.aspx (accessed Jan 12, 2018).
72. Compliant Armor: Body Armor that Complies with NIJ Stab Resistance Standard, 0115.00. https://nij.gov/topics/technology/body-armor/pages/compliant-stab-armor.aspx (accessed Jan 12, 2018).
73. Compliant Armor: Body Armor that Complies with NIJ Ballistic Resistance Standard, 0101.06. https://nij.gov/topics/technology/body-armor/pages/compliant-ballistic-armor.aspx (accessed Jan 12, 2018).
74. Croft, J.; Longhurst, D. HOSDB Body Armor Standards for UK Police (2007). Part 3: Knife and Spike Resistance. http://www.bodyarmornews.com/wp-content/uploads/2013/10/HOSDB__2007_-_part_3.pdf (accessed Jan 12, 2018).
75. Croft, D; Longhurst, D. HOSDB Body Armor Standards for UK Police (2007). Part 2: Ballistic Resistance. http://images.vestguard.co.uk/resources/1364486403_72.pdf (accessed Jan 12, 2018).

PART III

Protective Armor Design Aspects

CHAPTER 4

Stab Resistance of Flexible Protective Armor

ABSTRACT

In this chapter, the models for the penetration mechanism of a puncher blade cutting objects such as knives, spikes, and needles into a fibrous material in a form of mat or multi layers of fibrous mat and fabric are discussed. For the stabbing by knives, the anatomy of knives was examined. The mechanism of the stab resistance with a sharp object is found to differ from that of the punching mode. The mathematical analyses of the models for fibers, fiber mat, and fabrics for the different scenarios are given. The determination of the yarn and fabric stabbing resistance by sharp blade was discussed. The mechanism of fabric dynamic stab resistance, applying the energy balance principles, are theoretically supported. The experimental trials for different fabric structures are evaluated.

4.1 INTRODUCTION

The principal objective of the protective armors (PA) is to shield the wearer from the various threats that he or she may encounter and ensure comfort. However, there is an interaction between threat type and protective armor which have a reflection on the wearer, such as thermal stress, fatigue stress, movement disorder stress, deep injury, bruises, or trauma. Figure 4.1 represents the interaction between wearer's body and threat in the presence of protective armor. Obviously, protective armor system cannot protect against all levels of existing threats. The armor performance, usually, complies with a certain level of protection necessary for the purpose it is designed for. For stab resistance protective armor, materials, and armor system should be adapted to the standards on the level of knife or spike stabbing energy, such as NIJ Standard-0115.00, while the ballistic armors are referred to NIJ Standard-0101.06 for the bullet muzzle energy.

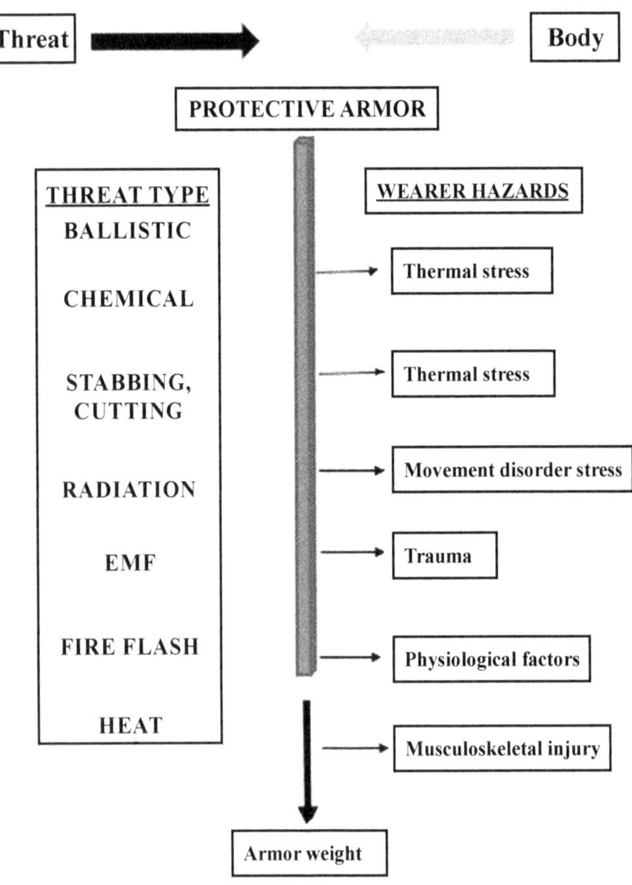

FIGURE 4.1 (See color insert.) The interaction between body and threat.

The PA designs are reliant on the fiber's type, structure of the fabric weave, number layers and their layout in the armor cross-section. Table 4.1 gives the SWOT analysis of the PA, their strength, weakness, and opportunities. The design of the PA may be one-piece or a conjunction armor system designed with a separate flexible armor panels. Some protective armors are designed to resist stab, spike, or cutting threats and others are designed as bulletproof. Protective armors must provide enough mobility and comfort to allow prolonged use.[1]

4.2 BEHAVIOR OF FABRIC UNDER PUNCHING FORCE

4.2.1 INTRODUCTION

Flexible protective fabrics (FPF) found their wide application in the industrial, medical, military, and civilian fields. However, continuous efforts have been applied to develop high-performance fibers for lightweight soft protective fabrics. Cutting and stabbing performances of the protective fabrics are essential parameters for the assessment of properties of the fabrics designed for protection against various types of mechanical impact. Soft armor should afford comfort for the wearer, lightweight, and cost-effectiveness. Stab resistant armor has several deficiencies, such as heavy weight, bulky shape, and uncomfortability,[2] especially female armor which may hinder her mobility.[3,4] Stab-resistant protective armor is designed to deal with edged blade tools, knife, or spike.[5] Within each of these categories, there are three levels of protection from low to high based on the energy that would impact the protective armor during an attack (NIJ Standard-0115.00). Cutting resistance is among the major mechanical properties required in protective clothing.[5]

Stab threats can be punching or cutting. Knife threats are, generally, more problematic to stop than punching ones due to the knife cutting edge continuous damage that initiated after the stab. The fabric punching strength is lower than the tensile strength or fabric biaxial tensile strength. The stabbing strength of the fabric is even much lower.[6] Consequently, for stab protective armor, different materials are used with high tenacity, tear, and cutting stress besides stiffness and shear. Many researchers[7–12] revealed that the fabrics of high cutting resistance and high strength forming multi-layer of different weave structure types enhanced the mechanism of stab resistance.

TABLE 4.1 SWOT Analysis of Protective Armors.

STRENGTHS	WEAKNESSES
• Global concern toward personal safety • High demand of protective armor • Increase of threats in different areas of daily life • Low percentage of injury • A strong community involvement with programs of safety of working personals • An increasing level of R&D and innovation within the companies • Strong encouragement for personal protective equipment	• Incomplete protection cost • Trauma effects • Thermal stress • Musculoskeletal injury • Alteration movement patterns of the wearer • Limitation of the weight of additional equipment wearers are required to carry • The ease of care
OPPORTUNITIES	**THREATS**
• Increase demand for protection of soldier in the battle fields • Progress of the HPF science and technology • Increase the awareness of the industrial protection in the work place • Progress of new armor designs • New high-performance fibers • Huge potential in innovations • Reduction in HPF cost • High research investment • Quick growth of technical textile production	• Increasing emphasis on recyclability • Cost • Weight–performance relation • Continuous increase of the capability of modern weapons • Increase of the weight of additional equipment soldiers are required to carry • High energy demand of HPF

The scenarios of penetration mechanism into textile material are:

- stabbing with sharp blade (knife) at high speed;
- penetration of sharp blade at low speed; and
- cutting with sharp blade.

Each case has the fabric resistance force for blade penetration and it relies on the blade speed of penetration and the construction of the punched textile. There are different types of fabric structures[13]:

- 2D structures (woven fabric, knitted fabric, nonwoven fabric, polymeric sheets, triaxial fabric, etc.)

- 3D fabric (3D woven fabric, 3D knitted fabric, multiaxial fabric, stitch bond fabric, Mali watt fabric, spacer fabric, and triaxial braided fabric).

Figure 4.2 illustrates the morphology of some textile structures.

For various weave structures, the punching manner by any shaped tool is different. The anti-stabbing protective fabric (ASPF) should possess the punching force greater than the force required to penetrate different tissues of the human, such as muscles, skin, soft tissue, cartilage, rib bone, and spine. The collagen fibers, which are almost inextensible, have strength between 147 and 343 MPa, and strain at rupture from 35% to 115%, depending on the age and protected area, and the modulus of elasticity from 15 to 150 MPa.[14] Thus, the forces to stab the skin have a very low value. It was revealed that the overarm stabbing speed is 10 m/s, depending on method of stabbing and the relative speed between walk-on and stab speeds. The peak punching force can reach a value of 1990–4740 N.[15] Manner in which the knife is held, directly influences the maximum speed generated during stabbing.[14] The protective armor should competently protect the wearer from the penetration of the sharp tools under such forces.

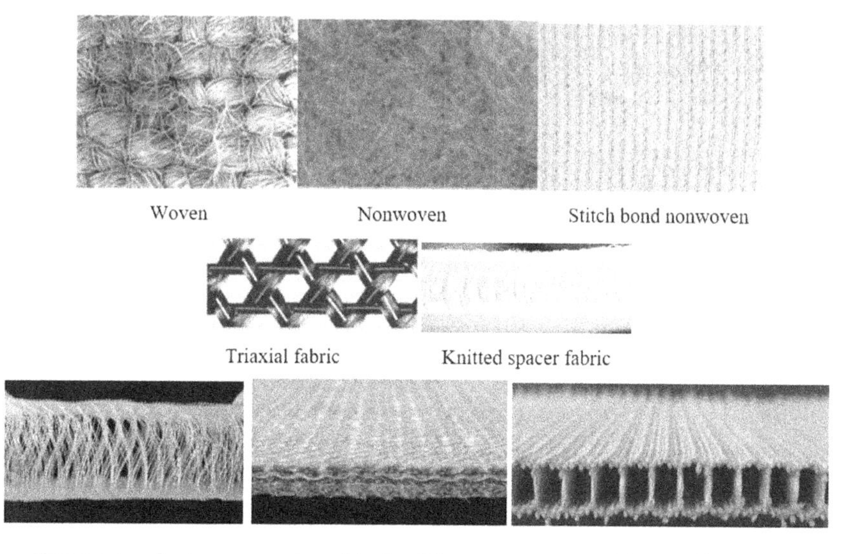

FIGURE 4.2 Fabric types.

4.2.2 ANALYSIS OF PENETRATION MECHANISM OF FIBROUS MATERIAL

4.2.2.1 PUNCHING OF FIBROUS NONWOVEN MAT

As it is clear from Figure 4.3, the structure of fibrous assemblies consists of fibers arranged in the 3D to fill the volume of the mat, in which the fibers may be oriented randomly or aligned in the specific form.

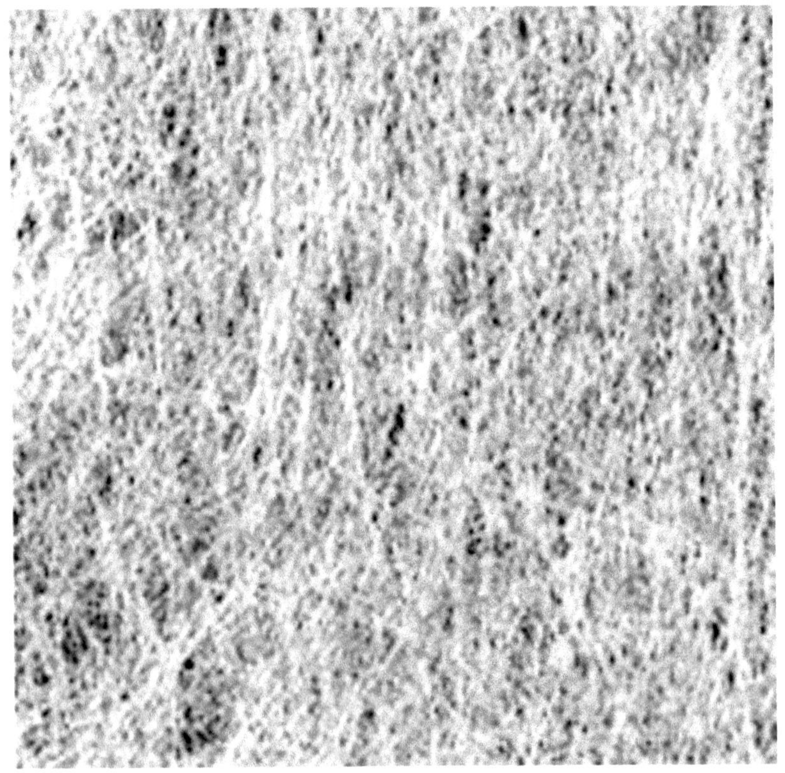

FIGURE 4.3 Fibrous nonwoven mat.

The packing density ϕ, which represents the fiber assembly compactness, is expressed as:

ϕ = (fiber volume/total fiber assembly volume)

The packing density for the yarns is 0.5–0.7, 0.03 for sliver, 0.15–0.3 for woven fabric, and 0.05–0.12 for nonwoven fabric.

The punching force of fibrous assembly is the force needed for the penetration of the puncher blade through the fibrous structure. Its components are as follows:

1. compactness resistance force;
2. puncher to fiber coefficient of friction;
3. fiber cutting force; and
4. fiber-to-fiber coefficient of friction.

The sketch of the blade penetration through fibrous assembly (Fig. 4.4) indicates that the fibrous structure compacted under the pressure of the puncher, increases the packing density of the mat under the compacting pressure till the thickness is almost constant (Fig. 4.5).

The fibrous mat thickness under the applied load relies on the type of fiber, the mat areal density, and the mat structure. Figure 4.6 represents this relation for different fibers. The puncher will continue its pressure to move the fiber out the fibrous structure. Mechanical failure begins when randomly aligned fibers structure moves at the cohesion points junctions between fibers in each layer and between the layers to be aligned into the applied force direction, the fibers are sliding over each other and some of the fibers are locking due to their entanglement with the surrounding fibers. The fibers packing density $\phi = V_f / V_c$ is changed. Moreover, the points of contacts of the fibers are tremendously high due to the low fibers diameters.[16] The penetration force of the fibrous structure is a function of fiber strength, fiber diameter, fiber surface morphology, fiber orientation factor, and the cohesion frictional forces. The number of fibers contacts is contingent with fiber length "l," fiber diameter "d_f," and packing density ϕ.[17] Theoretically, the contacts number of fibers in the structure will be:

$$\text{contact point per mm}^3\ (v) = 1.6227(\phi^2/d_f^3). \tag{4.1}$$

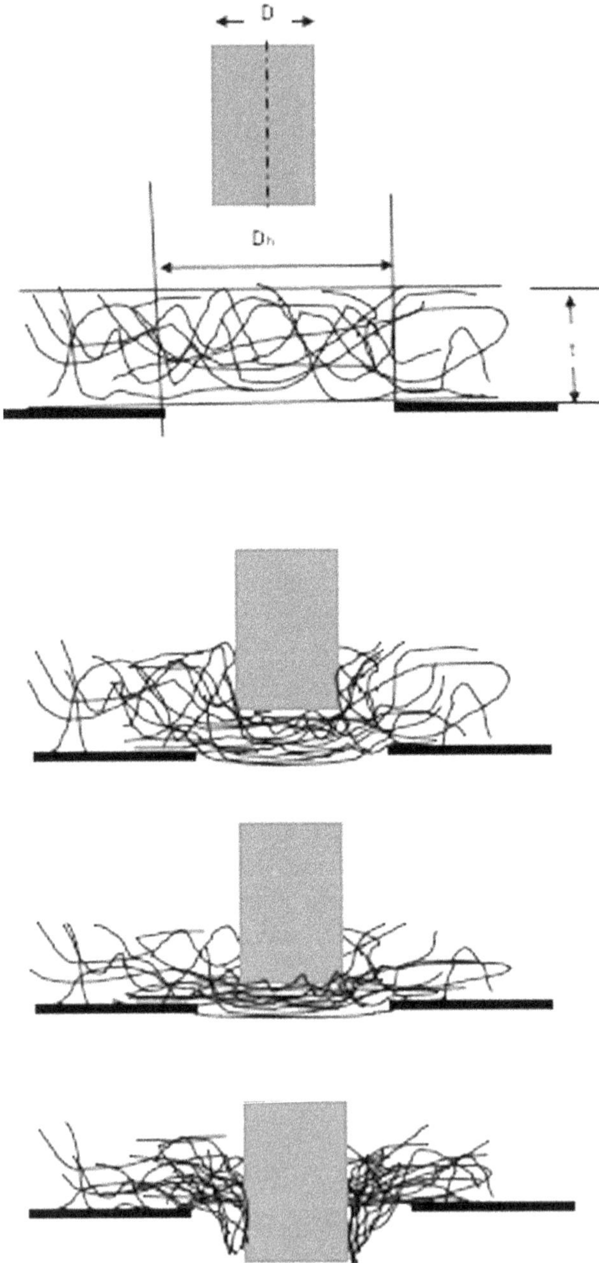

FIGURE 4.4 An illustration of the blade penetration through of the fibrous material.

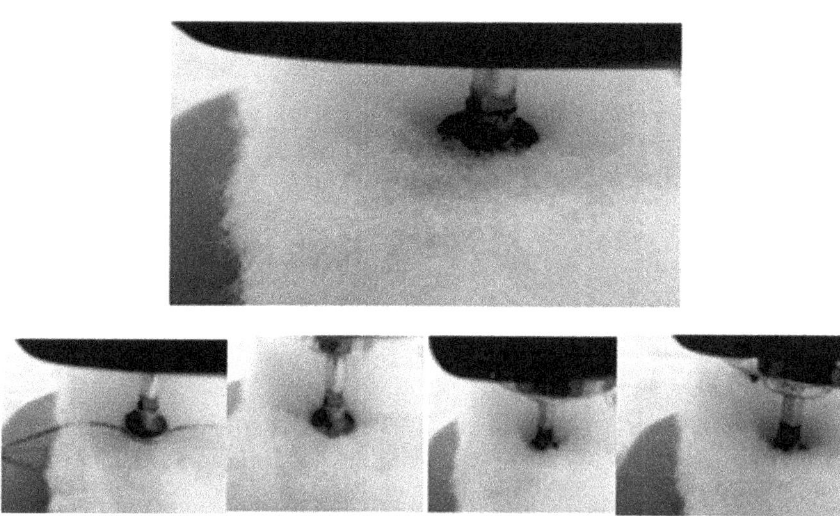

FIGURE 4.5 (See color insert.) Compactness of fibrous structure.

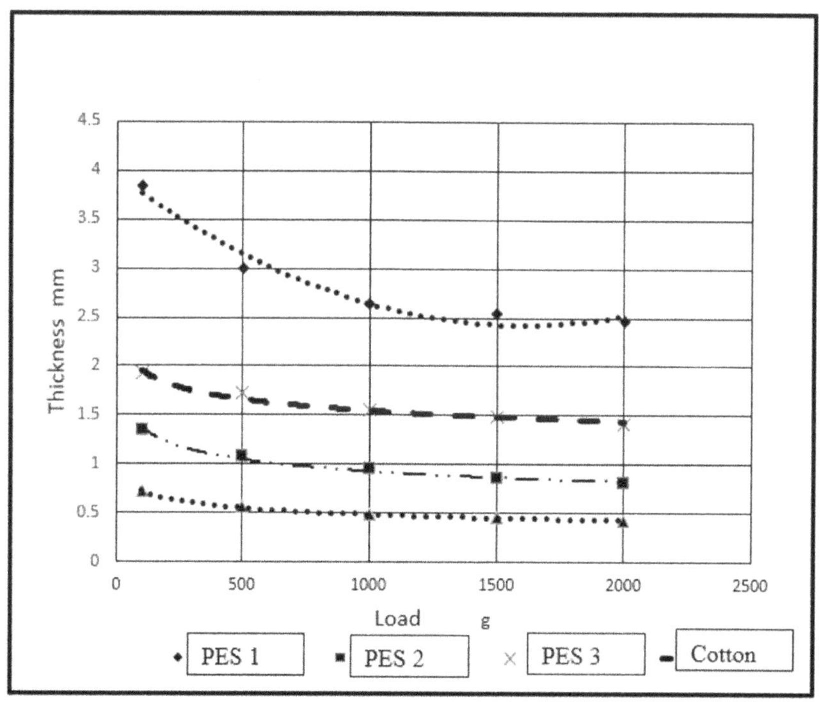

FIGURE 4.6 (See color insert.) Thickness–load curve of some types of fibrous structures.

For convenience, the symbols used throughout are listed in Table 4.2.

TABLE 4.2 List of Symbols and Abbreviations.

Symbol	Designation
HPF	high-performance fiber
Φ	packing density
V_f	fiber volume
Vc	fibrous mat volume
d_f	fiber diameter
P_{fiber}	fibers punching force
$W_{punched\ mat}$	weight of the mat
W_{fiber}	weight of a single fiber of length equal to the diameter of the puncher blade tip
H	fiber orientation factor
C_{fiber}	single fiber strength
K	percentage of failed fibers
$E_{punching}$	total energy of penetration
$E_{cutting}$	energy of fiber cutting
$E_{friction}$	friction energy
$E_{tension}$	energy to strain the fibers
$E_{compacting}$	energy to compact the fibers mat
P_{fabric}	fabric punching force
P_{fibers}	fibrous mat punching force
N_{warp}	number of warps/cm
N_{fill}	number of weft/cm
CT_{warp}	cutting force of warp yarn
CT_{fill}	cutting force of weft yarn
μ_1	blade–fiber coefficient of friction
μ_f	fiber coefficient of friction
N_1	normal force on the blade surface
N_f	normal force on the fibers
S_i	total path length of the fiber
Z_1	length of the blade path
σ_c	compressive stress
F_c	cutting force

TABLE 4.2 *(Continued)*

Symbol	Designation
d_{tip}	diameter of the blade tip
Z	geometrical factor
UT	tearing energy (lateral strain energy of the fiber)
UC	shear energy (associated with the blade cutting force)
FV	fiber cutting force
TA	axial tension force
G_f	fiber modulus of rigidity
E_f	fiber modulus of elasticity
A_f	fiber cross section
D_h	diameter of the gripped fabric area
N_{fill}	number of the weft yarns in contact with the plunger
N_{warp}	number of the weft yarns in contact with the plunger
W_1	number warp per cm
W_2	number weft per cm
D	diameter of the plunger
ε_{warpi}	strain of *i*th warp yarn
E_{warp}	warp yarn Young's modulus
E_{weft}	weft yarn Young's modulus
A_{warp}	area of the warp yarn cross section
A_{fill}	area of the weft yarn cross section
m, n	number of warp and weft yarns involved in the resistance of punching
θ_i	angle of contact between warp and weft yarns at the intersection points
FS	fabric deformation force under puncher's downwards movement
$F_{p\,max}$	fabric punching force
I	number of failed intersections
F_{fric}	friction force between the yarns and surface of the puncher tip
D_w	yarn width
H	spaces between the yarns (1/number of yarns per cm)
σ_y	yarn stress
ε_y	yarn strain
EY_s	yarn strain energy to deform all other yarns due to the penetration movement of blade
EY_c	yarns cutting energy

TABLE 4.2 *(Continued)*

Symbol	Designation
EY_f	energy to overcome friction between blade and yarns
EF_d	energy to move the fabric during blade penetration
$EF_{cutting}$	energy fabric cutting
KE	kinetic energy
E_y	yarn Young's modulus
ρ_y	yarn density
$EY_{cutting}$	energy of yarn cutting
$EY_{friction}$	energy to overcome friction between blade and yarns
$EY_{tension}$	energy of tensile strain of yarns at the tip of the blade (yarns pull-out force)
EF_{cone}	energy to move the fabric during blade penetration (cone forming)
$EY_{deformation}$	energy to deform the yarns due to the penetration movement of blade
$EF_{cutting}$	fabric cutting energy
FCI	fibers energy absorption capacity index
T_{fiber}	fiber tenacity (GPa)
T_{yarn}	yarn tenacity (GPa)
V_s	longitudinal wave speed (m/s)
E_{armor}	armor absorbed energy
W_{fabric}	weight of fabric (GMS)
N_{fill}	number of wefts ends in one meter
N_{warp}	number of warps ends in one meter
tex_{fill}	weft count
tex_{warp}	warp count
C_{fill}	weft crimp %
C_{warp}	warp crimp %
P_{fabric}	average stabbing resistance force per fabric layer
$P_{standard}$	stabbing force according to standards
E_{fabric}	average stabbing resistance energy per fabric layer

Figure 4.7a,b illustrates the effect of value of (d) and the packing density ϕ on contact point per mm^3 (v). The punching of the fibrous structure is quite different from metal punching. A cylindrical puncher tool pierces the sheet metal by applying enough shearing force. The punching force of a fibrous mat (Fig. 4.8) specifies the presence of three different distinguished zones.

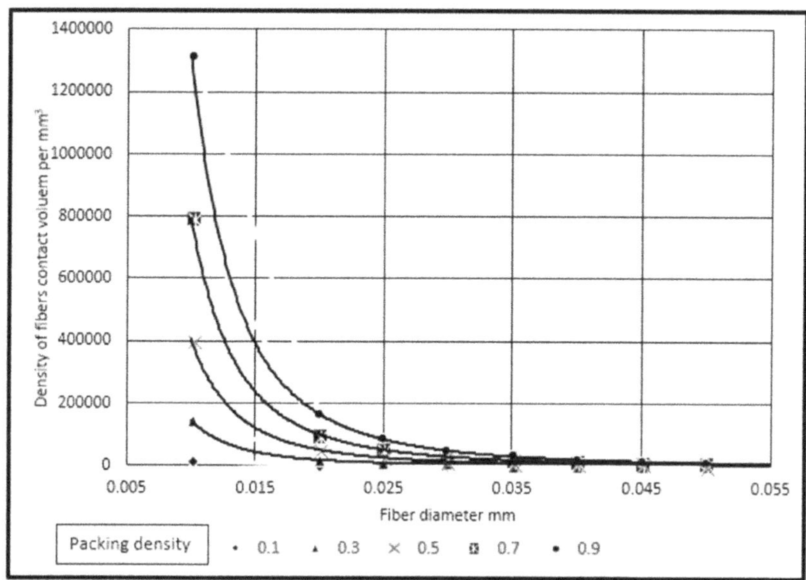

(a) Effect of the fiber diameter

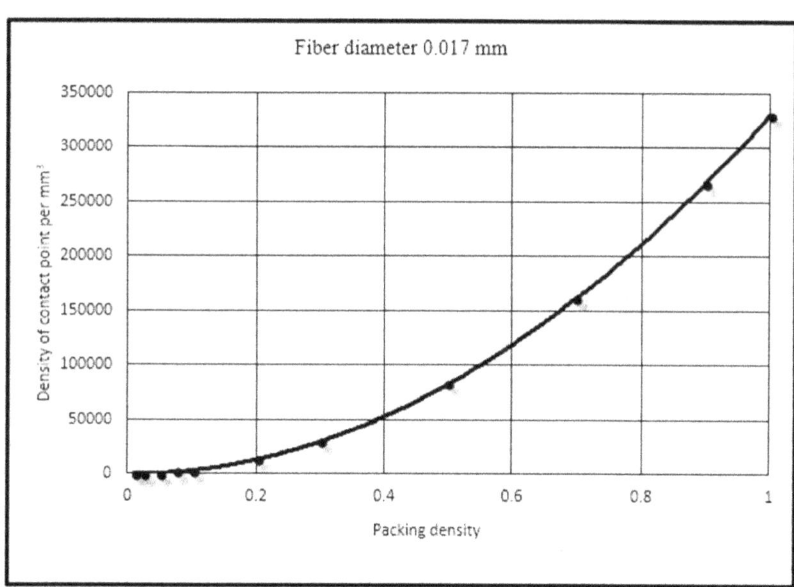

(b) Effect of packing density

FIGURE 4.7 Density of contact point per mm^3 versus (a) fiber diameter and (b) packing density.

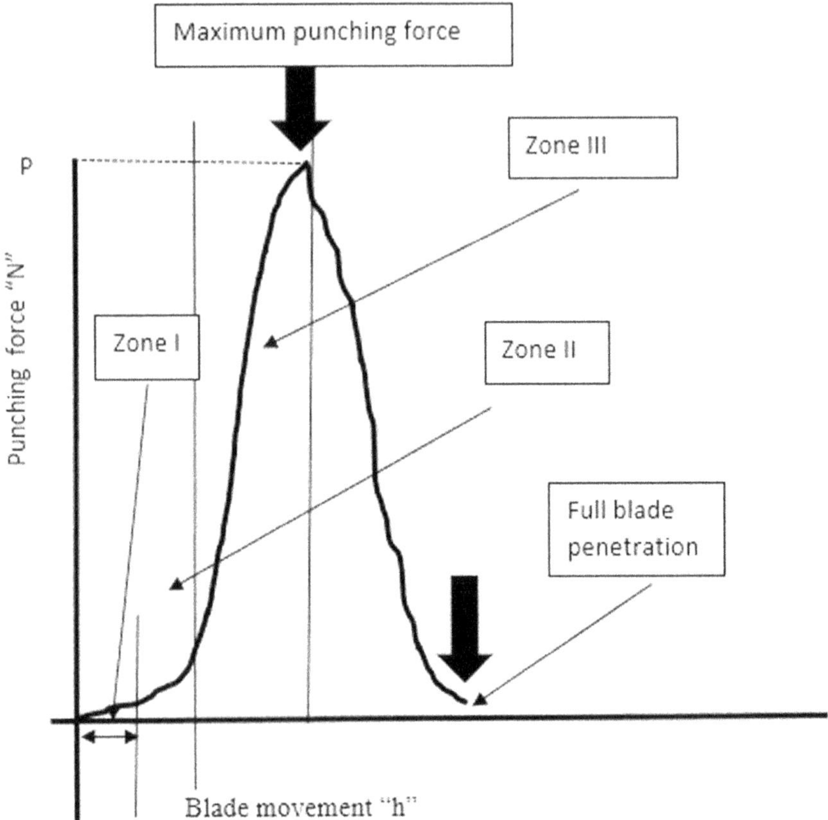

FIGURE 4.8 The punching force–blade displacement curve.

Zone I: Compactness zone—the mechanism of punching of fibrous bulk, such as web structure, differed from the woven or knitted fabrics hence the low value of structure compactness makes a critical change in the packing density (υ) under the punching blade and the area around it. During the punching process, the fiber gets stretched and entangled with the punching blade. The resisting force is the compacting force on puncher blade tip.

Zone II: with further movement of the puncher blade, the fibrous structure becomes more compressed and the fibers more strained to resist the punching. The fiber strain relies on friction force with the puncher surface and the friction with the other surrounding fibers. Fiber movement during puncher penetration is largely responsible for the difference in the punching behavior of the fibrous structure.[18] The punching force is the resultant of the fiber friction at the contact points between the fibers and the compactness force on puncher blade tip.

Zone III—(mechanisms controlled by the puncture and cutting); with further puncher penetration, the fibers involved with the puncher tip will start to fail and/or slip. The total resisting force is resultant of the straining force for all the fibers involved with the puncher blade and the frictional forces between the contacted fibers at the blade surface and the compactness force on puncher blade tip for a penetration height "*h*." The subsequent failure of the adjacent fibers occurs in all directions. If the fibrous mat has a high value of packing density and the fibers are long and well entangled, the punching resistance will be highest because most of the fibers under the punching blade will be cut.

Fibers punching force = number of fibers under the puncher blade × the single fiber cutting force

$$P_{\text{fiber}} = k\,(W_{\text{punched mat}}/W_{\text{fiber}})\,\varphi\,\eta\,C_{\text{fiber}}. \tag{4.2}$$

The fiber orientation factor $0 \leq \eta \leq 1$, which is equal to 1 when all the fibers are oriented perpendicularly to axis of the punching and zero when all fibers are parallel to axis of the puncher, the percentage of failed fibers for long continuous fibers (k) is equal to 1.

The total energy of penetration E_{punching} is given as follows:

$$E_{\text{punching}} = E_{\text{cutting}} + E_{\text{friction}} + E_{\text{tension}} + E_{\text{compacting}}$$

$$E_{\text{fiber}} = \sum (E_{\text{friction}} + E_{\text{tension}} + E_{\text{compacting}})$$

$$E_{\text{punching}} = E_{\text{cutting}} + E_{\text{fiber}}. \tag{4.3}$$

The energy of friction forces is to overcome the friction between the fibers during compactness and between the surface of the puncher and the fibers during the puncher penetration.

4.2.2.2 PUNCHING MECHANISM OF MULTILAYER FIBROUS MATERIAL

In most designs, armor has multilayers of fabrics and other materials. Using fibrous web over fabric, the punching mechanism is dissimilar to that described above. The fabric stiffness will support the fibrous assembly and elevate its compactness force. Consequently, fiber packing density will be higher, increasing the punching resistance force. Figure 4.9a illustrates the punching force–blade displacement curve. The puncher penetration through high packing density fibrous web is ended by a fabric failure through cutting and straining of the yarns, higher punching resistance is achieved.

$$\text{Punching resistance force} = P_{\text{fabric}} + P_{\text{fibers}}, \tag{4.4}$$

$$\text{Fabric punching force } (P_{\text{fabric}}) = 2\,D_{\text{puncher}}\,(N_{\text{warp}}\,C_{\text{warp}} + N_{\text{fill}}\,C_{\text{fill}}). \tag{4.5}$$

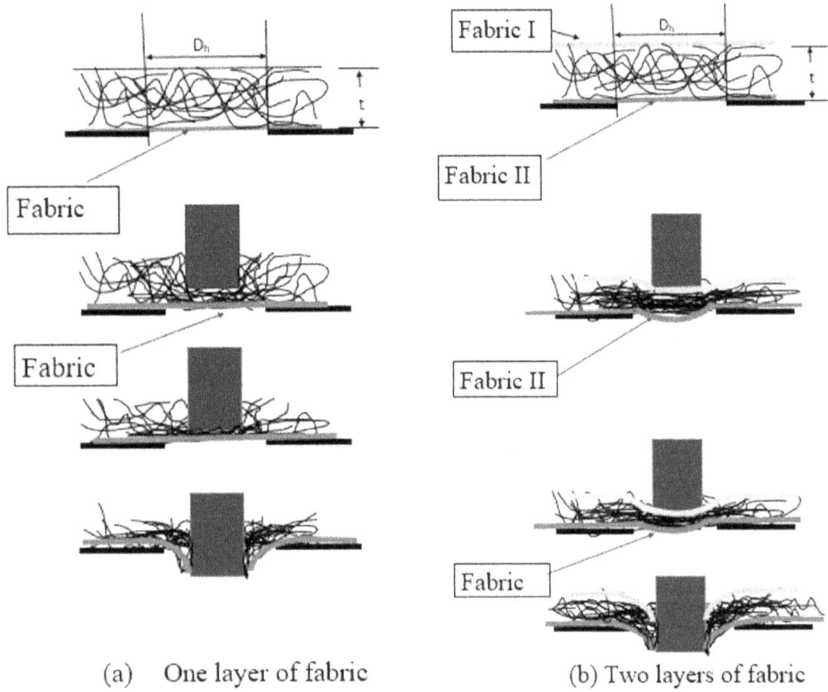

FIGURE 4.9 Sketch of the punching mechanism of the multilayer fibrous structure.

If the fibrous material is laid between two fabric layers, the top fabric layer will raise the compactness of the fibrous mat. More fibers will participate in the cutting resistance of the multilayer structure, as demonstrated in Figure 4.9b. Because of the deflection in the first and second fabric layers, the pressure on the fibers will extend on the sides of the puncher, increasing the number of the fibers participated in resisting puncher movement. The total resisting forces will act on the puncher blade more effectively. The multilayer structure of fabric and fibrous mat gives a higher specific punching force.[19,20] Using multilayer structure, the punching resistance of the upper layer fabric is higher than the cutting resistance of the fibrous mat alone to allow utmost compactness force by deflection of fabric under the punching stress. The use of the fibrous mat as a layer in the anti-stabbing armor improves thermal comfort. Fabric individual layers punching force preferably be higher in the direction of punching force.[19,21] For puncher with tip of a conical shape, a friction force between the tip's surface and surrounding fibers growths as the puncher penetrates through the fibrous mat. Higher packing density will change described above fibrous mat penetration mechanism. Punching force-displacement curve of conical puncher (Fig. 4.10) indicates the mechanism of conical puncher. Fibrous material packing density expresses the punching force of the penetrating puncher.

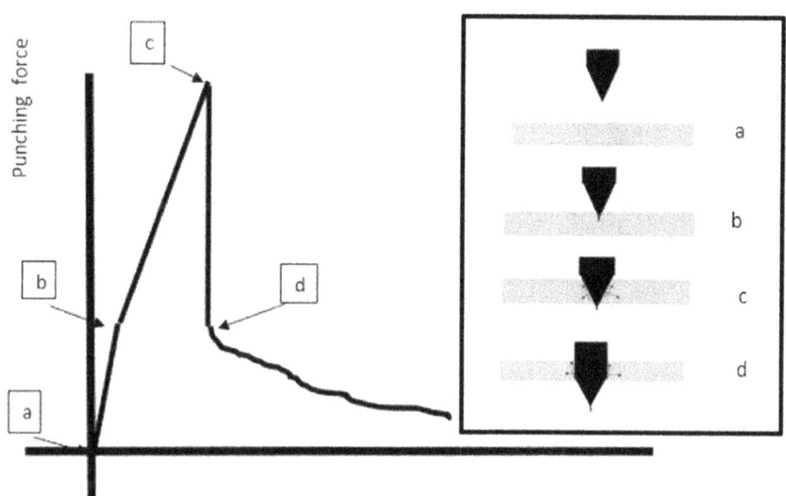

a- indentation, b- penetration, c- failure, d- perforation

FIGURE 4.10 Punching force-displacement curve of conical puncher.

4.2.2.3 STAB RESISTANCE OF FIBROUS STRUCTURES

Stab resistance of armor is the protection against cutting objects, such as knives, spikes, blades, and needles. Stab resistance is contrasted from the punching resistance hence the punching tools have no cutting edges. To describe the stabbing by knives, the anatomy of knives will be described. There are enormous shapes of knife's blade designed along with the evolution of humanities hence the knife was considered as an essential tool for daily activities, such as hunting, self-defense. Some knife cutting edge shapes are exemplified in Figure 4.11. Edge sharpness determines the efficiency of the knife to penetrate through a material.[22]

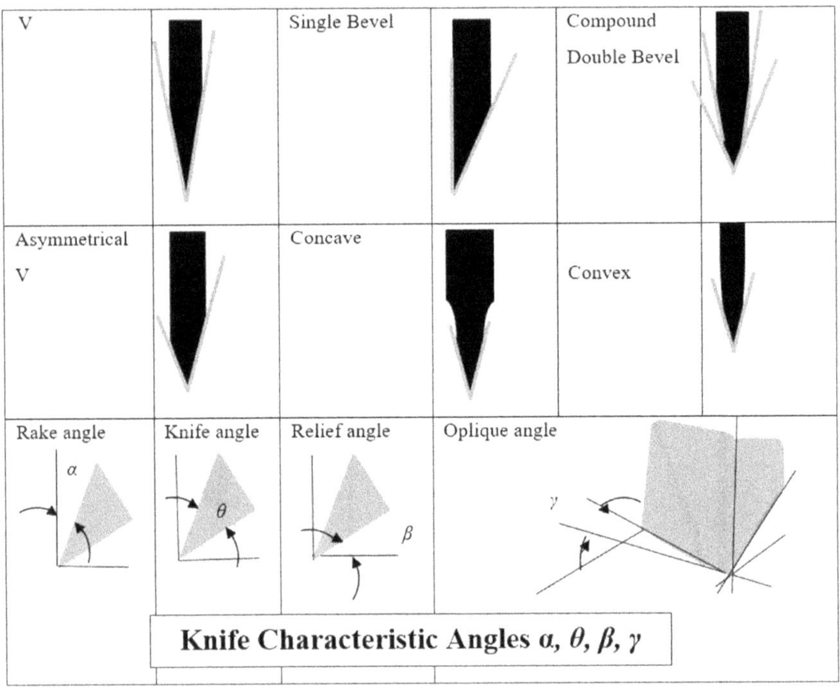

FIGURE 4.11 Classification of the knife types.

Figure 4.11 illustrates the knife cutting edge shapes. The edge of the knife can be sharped from one side "single bevel" or "double bevel." Its edge may be straight, concave, or convex. The knife edge tip has different shaped parts: simple with single edge angle or it may have primary edge at the tip followed with secondary edge.[22] Figure 4.12a, b illustrates the sharping angles of the V-shaped and double edge knifes. The sharping angle varies between 8° and 32° depending on the end use.[22–25] The cutting force relies on the blade angles α, θ, β, γ.

Double edge knife secondary edge is the cutting part of the knife while primary edge helps in the knife penetration. The combinations of angles θ, γ, and α and lengths l_1, l_2, and l_3 will determine the cutting force of the knife.[26]

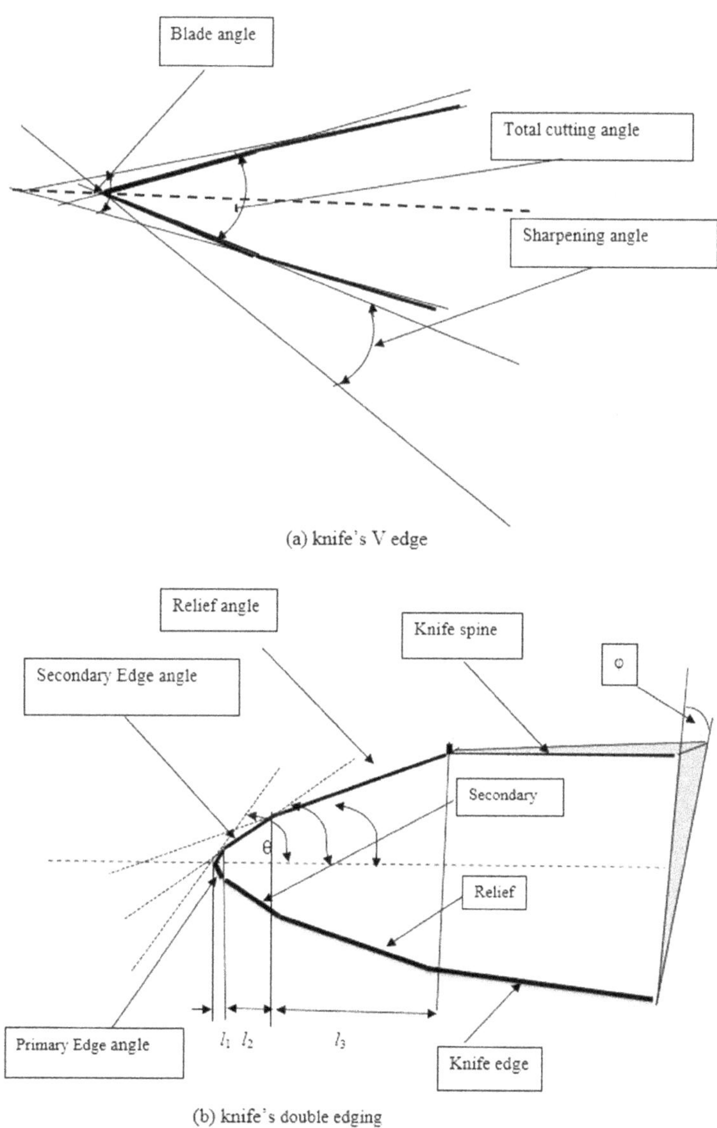

FIGURE 4.12 Different edge and cross sections of knife blade.

The depth of blade indentation required to initiate a cut or crack in the target material is a function of the condition or sharpness of the blade's tip, and this property is used to formulate a so-called "blade sharpness index" (BSI).[27] It was revealed that the blade edge radius has a critical effect on the cutting force hence the contact pressure at the indentation area is very high as the radius of blade edge becomes smaller.[28] Moreover, the grade of the steel used for the knife blade is the determining factor of the knife cutting capacity. Also, the cutting speed and knife-edge angle significantly influence the peak force and specific cutting energy.[15,29–32] In the stab resistance test of the protective armor, NIJ Standard–0115.00,[33] two shapes of knives are used and one shape of spike with dimensions as given in Figure 4.13.

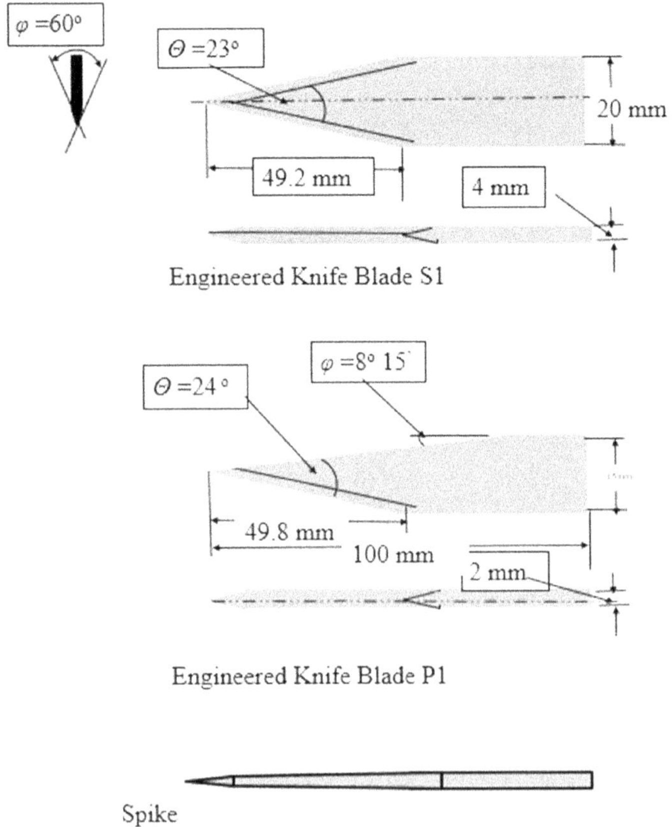

FIGURE 4.13 Engineered knife blade and spike NIJ Standard–0115.00.

4.2.2.4 STABBING RESISTANCE OF FIBROUS MAT

The stabbing and slashing have their specific features, depending on the design parameters of the fibrous material[34] of the protective vest designed to secure shield, comfort, and mobility. The stabbing mechanism is a combination of cutting, straining, and friction mechanisms. Contribution of friction in stabbing plays a decisive role as it appears all the time during the blade penetration:

- fiber-to-fiber friction throughout the compactness stage;
- between fibers due to their sliding over each other during their movement under the effect of blade penetration; and
- between the fiber and the blade surface during its movement through the fibrous structure.

The friction energy is:

$$E_{\text{friction}} = \mu_1 N_1 Z_1 + \sum \mu_f N_f S_i. \tag{4.6}$$

The scenario of the puncture/cutting mechanism is given in Figure 4.14 to provide a better explanation with respect to the penetration mechanisms associated with a puncture/cutting by pointed blades.[35] The penetration force includes frictional forces (friction forces on the side of the blade ($F_{\text{sides friction}}$), friction force on the edge of the blade ($F_{\text{edge friction}}$), compactness force that represents the reaction of the fibrous material on the penetration of the blade ($F_{\text{compacting}}$) and cutting force due to the shearing of the compacted fibers (F_{cutting}). The analysis of the distribution forces clearly confirms that the friction force increases, depending on the packing density, mat thickness, and as well as the coefficient of friction between the blade material and fibers.[36] The penetration force will increase as the tip angle of the blade increases. However, an increase of the friction, associated with the lateral pressure of material on both sides of a blade because of high value of packing density, makes the friction resistance related to puncture/cutting mechanism the highest. The penetration force in equilibrium can be expressed by:

$$\overrightarrow{\boldsymbol{F}_{\textbf{stabbing}}} = \overrightarrow{\boldsymbol{F}_{\textbf{cutting}}} + \overrightarrow{\boldsymbol{F}_{\textbf{compacting}}} + \overrightarrow{\boldsymbol{F}_{\textbf{sides friction}}} + \overrightarrow{\boldsymbol{F}_{\textbf{edge friction}}} \tag{4.7}$$

The stabbing force will be a function of the stabbing blade angles α, θ, β, and γ (Fig. 4.11).

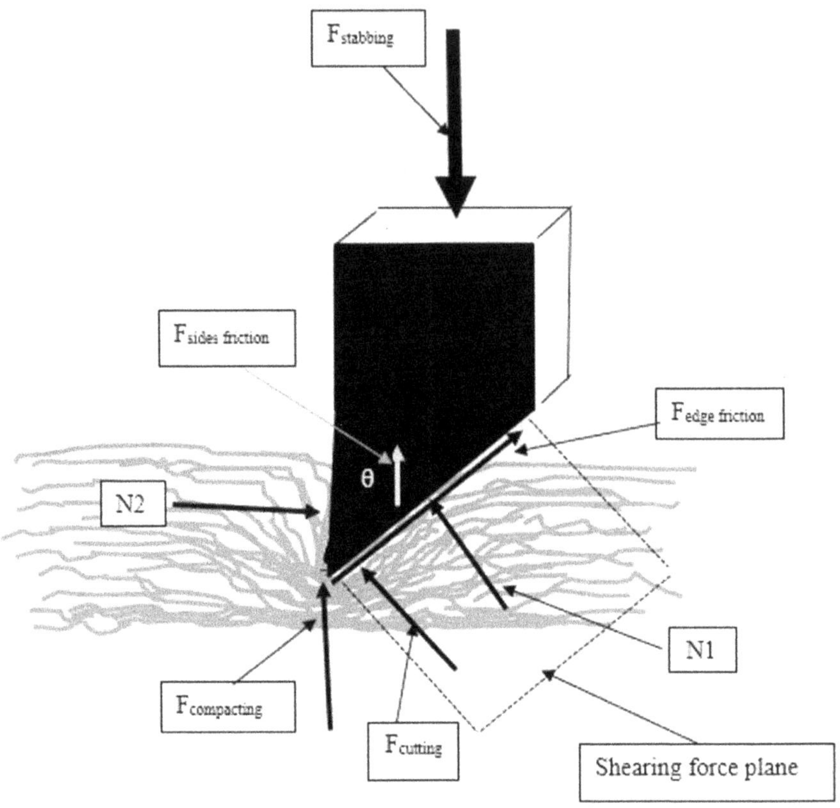

FIGURE 4.14 The forces acting on a knife in initial cutting of the fibrous mat after compactness.

4.3 MECHANISM OF FABRIC STABBING

4.3.1 INTRODUCTION

The penetration of sharp edge blade through the fibrous structure is different from the model discussed above due to the fact that the stabbing blade has a sharp edge capable to cut the material's withstanding to the penetration. The material cutting resistance is the force required by the sharp object to slide through a certain material and therefore strongly rely on material shear properties and coefficient of friction between the cutting edge and the material.[36] The actual fiber failure is either due to tension or shear.

In some situations, the textile material is slashed by knife or other sharp tools, and then the mechanism of failure, in this case, will be different from that of knife penetration. The mechanism of stabbing has several scenarios:

1. when the knife is penetrating vertically and
2. the knife moving horizontally.

The fibers under the attack of the knife will be cut to allow the knife to pass through the fibrous structure and move forward.

Generally, the cutting force of fabric relies on:

1. cutting force of the fibers and yarns,
2. fabric structure,
3. fabric thickness,
4. fabric tightness,
5. knife sharpness,
6. knife edge angle,
7. knife geometrical parameters,
8. knife penetration angle,
9. speed of punching,
10. friction properties,
11. knife velocity,
12. fabric tearing strength, and
13. direction of cutting.

Usually, different situations occur as a knife penetrates a fabric: pushing yarns, cutting yarns with shearing due to blunt transverse pressure on the fibers, and tearing the yarns when strained to break.[21,37]

4.3.2 MECHANISM OF FIBERS CUTTING

Before analyzing the fabric stabbing, the cutting of the fibers by blade moving crosswise it at varying blade and fiber angles should be highlighted.[37,38] It was revealed that, based on blade stress normalization, inorganic fibers demonstrate higher cut resistance than organic fibers. The analyses of the cutting mechanism indicated that during the cutting processes the fiber will be deflected under the cutter load, then the cutter starts to penetrate through the fiber with the increasing force till the cross section of fiber is reduced to the limit, and it fails under tension. Consequently, the fibers cutting force depends on their mechanical properties. The cutting of the fiber cross-section may have four distinct zones: (Fig. 4.15) indentation zone, deformation of the material under the pressure at the blade tip; cutting zone, shearing of the material by the sharp edge of the blade; shear–tension zone; and tear failure zone.[39,40] The area of each zone relies on the hardness, shear strength, and the fibers tenacity. The roundness of the tip affects the indentation zone and develops a local transverse compressive stress on blade tip σ_c:

$$\sigma_c = F_c/(d_{tip}\ \zeta\ d_f). \tag{4.8}$$

The higher diameter reduces the applied local stress on blade tip, and more force is needed to create indentation zone.[40]

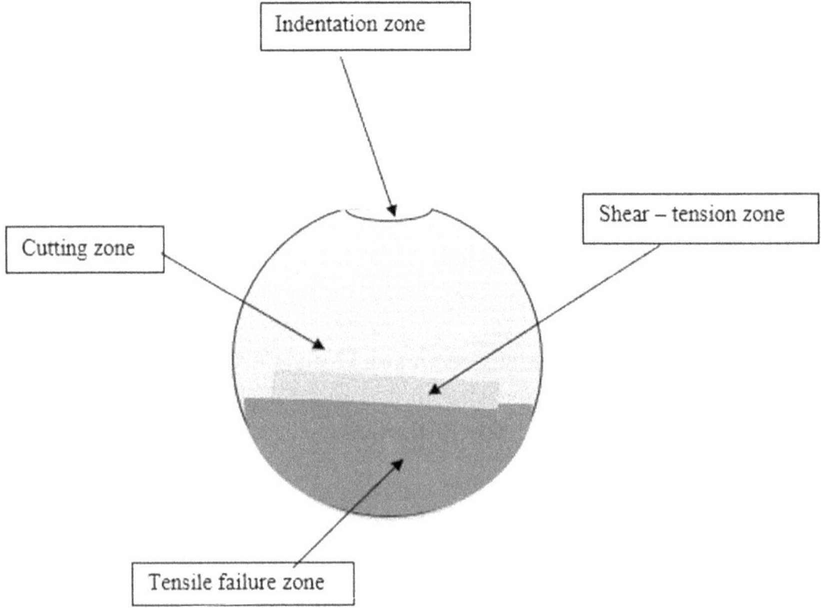

FIGURE 4.15 Fiber fracture zones.

The cutting force–cutter displacement curve for organic and inorganic fibers is given in Figure 4.16 and demonstrates the blade cutting mechanism of the fiber. As the effective area of the fibers' cross section decreases, the fiber strain increases by the continuous downwards displacement of the blade till the fiber failure. The cutting resistance relies on the relation of the fiber shear stress and the tenacity properties. The inorganic fibers have higher indentation stress and elastic modulus than the organic fibers, thus the increased force is needed to start the cutting processes.

The cutting force of the fibers is decreased by the increase of the sharpness of the blade and its area. Dyneema fiber shows the highest cutting resistance force, while Zylon fiber gives the lowest due to its lower coefficient of friction between cutting blade and the Zylon fiber.[41]

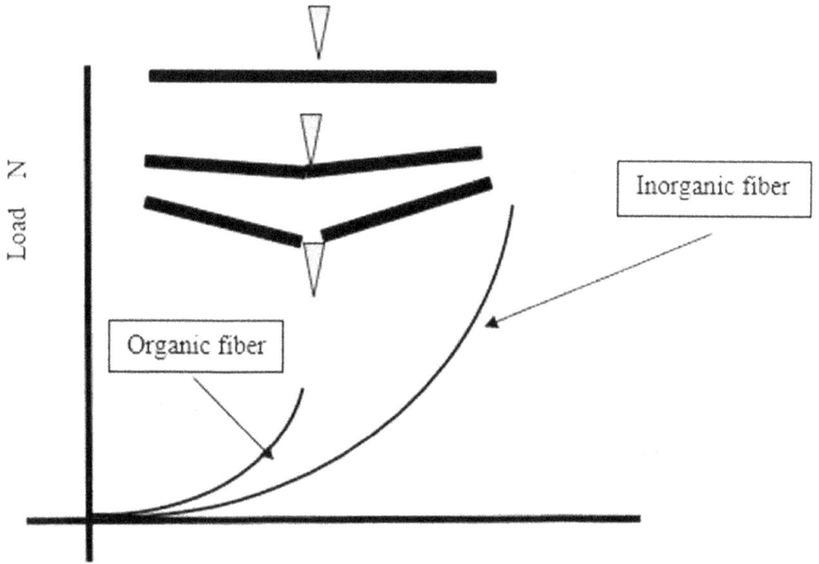

FIGURE 4.16 Cutting load–displacement curve of fibers.

The cutter angle of inclination, cutting in longitudinal cut or transverse cut, expressively influences the fiber cutting force, the later gives higher cutting values. Carbon fibers possess the highest cutting force, followed by Technora, Twaron, S-glass, Kevlar, Zylon, Dyneema, and Vectran. Fiber cutting is a mixture of fiber shearing and the fiber tension straining under the applied cutting force. For fiber material with lower cutting force, the failure under tension forces predominates.

4.3.2.1 CUTTING RESISTANCE FOR STRETCHED FIBER

Figure 4.17 shows that a fiber during cutting by a sharp blade is under tension at both ends. The cutting process of the fiber will be under two effects: First, shearing the outer fiber's layer after indentation. With the continuous penetration, the fiber cross-section will be sheared reducing its area. Second, the cross section area (A) is strained under tension. The total energy is shear and strain energy under cutting and tension load:

Total energy = UT + UC

$$\text{Total energy} = 0.5 \int (k\, FV^2/A_f G_f)\, dz + 0.5 \int (TA^2/A_f\, E_f)\, dx, \qquad (4.9)$$

where k is form factor which relies on the shape of the fiber cross section (for circular cross section $k = 1.1$).

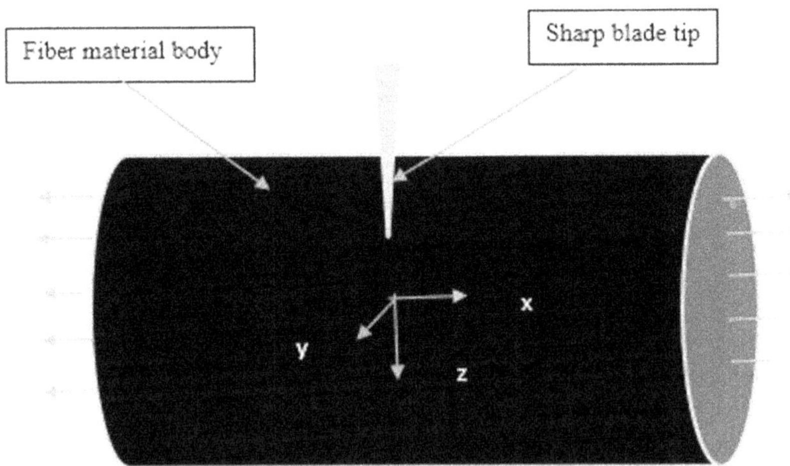

FIGURE 4.17 Fiber undercutting–stretching by a sharp blade.

However, the TA, FV, and A_f vary as the cutting blade penetrates thus reducing the fiber cutting resisting energy, resulting in early failure undercutting–tension mechanism.

It was revealed that, when the fiber is tensioned and subjected to stabbing, its cutting force is low, as demonstrated in Figure 4.18. This explains while testing a skin samples under tension, the depth of out of plane displacement of the skin before the knife penetration and larger levels of in-plane tension in the skin are associated with lower penetration forces.[42]

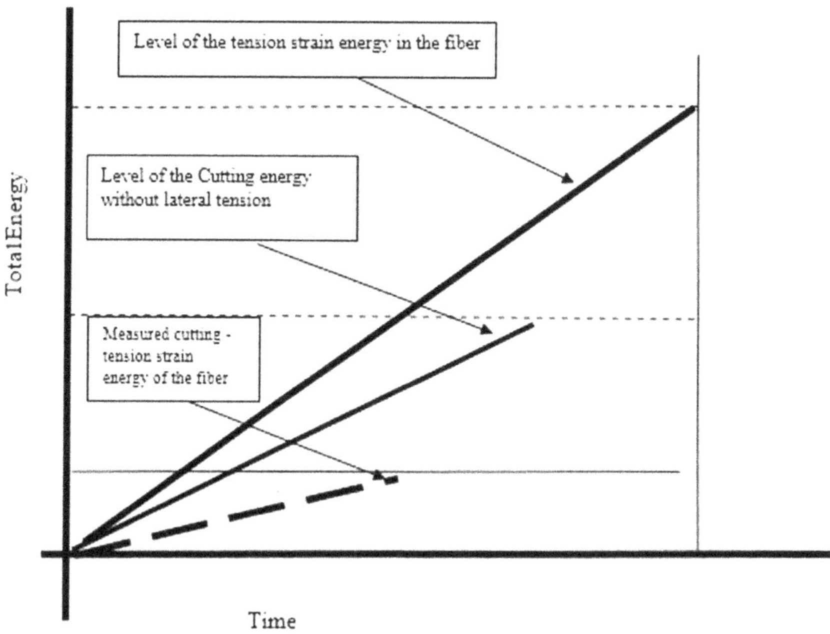

FIGURE 4.18 The energy–time curve of fiber under cutting–tension mechanism.

The component of the shear energy for blade with cutting angle θ is:

$$\text{The shear energy} = 0.5 \int k\,(FV_\theta/\cos\theta)^2/A_f G_f)dz \qquad (4.10)$$

FV_θ is fiber cutting force inclined by angle θ.

The cutting force of the fiber's material will vary according to the cutting-edge inclination.[39]

Then:

$$F_C = FV_\theta/(\cos\theta)^2. \qquad (4.11)$$

The fiber cutting force is a function of fiber shear angle θ. The higher the shear angle, the lower is shear strength, which contributes to the resistance of a material. The cutting force of fiber will increase as the knife edge angle increases,[31] and more cutting energy is required. Experimentally, the cutting force and cutting energy were expressed by various equations, based on the properties of the fiber material and design of the cutting tool.[21,42]

The shear force in the section inclined by angle α to the fiber axis (Fig. 4.19) is usually less than its value when cutting angle of inclination is perpendicular to the fiber axis. The required shear force is the smallest when the blade is cutting parallel to the fiber longitudinal axis.[39] The minimum cutting force is when angle α equals to 90°. Less cutting force and energy are needed to cut the material when applying a high tension.

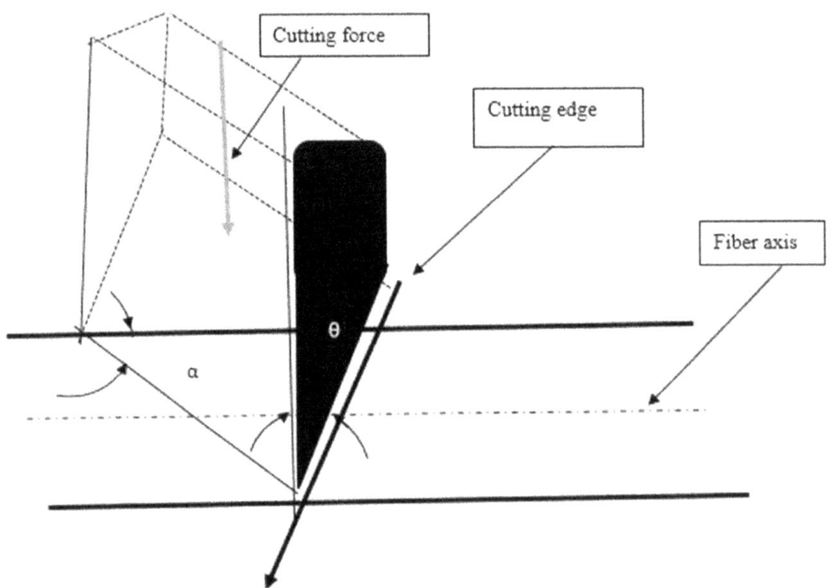

FIGURE 4.19 Blade cutting angles.

4.4 MECHANISM OF PUNCHING/CUTTING OF FABRIC MATERIAL

4.4.1 MECHANISM OF PUNCHING OF THE FABRIC

The material punching resistance is the force acting on the blade penetrating through a material.[37] The actual fiber failure is either due to tension or shear. The mechanism of the punching of the fabrics is dissimilar than that of a fibrous mat. The morphology of fabric (Fig. 4.20) demonstrates the yarns interlacements and pores between them. The behavior of the fabric under the punching tool is contingent on the tip configuration, material, and fabric design: woven, knitted, triaxial, nonwoven, or combined structure.[13]

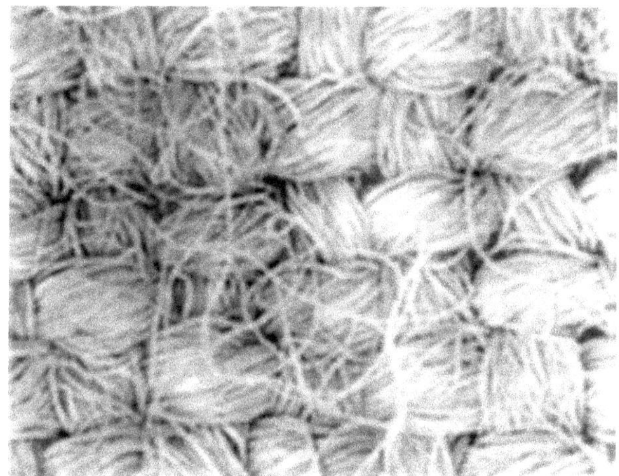

FIGURE 4.20 Plain weave.

The fabric is the interlacements of warp and fill yarns in a certain manner to form 2D or 3D fabric. The punching force reorients the yarns in a fabric since they tend to get the position of the minimum energy level. The punching force straightens the crimped yarns. At high values of a punching load, fabric loses its integrity initiating the yarn strain and failure of the warp or fill yarns or both under the critical force. The punching loading has the destructive effect only when it causes the punching stress higher than the bursting stress.

4.4.1.1 CIRCULAR ROD PUNCHER

The punching by a round rod puncher strains the warp and fill yarns in contact with the puncher. Figure 4.21 shows the yarns deformation caused by punching with circular rod puncher. The yarn deflection fluctuates according to its location. The fabric punching force can be calculated neglecting friction resistance and fabric stiffness. The puncher diameter should verify: $D1 > p_2 + d_{warp}$ and $D2 > p_1 + d_{fill}$, otherwise it will pass freely through the fabric pores ($N_{fill} = D/p_1$ and $N_{warp} = D/p_2$).

The warp and fill yarns will be strained differently, depending on their position in contact with the puncher.

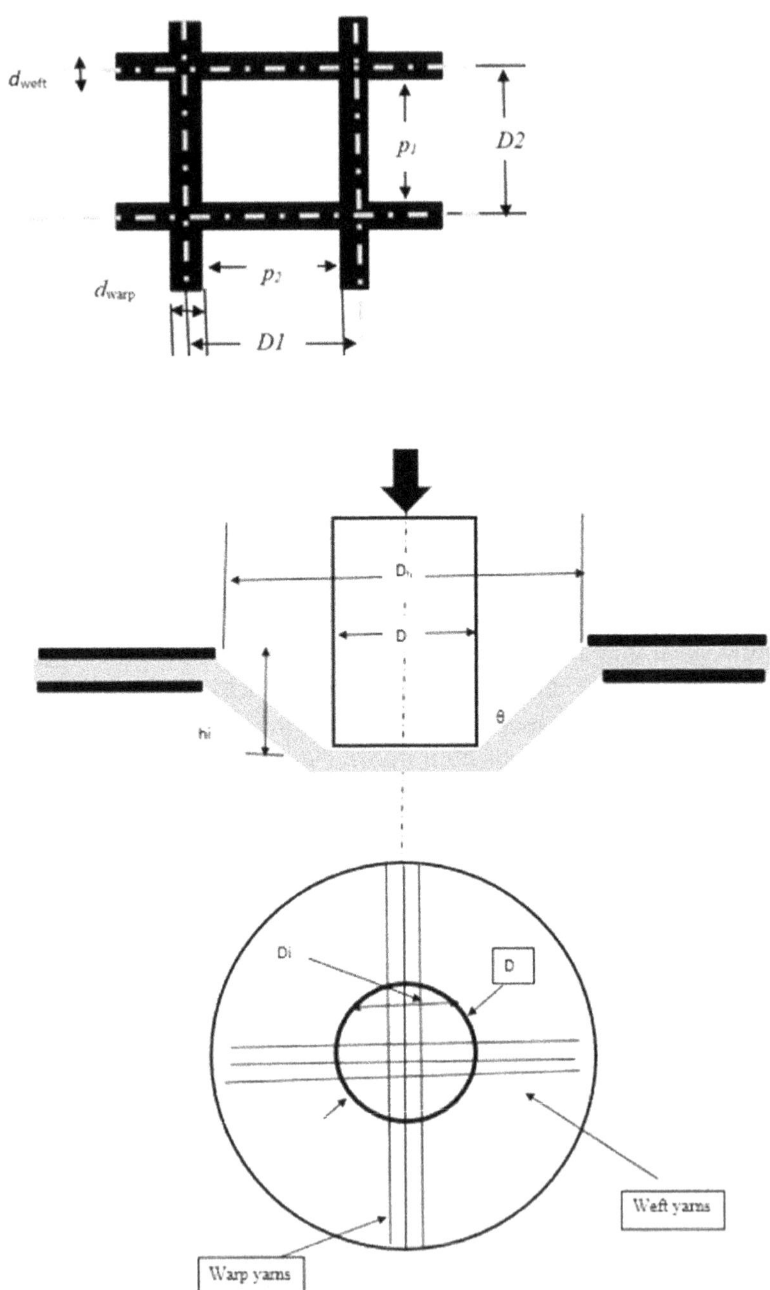

FIGURE 4.21 Mechanism of circular rod punching.

The forces straining the warp or fill yarns under the puncher (Fig. 4.22) can be calculated as:

the strain of the yarn that passes under the puncher center:

$$\mathcal{E} = (L - D_h)/D_h, \tag{4.12}$$

where

$$L = D + 2\,((h_k^2 + ((D_h - D)/2)^2))^{0.5}. \tag{4.13}$$

The yarn strain at distance (ip_1) of contact length D_i (Fig. 4.22) is:

$$\mathcal{E}_i = (L_i - D_h)/D_h \tag{4.14}$$

$$L_i = D_i + 2\,(h_k^2 + ((D_h - D_i)/2)^2)^{0.5}, \tag{4.15}$$

where $D_i = (D^2 - (ip_1)^2)^{0.5}$

$D_j = (D^2 - (j\,p_2)^2)^{0.5}$

i is 1, 2, 3..., n, $n = D/W_1$, W_1 the warp per cm, j is 1, 2, 3…, m, $m = D/W_2$, W_2 the fill per cm.

Then force on the ith warp yarn is:

$$T_{\text{warp i}} = (\mathcal{E}_{\text{warp i}}\,E_{\text{warp}})\,A_{\text{warp}}.$$

While the force on the jth fill yarn is:

$$T_{warp\ \mathrm{i}} = (\mathcal{E}_{fill\ \mathrm{j}}\,E_{fill})\,A_{fill}.$$

The punching force on the yarns passing under the puncher tip is:

$$\text{Punching force } P = (\textstyle\sum \cos(\theta_i)\,T_{\text{warp i}} + \sum \cos(\theta_i)\,T_{\text{fill j}}) + FS. \tag{4.16}$$

The fabric failure begins when the strain in any of the yarns under the puncher tip reaches its failure strain value. Assuming $W_1 = W_2$, then the punching force is:

$$P = 2(\textstyle\sum \cos(\theta_i)\,T_{\text{warp i}}) + FS. \tag{4.17}$$

FS is fabric resistance force of the deformation under puncher's downwards movement.

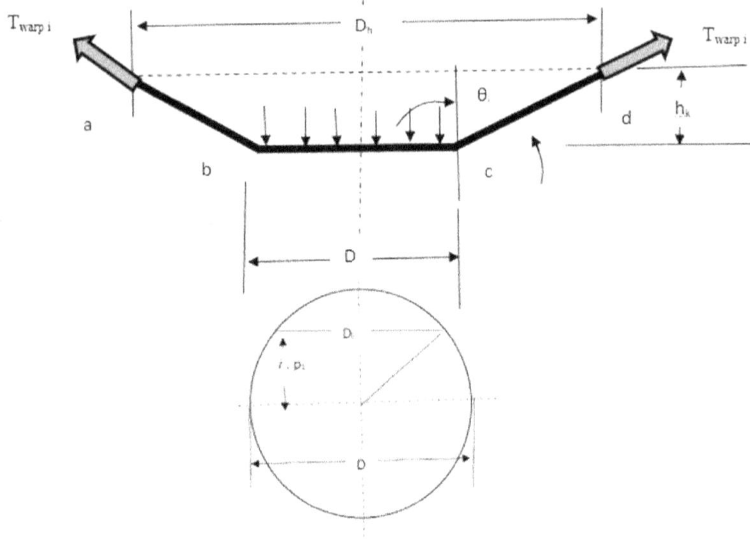

FIGURE 4.22 Tension of the yarns during punching.

4.4.1.2 SPHERICAL PUNCHER TIP

The yarns in contact with puncher's round tip will be strained under the punching force, as illustrated in Figure 4.23.

The forces applied on the yarns at the point of intersection are as follows[43,44]:

- punching force F_p,
- packing force P_p,
- friction force F_f,
- friction force F_c,
- bending moment M_b, and
- tension force T_i.

For flexible yarns with low coefficient of friction μ (yarns to metal of the sphere), the main forces acting on the yarns are T_i and yarn to yarn friction at the point of intersections.[45–47]

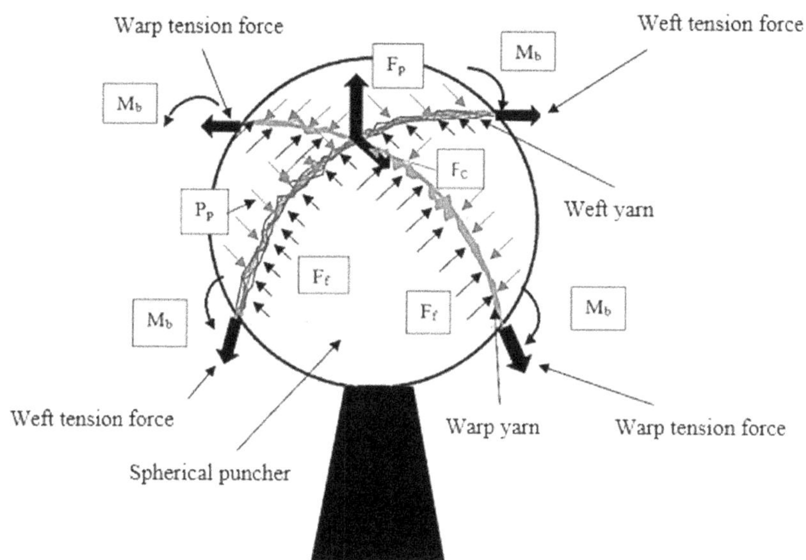

FIGURE 4.23 Forces acting on the yarns elements under punching force F_p.

The yarns tension relies on its location on the surface of the puncher tip. As the punching force increases, these components increased, too. Then, the total warp yarn stress, neglecting the adhesive force, is:

$$T_{\text{warp i}} = (\varepsilon_{\text{warpt i}} E_{\text{warp}}) A_{\text{warp}}.$$

Tension in the *j*th fill yarn:

$$T_{\text{fill j}} = (\varepsilon_{\text{fill j}} E_{\text{fill}}) A_{\text{fill}},$$

$$\text{punching force } P = (\textstyle\sum^{m} \cos(\theta_i)\, T_{\text{warp i}} + \sum^{n} \cos(\theta_i)\, T_{\text{fill j}}). \tag{4.18}$$

Fabric punching force is:

$$F_{\text{fabric punching}} = \textstyle\sum^{n1} (\varepsilon_{\text{warpi}} E_{\text{warp}} A_{\text{warp}}) + \sum^{m1} (\varepsilon_{\text{filli}} E_{\text{fill}} A_{\text{fill}}) + FS, \tag{4.19}$$

n_1 and m_1 are number of failed warp and filling yarns, respectively.
For round puncher tip, punching force relies on:

- yarns properties,
- yarn-puncher frictional properties,

- Young's modulus,
- friction between warp and fill yarns,
- fabric tightness,
- number of yarn's interlacements,
- fabric bursting force, and
- fabric shear modulus.

In the case of $n_1 = m_1$ and the number of interlacements to be punched to pass the puncher tip through the fabric is I, with small angle θ, then:

$$F_{\text{fabric punching}} = 2\,I\,(\mathcal{E}_{\text{warpi}} E_{\text{warp}} A_{\text{warp}}) + FS. \tag{4.20}$$

4.4.1.3 CONICAL TIP PUNCHER

For conical tip puncher (Fig. 4.24) the tip of the puncher either strikes the fabric at the point of the interlacement of the yarns or at the spaces between them, trying to push the yarns aside and proceeding downwards till the yarns are jammed. The yarns will be strained on further penetration of the puncher. The friction force between the puncher tip and the adjacent yarns strains them and pull off some of the yarns (Fig. 4.25) and/or the puncher tip pushes some yarns out the fabric plane, thus the failure of the yarns relies on their pull-off force. The punching force depends on the cone apex angle 2α.

Thus, the fabric ultimate punching resistance is a function of fabric design, geometrical characteristics of puncher's tip and yarns mechanical properties.[47] As the fabric deformed under the punching force, the yarns in contact with puncher's tip will take angle α_i and α_j. The maximum punching force F_p arises once the puncher pushes the yarns at the point of intersections. Consequently, for balanced fabric:

$$F_p = 1.55\,(D_{\text{blade tip}})^2\,\varepsilon_y E_y \cos(\alpha).$$

For I number of failed intersections, the total punching force F_{pt} is (IF_p).

F_{pt} is a function of the time, starts at zero value and continuously increases till the puncher succeeds to enter through the fabric:

$$F_{p\,max} = IF_p + F_{fric}\cos(\alpha) + FS. \tag{4.21}$$

Figure 4.26 illustrates the force–displacement curve of the punching.

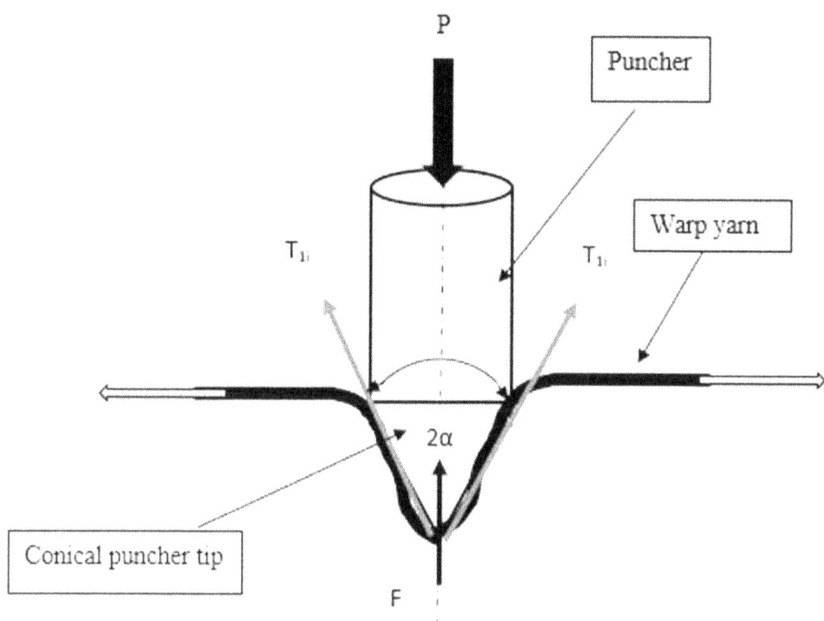

FIGURE 4.24 Conical puncher tip.

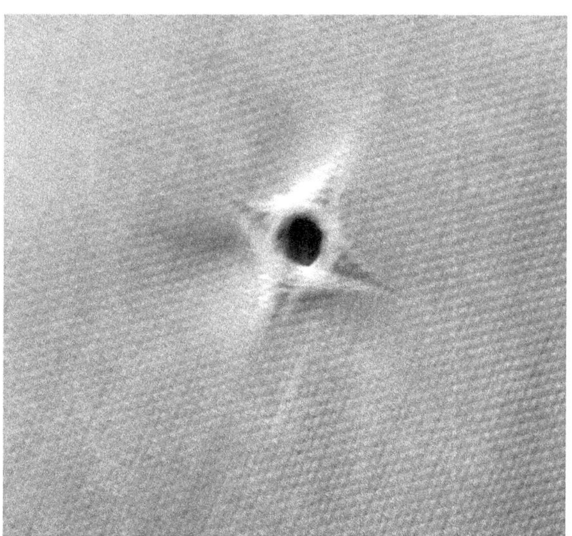

FIGURE 4.25 Punched fabric.

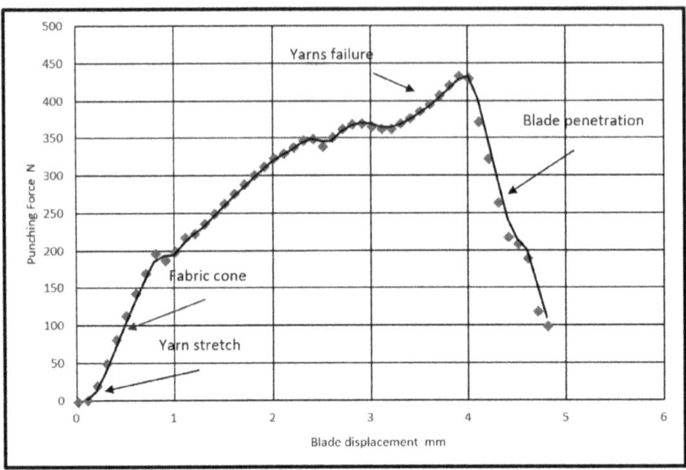

FIGURE 4.26 Punching force-displacement diagram.

The conical tip puncher is more effective punching tool since it makes the fabric to lose its punching resistance.

4.4.2 PUNCHING OF THE TRIAXIAL FABRIC

The triaxial fabric differs from the woven fabric in structure and is formed by three sets of the yarns (Fig. 4.27). The opening of hexagonal shapes is formed between yarns and depends on the number of yarns/cm in each direction. Consequently, F_p is a function of:

- size of the puncher tip diameter,
- size of hexagonal opening between the yarns, and
- yarn properties.

To be sure that the triaxial fabric will resist the punching (Fig. 4.27) the following should be met[49]:

$$D_{\text{blade tip}} \geq 0.87\,h + 2dw \tag{4.22}$$

Figure 4.28 shows the forces applied to the yarns under the pressing of the puncher. Consequently, the punching force[49] will be:

$$F_{\text{p max}} = 2.355\,(D_{\text{blade tip}})^2\,\sigma_y \cos(\theta)\,I + F_{\text{fric}} \cos(\theta) + FS \tag{4.23}$$

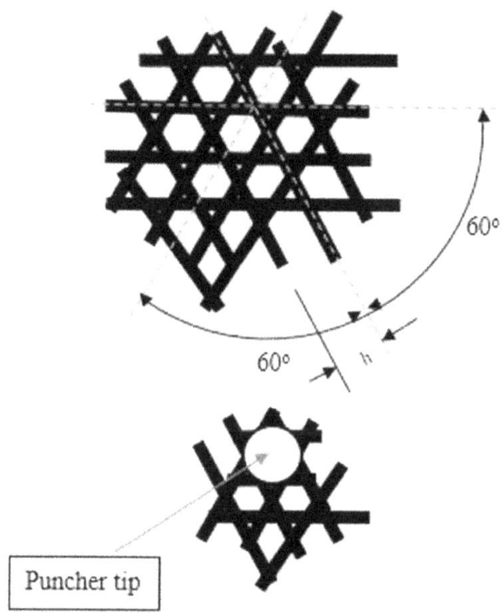

FIGURE 4.27 Triaxial fabric structure.

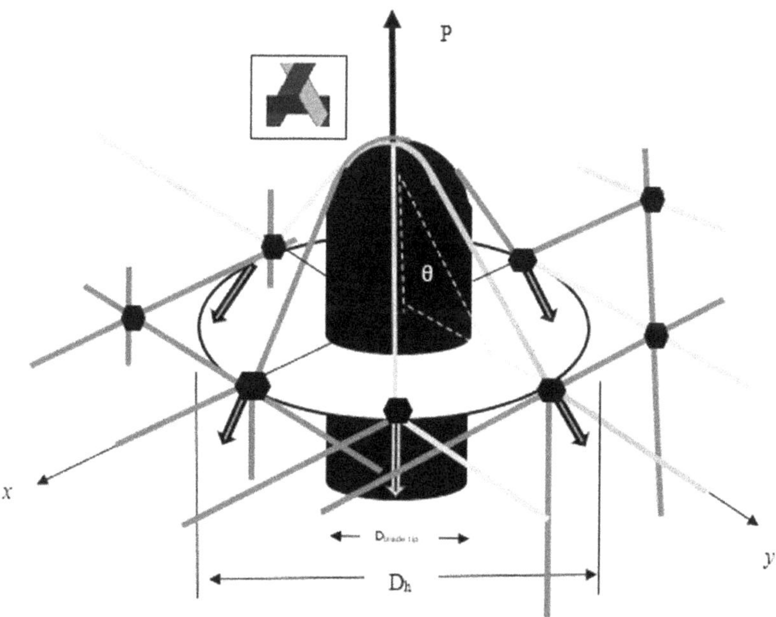

FIGURE 4.28 Forces acting at yarns intersection.

The structure of 3D fabric is different than 2D, the yarns take geometry structures in XYZ; highest dimension of fabric is in Z direction. The punching of the yarns in *Z* direction raises fabric punching resistance.

4.5 FABRIC STABBING RESISTANCE

4.5.1 INTRODUCTION

Fabric stabbing or penetration by sharp blade has the following scenarios: cutting, quasi-static puncture, and dynamic puncture.

Weave designs represent the yarns interlacement, as shown in Figure 4.29. The knife cutting tip faces the interlacement of yarns or between the yarns, so the resisting force will be different. During penetration of the knife tip through a material, yarns in contact with the knife edge have high stress and tend to be stretched and spliced and will move in the direction of fill and warp to allow the edge of the knife to pass through. The fabric will be deformed under the build-up force; furthermore, the edge of knife will start to cut the yarn.

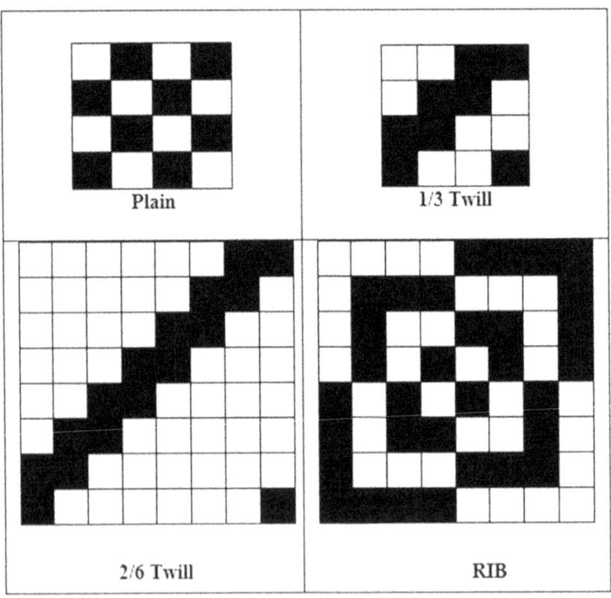

FIGURE 4.29 Fabric designs.

The fabric stabbing force represents a function of the yarn properties and their arrangement in the fabric (fabric design and specifications). Moreover, if the angle between the blade surface and the warp direction (Fig. 4.19) more than 0°, both warp and fill yarns will be cut. For the cutting angle $\alpha = 90°$, all cut yarns are warp, if $\alpha = 0°$, all cut yarns are fill, while several fill and warp yarns will be cut if $0 < \alpha < 90°$, depending on the blade width and warp and fill densities (Fig. 4.30).

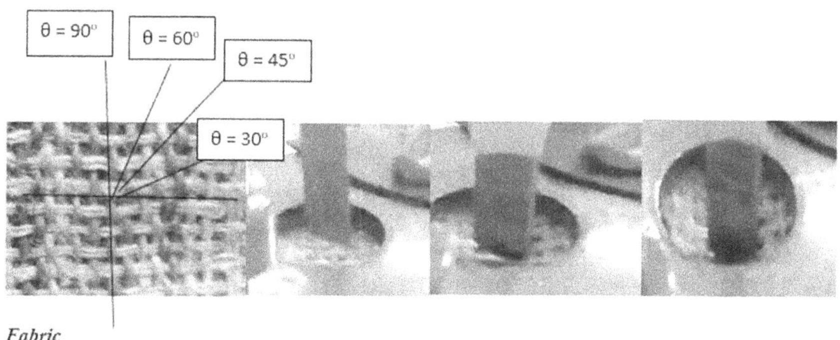

FIGURE 4.30 The cutter penetration through the fabric at different angles θ.

4.5.2 ENERGY BALANCE FOR FABRIC QUASI-STATIC STABBING RESISTANCE

The total energy required to penetrate strongly relies on the stabbing energy EF_{cutting} to cut the yarns by the edge of the blade and it consists of several components: yarns cutting energy, yarn strain energy to deform all other yarns due to the penetration movement of blade, energy to overcome friction between blade and yarns, energy to move the fabric during blade penetration

$$EF_{\text{cutting}} = EY_c + EYs + EY_f + EF_d \tag{4.24}$$

The fabric cut force was found to be function of fiber and yarns properties, the yarn cross-section structure and shape, the number of yarns interlacements in the fabric design, the fabric tightness and fabric cover factor. Fabric stabbing resistance is higher if the friction between the yarns and the sliding blade, fabric tightness, yarn cutting force, yarn strength, yarn to yarn friction, and fabric stiffness are having a high value.

The cutting resistance of fabric is contingent on the incident angle of the blade. For vertical stabbing $\theta = 90°$, the yarns in the fabric are cut under shear and tension stresses. If fabric is placed on a horizontal surface and the blade is drawn over it $\theta = 0°$, the fabric failure is under shear and compression stresses. If the fabric is slashed with a knife, its failure will be in tension, shear and compression stresses.

The shapes of the failure of the fabric under stabbing force by knife or spike are diverse (Fig. 4.31).

FIGURE 4.31 (See color insert.) Different examples of stabbing cut shapes.

4.5.3 CUT RESISTANCE OF YARNS

The yarn cutting force plays a critical role in predicting the protective performance of the fabric, the failure of the yarn under cutting blade needs to be carefully analyzed.[50] The yarn failure relies on the structure of the yarn, either staple or filament yarns.

4.5.3.1 YARN STRUCTURE

The macrostructure of yarn is divergent owing to fiber properties, spinning process, value of the inserted twist. The yarn structure may be voluminous or compact; smooth or rough or hairy; soft or hard; round or flat; thin or thick; even or irregular.[51] Figure 4.32a illustrates the longitudinal section of the spun yarn. Generally, there are several types of the yarns classification: spun or multifilament yarn, single or plied, low or high twisted yarns. The yarns from short staple fibers utilize the twist to lay the fibers in a helix form around their axis. The integrity of yarn is based on the fact that the coaxial helical forms of the fibers will insert pressure on the fibers layers in the yarn cross section with increasing intensity from outside layers further toward the maximum value at the yarn center. Spun yarns under tension will fail when 55–62% of the fibers in their cross section failed, on the contrary to multifilament yarn, 100% of its filament will be broken upon its failure. Consequently, the yarn structure determines the yarns failure during cutting owing to:

- The cutting of the fiber at the outer layer will reduce the pressure on the inner layer and therefore reducing yarn tenacity.
- The cutting of the fibers will take place at inclined angle to fiber axis which is varied from layer to layer, the slicing angle relies on the value of the twist.
- The pre-tension affects the number of fibers associated with the yarn cutting force.
- The possibility of compactness of the yarns undercutting force depends upon their packing density.
- The distribution of the radial packing density varies from the yarn center to its surface.

4.5.3.2 YARNS CUTTING RESISTANCE

The yarn in cutting test is fixed at both ends after application of pre-tension (Fig. 4.33). The blade is moving toward the yarn, creating a stress on the yarn's surface, which continuously is increased by the downward blade displacement. When the blade initially contacts the fibers, the yarn tension increases the stress between the outer layer fibers and the blade edge, causing a shear of the fibers, and the blade further penetrates in the successive layers of the fibers in the yarn cross-section. The fibers in all layers are subjected to tension so their failure will follow the same mechanism discussed above. As mentioned before, the packing density of the spun yarns varies across the yarn diameter (Fig. 4.32b). Consequently, the yarn cutting resistance surges as the blade penetrates toward the yarn axis till it passes through it, then the cutting force drops till the cutting of all the gripped fibers under shear and strain forces. The spun yarns have hairiness and have mostly a free end that cannot be cut by the blade; that reduces the cutting yarn force. At the same time, the cutting of the outer layer fibers will reduce the pressure on the other fibers in the inner layers, and slippage occurs. As the cutting blade penetrates further, the yarn loses its integrity, accelerating its failure.

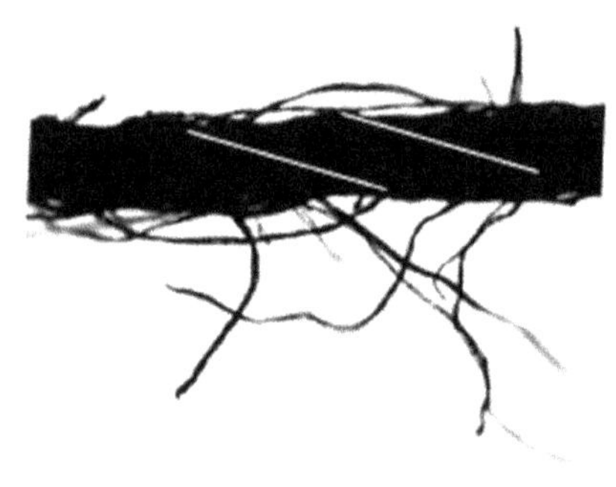

(a) Ring spun

Continuous multifilament yarn

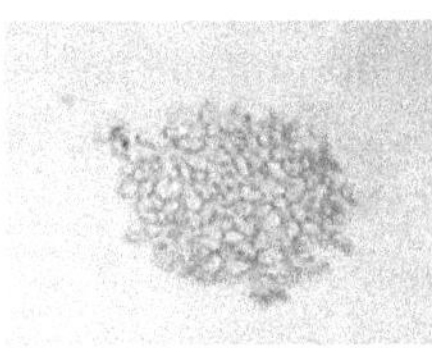

Ring spun yarns

(b)

FIGURE 4.32a,b Yarn structures (a) longitudinal and (b) cross sectional.

Continuous multifilament yarns have high packing density, thus all filaments will participate in the yarn cutting force. It was revealed that the cutting force of Zylon yarns has higher value than that of Kevlar and Spectra at all values of slice angles, blade sharpness, and pre-tension loads.[52] The Vectran HT and NT show higher cutting resistance than Aramid or HMPE yarns.[53]

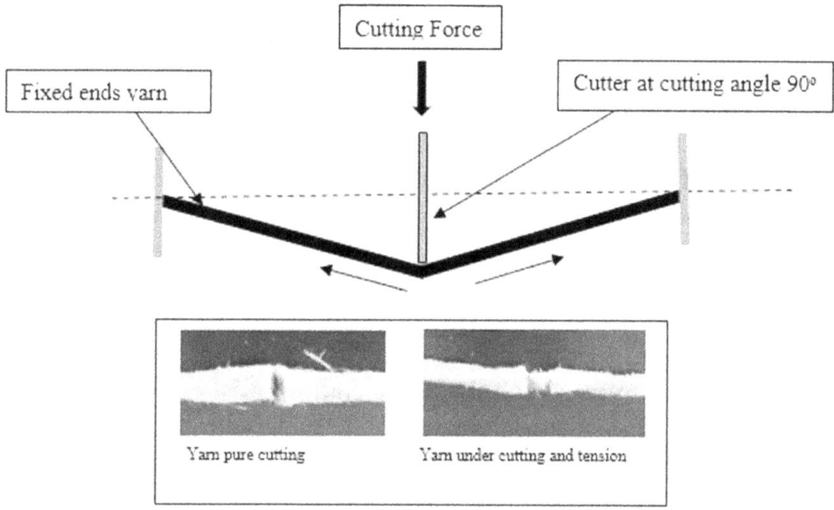

FIGURE 4.33 Mechanism of yarn cutting.

The effect of blade of slice angle is dramatic; the cutting energy falls steeply as the cutting angle is reduced. It was revealed that cutting energy of Zylon yarn, 500 *den,* was reduced by 75% as the cutting angle decreased from 90° to 85°, and reached 95% at a slice angle of 45° (Fig. 4.34).[50]

4.5.4 CLASSIFICATION OF FABRIC CUT PROTECTION PERFORMANCE

Several standards address the material cut resistance when the fabric is fixed on a mandrel and subjected to a cutting edge under a specified load related to a cutting action by a smooth sharp edge across the surface of the material, ASTM F1790-97, ASTMF1790-05. The cut protection performance (CPP) tester is widely used for the classification of cut resistance level, weight in grams can be applied to a razor blade, while moving the blade over the fabric without cutting through the fabric more than 20 mm. The ranking of CPP has five levels: 0 < 200, I ≥ 200, II ≥ 500, III ≥ 1000, IV ≥ 1500, V ≥ 3500 g. For EN 388 standards, CPP level for cut resistance is measured using circular, free-rotating blade under pressure of a standard weigh 5 N, which moves backward and forwards over the surface of the tested material over a fixed stroke length. The test's result is the number of cycles taken for the blade to cut through the material. To take the sharpness of the blade into the account, the test is performed using a standard material before and after testing the sample, the mean of these two tests on the standard material is defined as blade cut index 1. The test's result is the ratio of the number of cycles, required to cut through the sample, to the number of cycles, required to give blade cut index 1. The fabrics with high warp and fill density, high tightness plain weave, higher thickness show higher cutting resistance than the other fabric designs.[54] Less force and energy are required to puncture the fabric when the plane of the blade is parallel to a direction of tensioned fabric than when perpendicular.[15] In all cases, the fabric's in-plane tension in the direction of the warp and fill has a significant influence on the fabric cutting resistance force. For the ASTM F 2992-15 standard, CPP levels of the material expanded up to nine levels: 0 ≥ 200, I ≥ 500, II ≥ 1000, III ≥ 1500, IV ≥ 2200, V ≥ 3000, VI ≥ 4000, VII ≥ 5000, and VIII ≥ 6000 g. The cut through distance is 25.4 mm.

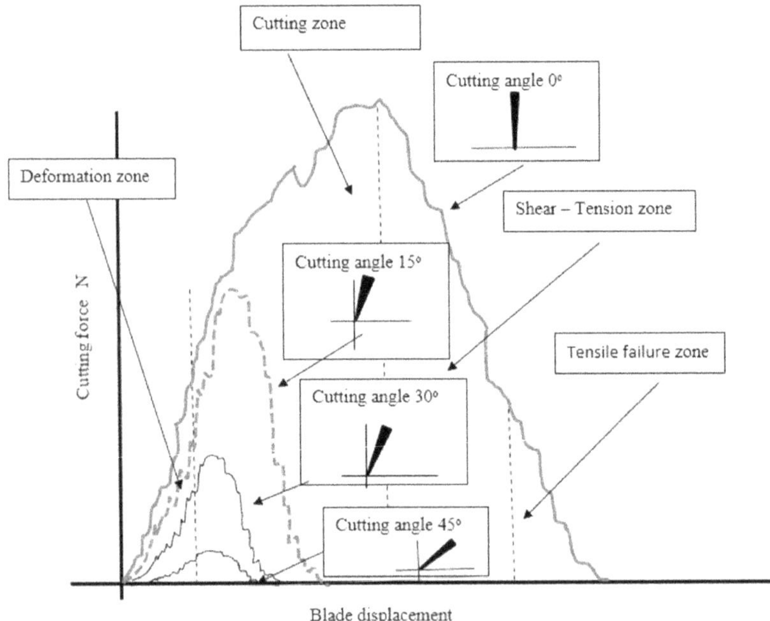

FIGURE 4.34 An illustration of yarn cutting force diagram at different cutting angles.

4.6 ANALYSIS OF THE MECHANISM OF FABRIC DYNAMIC STAB RESISTANCE

The total energy required to penetrate a fabric intensely relies on the lost energy to cut the yarns, the yarns straining around the edge of the blade, energy to overcome the friction between the blade surface and the surrounding surface of a fabric as well as the fabric deformation energy. Flexible stab-resistant fabrics have been widely used for military and civilian purposes. However, numerous efforts have been taken to construct better lightweight soft protective armors, for instance using high-performance fibers. Recently, the studies on the protection against impact by sharp blade or spike have been studied.[2,20,55,56] Resistance to the penetration is the most essential parameter considered for estimation of useful properties of the fabrics designed for protection against various types of mechanical impact, moreover, a soft vest should also provide comfort for the user, a lightweight, and cost-effectiveness.[55] Nevertheless, a deep understanding of the design of the protective stab resistant clothes to minimize the weak points, such as heavyweight, bulky shape, and uncomfortable characteristics, should be exposed.

During the interval of blade penetration time, the cutting force will increase till full penetration, and then it will gradually decline.[21] To increase the cutting performance of the fabric, it is expected to involve each component of the total fabric energy $EF_{cutting}$: use yarns having high cutting force, increase the friction between the blade and the yarns, increase the resistance to moving of the yarns in warp and fill directions due to the blade penetration, and increase the yarn pulling-off force.[21]

The fabric stabbing with sharp blade of mass (m) moving with velocity (V) will start when its tip touches the surface of the fabric; since the textile material has enough elasticity, it starts to bend under the dynamic force. The blade kinetic energy:

$$KE = 0.5\, m\, V^2. \tag{4.25}$$

There is a substantial difference in the behavior of the fabric under stabbing and bullet shooting, bullet's mass is very small, however, the impact velocity is very high, whereas a knife mass is larger than the bullet, but impacts are at very low velocity. Nevertheless, because of the pointed nature of a knife or spike tip, the impact energy density is actually higher.[34] The impact force of the blade will strain the fabric at the area of contact.

The strain will propagate through the fabric creating a resistance force to the blade to move forward. The strain energy of the yarn per unit volume:

$$EY_s = 0.5\, \sigma_y(t)\, \mathcal{E}_y(t), \tag{4.26}$$

where $\sigma(t)$ is yarn stress as a function of time, $\mathcal{E}(t)$ is yarn strain as a function of time.

The velocity of strain propagation V_s can be expressed by the following equation[57]:

$$V_s = (E_y/\rho_y)^{0.5}. \tag{4.27}$$

In the impact, faster moving longitudinal waves help to scatter the impact energy through the yarns intricate at the impact area.[58] The investigation of the relation of the yarns acoustic dynamic modulus revealed that it relies on their twist factor and fibers orientations. Sonic modulus was found to be affected by the blending ratio of the different components.[59–61] Fabric dynamic moduli of elasticity are dependent on the arrangement of the fibers within a fabric.[61] Also, it was specified that the transformation of fiber into yarn results in a drop in the level of sonic modulus, it represents 0.05 to 0.6 of the fiber modulus, while the weaving of yarn into fabric reduces the sonic modulus of fabric to from about 0.017 to 0.25 that of the fibers. The interrelation between sound velocity, fibers alignment in a fabric, material anisotropy, strength, stiffness, elasticity, resilience, and fatigue properties was verified.[62,63] As the rate of strain propagation becomes low, the cutting mechanism starts earlier, and less area of material will be stained under dynamic stabbing. Figure 4.35 shows the parameters affecting the strain propagation velocity of the fabric.[63]

4.6.1 ENERGY BALANCE FOR FABRIC PUNCHING

As the yarns are fully strained, the blade starts to cut yarns in contact with till it fully passed through the fabric. Figure 4.32 exemplifies several cases that characterize the blade penetration through the different types of the fabrics.[21] The energy balance during blade penetration, to absorb the kinetic energy *KE* before full penetration of the blade, will be equal to the sum of the fabric cutting energy "EF_{cutting}."

The cutting energy = blade kinetic energy − fabric energy resistance *ER*.

Kinetic energy absorbed by the fabric "*ER*" is defined by five different components.

Therefore, the total cutting energy of the fabric can be expressed as:

$$EF_{\text{cutting}} = EY_{\text{cutting}} + EY_{\text{friction}} + EY_{\text{tension}} + EY_{\text{deformation}} + EF_{\text{cone}}.$$

All components of the fabric cutting energy depend on fabric structure and specifications as well as fiber, yarn, and fabric properties.[58]

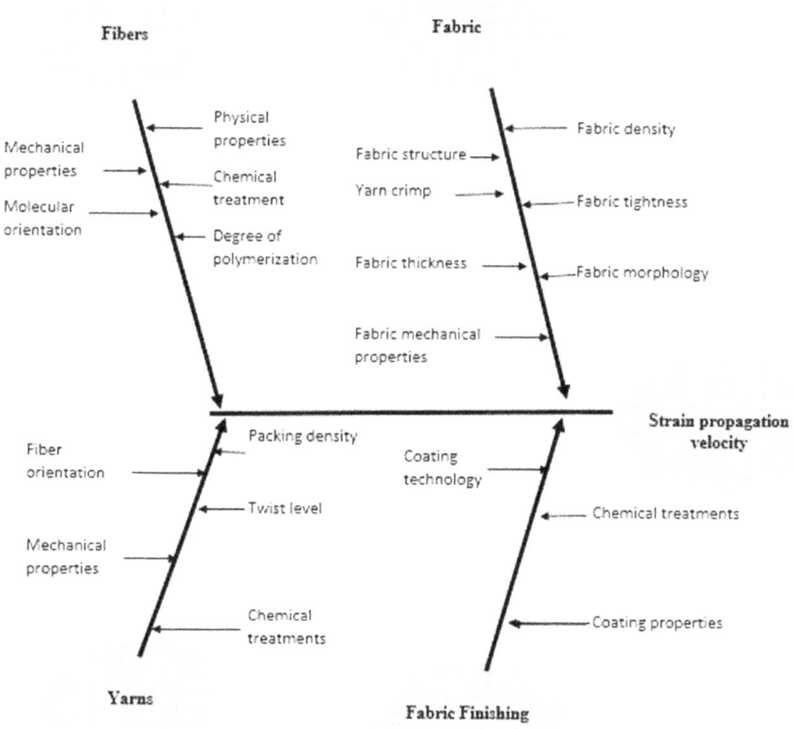

FIGURE 4.35 Analysis of main factors affecting strain propagation velocity of fabric.

4.6.2 *EVALUATION OF FIBER RESISTANCE CAPACITY*

Principally, the fabric stabbing resistance is a fiber property dependent. An index for ranking the fiber stabbing performance was introduced (*FCI*)[21]:

$$FCI = (TS\ V_s)\ \text{(N/ms)}. \tag{4.28}$$

The punching force measured on drop mass tester may be expressed as:

$$\text{Fiber punching force (N)} = k1\ (FCI)^{0.46} \tag{4.29}$$

$k1 = 2.8729^{21}$.

The specific stabbing force:

$$\text{specific punching force single layer (N/g/m}^2\text{)} = k2\ (FCI)^{0.46}, \tag{4.30}$$

$k2$ is a constant depending on the specifications of the fiber material and fabric and sharpness of the impactor.

The multilayer fabric performance is better than of one layer with the same areal density, since its stiffness will be more than multilayer fabric. Fabric weaved from finer filament has less fabric stiffness.

Figure 4.36 shows the specific stabbing force N/g/cm^2 according to *FCI* values.

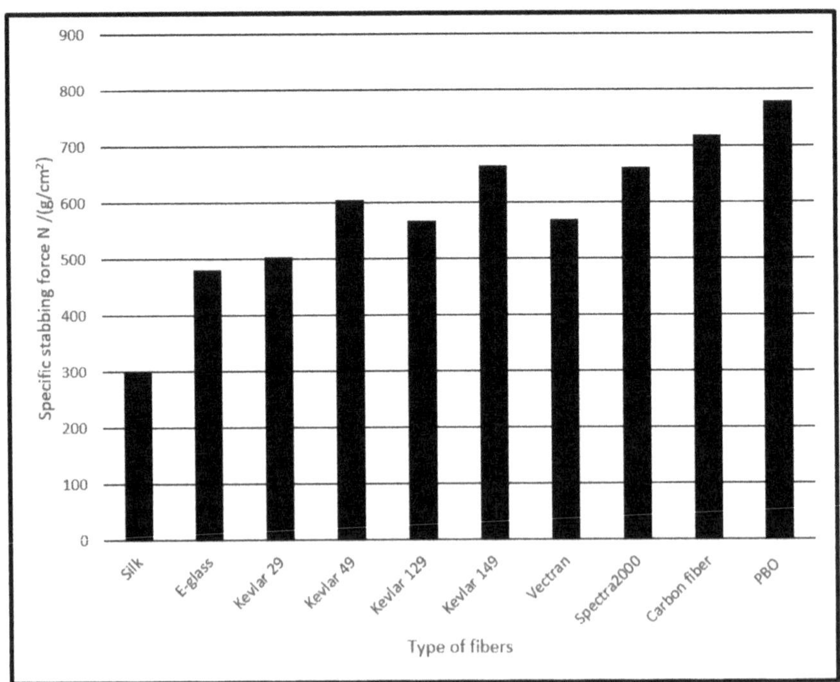

FIGURE 4.36 Specific stabbing force versus FCI.

4.7 STAB RESISTANCE OF NATURAL FIBERS FOR SOFT PROTECTIVE ARMOR

4.7.1 INTRODUCTION

Cutting and stab performance of the protective fabrics is the main quality to be evaluated for the lots of the industrial applications to shield against various mechanical threats, stabbing, punching, cutting, and needle piercing.[20] Studies pursuit the capability to increase the stabbing energy absorption by the use of multilayer compositions and optimizing them for applying into suitable stabbing protective fabrics from natural or synthetic fibers and their blends.[20,64] The stab resistance is often calculated by taking an average value between the cutting and tearing or tensile strength. Material tear resistance relies on fabric strength, elongation, and packing density. More loosely packed fabric can be an advantage for tear resistance, but it inadequately resists the stabbing. One of strongest natural fibers is silk, it was revealed that as protective fabric, silk multilayer woven fabrics have high penetration resistance and provide more comfort in use. Additional fabric layers contribute to the increment of anti-stabbing fabric with the areal density adequate to the degree of protection required. The stabbing force of silk fabric is linearly related to its weight/m^2 (Fig. 4.37).

4.7.2 IMPROVING THE STABBING PERFORMANCE OF PROTECTIVE ARMOR

To improve the stabbing resistance of a material for protective armor, several suggestions can be considered[13,64–66]:

- Use the Nitinol fibers as the reinforcement for fabrics made from cotton/nylon and cotton/polyester to give the fabrics increased cut, puncture, and tear resistance without decreasing the flexibility or comfort of the fabric.
- Use multilayer silk fabric supported on a net of high-performance material, either plain weave or triaxial fabric made of para-aramid or polyester.

- Use composite structure of natural fabrics layers, fabric pads, and layers of high-performance fabric.
- Use shear thickening fluid (STF) structures.
- Use high-density nonwoven structures.
- Use hybrid fabric layers from natural fibers with layers with other HPF fabrics.

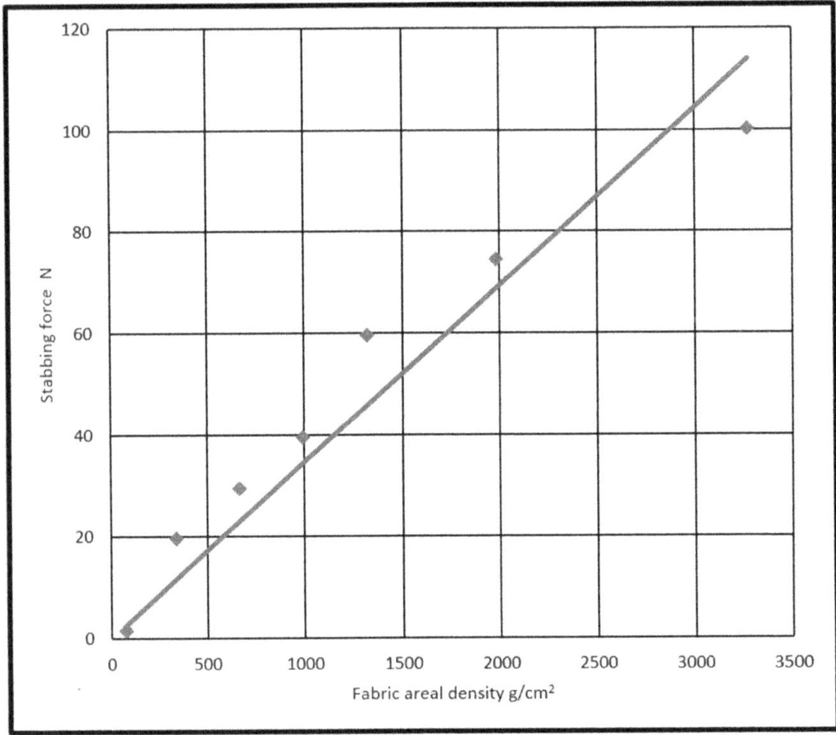

FIGURE 4.37 Stabbing force versus areal density of silk fabric.

4.7.2.1 TRIALS

4.7.2.1.1 Natural Fiber–HPF Composite

Figure 4.38 illustrates that the composite structure of silk fabric, supported on one layer of Kevlar triaxial fabric (TAF), increases the stabbing resistance force to knife penetration to perforate the structure by more than

100%. The fabric is deformed in three directions of the yarns, dissipating more energy before the blade can pass through. The use of a high tenacity polyester (HTPES) net under the silk fabric was effective in expanding the resistance to stabbing knife to penetrate.[20] The same effect was noticed when wool/polyester fabric and HTPES weave were used (Fig. 4.39). The stabbing resistance force of the composite structure from silk fabric supported on one layer of HTPES fabric may increase by 60%.

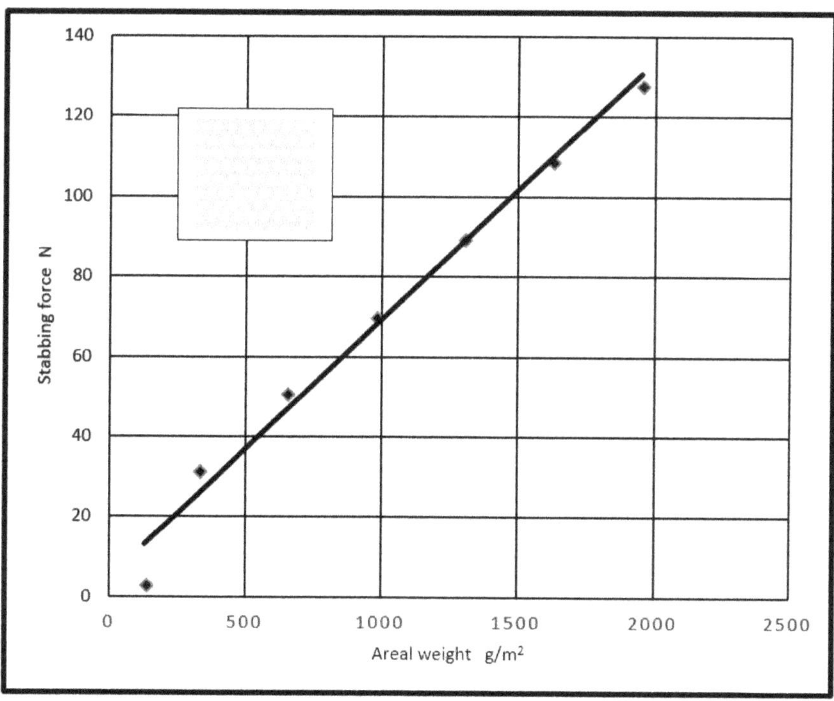

FIGURE 4.38 (See color insert.) The stabbing force of the composite structure of silk fabric supported by layer of Kevlar fabric.

4.7.2.1.2 Knitted-Triaxial Composite Structure

The application of triaxial fabric with other materials can improve its stabbing resistance performance. The triaxial fabric is different in structure from knitted or woven fabric, as mentioned by several investigators.[12,14,20,64] Figure 4.40 shows that the stabbing resistance force of the triaxial fabric linearly increases with number of layers. The combination of layers from polyester knitted fabric with 16% of Lycra yarn and HPF triaxial fabric improves the stabbing resistance performance of the structure,[21] especially, with higher number of layers of the knitted fabric as demonstrated in Figure 4.41.

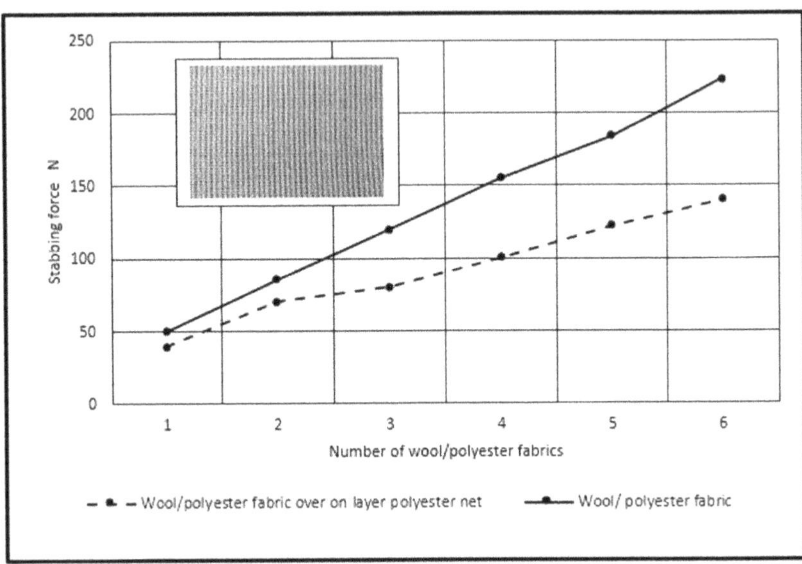

FIGURE 4.39 The stabbing force of the composite structure of wool/polyester fabric supported by layer of HTPES net.

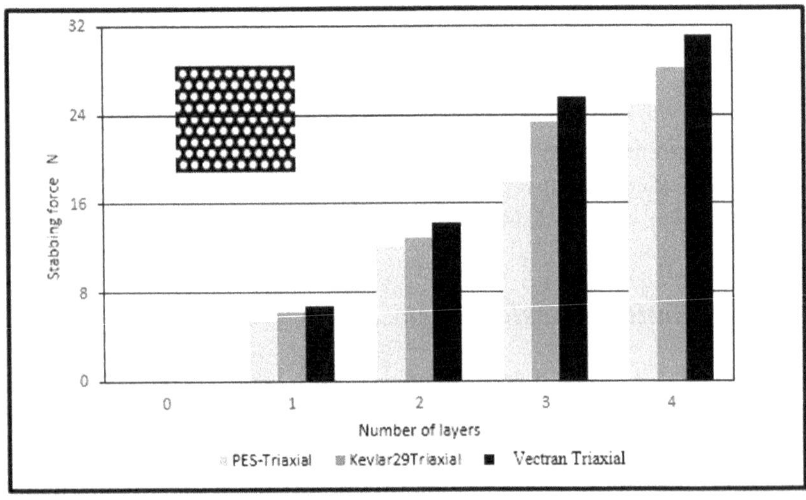

FIGURE 4.40 The stabbing force of triaxial fabric.

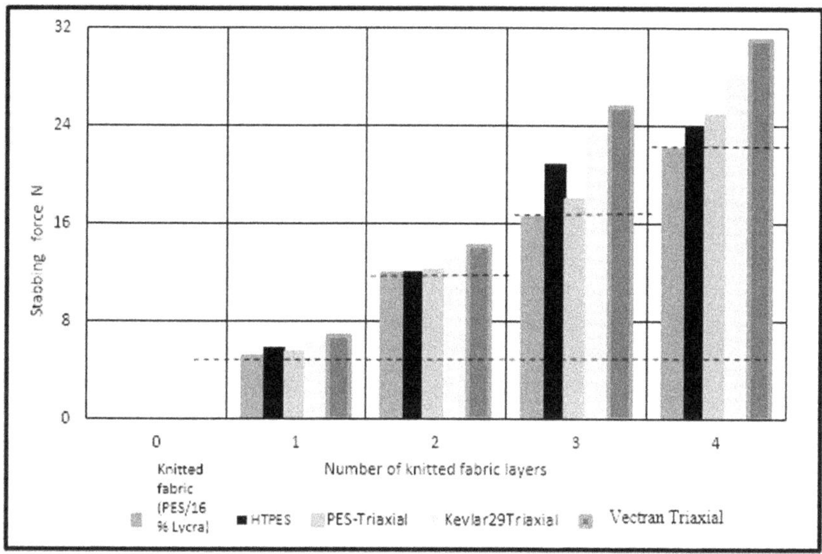

FIGURE 4.41 Stab resistance of multilayer knitted fabrics with HPF triaxial fabric.

4.7.2.1.3 Sandwich Pad Structure

To increase the armor stabbing resistance, the pads are made from fibers and covered by HPF fabric layers to form a composite sandwich pad structure[20]. Polyester fibers structures of several weights, supported by different types of high-performance fabrics with various areal densities, and covered with PES-Lycra fabric, were suggested.[9,20] The cutting energy required to cut through a pad material involves some other energy components: a lost energy dissipated by the lateral force exerted by the pad material on the blade sides and a cutting energy of entangled fibers at the tip edge of the blade. These energies affect the measured cut resistance force; the lateral force contributes to the increased material cut resistance, while the one at the tip of the blade corresponds to the necessary energy to cut through the pad material. The blade is entangled with the fibers, resisting the knife motion that increases the total stabbing resistance force. The usage of the designed pads will upsurge the stabbing resistance force controlled by the friction distribution between the fibers of the pad. The tip of the knife will be resisted by entangled fibers that will move when the force is applied to overcome the movement of the knife. The stabbing resistance energy of the pad sandwich structure for different weights is given in Figure 4.42. The packing density of the pad is the controlling factor in this mechanism of stabbing.

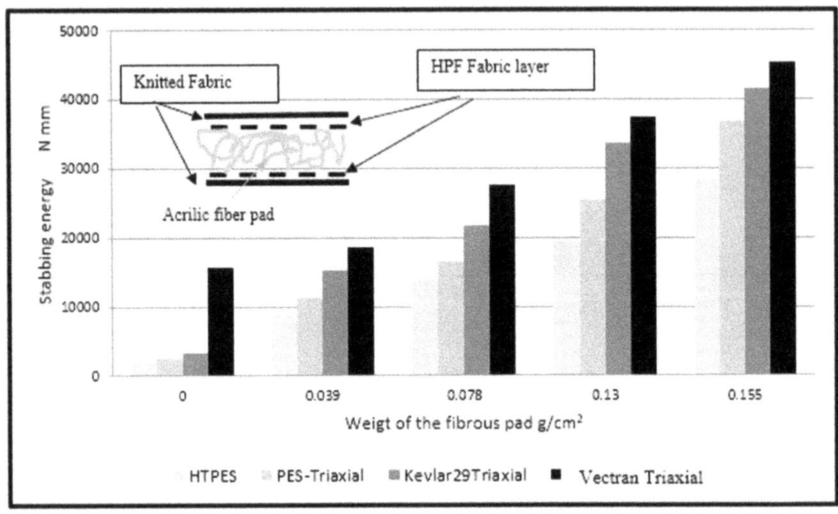

FIGURE 4.42 Stabbing energy for different sandwich pad structures.

The specific resistance stabbing force (N/g/m^2) of fabrics in combination of composite structures, for instance, silk fabric, silk pad on supporting net of HTPES, Kevlar fabric, improves their stabbing resistance force. The values of the specific resistance stabbing force depend not only on the material and structure but also on the fabric design and specifications, speed of stabbing, and sharpness of the knife.

4.7.2.1.4 High-Density Nonwoven of HPF

The application of high-density nonwoven as anti-stabbing structure enhances the stab resistance of armors.[67] The fabric is prepared using high-performance fibers or blend with other thermal fibers under hot press to reach the fabric density up to 0.46 g/cm^3. The stabbing resistance force of the high-density nonwoven fabric can withstand high stabbing energy however, it is mostly appropriate for anti-trauma plates. For stab resistance armor, the nonwoven structures were improved through the embedded layers of high-performance fibers within the nonwoven layers of the needle punched or thermal bonded.[13]

4.7.2.1.5 HPF Reinforced Polymer Composite

The high-performance fiber reinforced polymer composite structures are a field for improving the stab resistance of the material, either for flexible or hard composites. This structure is a good candidate to construct hard anti-stabbing plates.[49,68] The analysis of the failure of such composites indicates the appearance of several cracks, as the blade penetrating the composite till the full penetration, all the yarns and the matrix were completely engaged under the punching load of the blade. Figure 4.43 indicates the mechanism of failure of the composite under punching force.

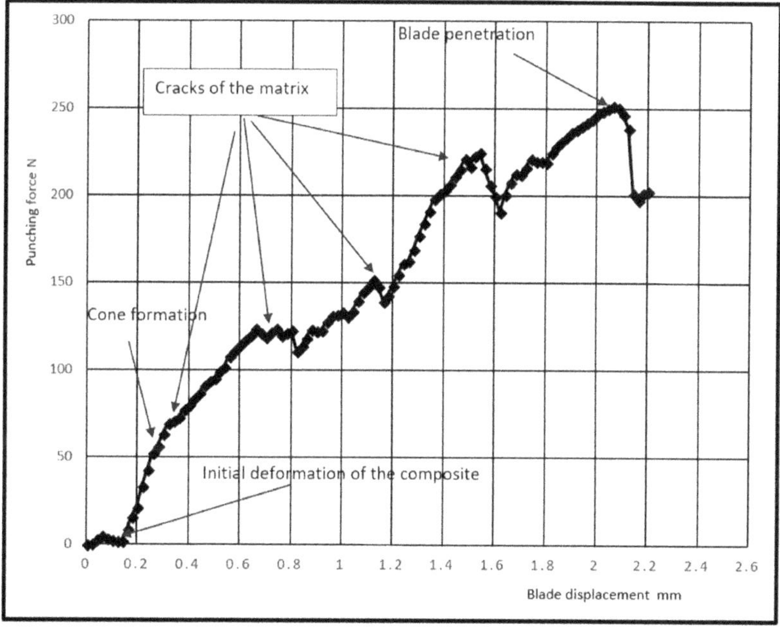

FIGURE 4.43 (See color insert.) Punching resistance of the triaxial fabrics reinforced polymer composite.

The triaxial fabric reinforced polyester polymer with 2% nano glass particles composite improves the punching resistance of the triaxial fabric.[49,68]

4.7.2.1.6 *Shear Thickening Fluid Structures*

The shear thickening fluid (STF) is a combination of hard metal oxide particles suspended in a liquid polymer. This mixture of flowable and hard components at a composition results in a material with the remarkable properties. The results showed that STF impregnated fabrics have better penetration resistance without affecting the fabric flexibility.[66,69,70] The stab resistance of STF was found to have significant improvement over fabric targets of the equivalent areal density, especially for spike punching threat,[71] due to the improvement in the energy dissipation in the penetration zone. This behavior could also explain the high elongation of the fabrics, combined with the slow loading rates, that may have allowed the fabrics to stretch rather than to be cut or punctured.[72] The knife stab resistance performance of the STF strengthened fabric was dominated by hardness of the STF particles. The fabric strengthened by the SiO_2-based STF exhibited the highest knife stab resistance.[66] The quasi-static stab-resistance of the STF/glass fabric composites offers the superior stab protection as compared to the glass fabric targets for knife threats.[73] Nano-silica-based STF is used with Kevlar fabric to design soft flexible composite fabric for stab protection; the fabric composite shear modulus will change due to the presence of the thickening material. The stab resistance properties of UHMWPE, Twaron, Nylon fabrics were improved by the impregnation of STF. Critical increase in stab resistance is reached as the mass fraction of Nano-silica in the STF was increased. It was suggested that STF is the optimal for knife and stab protection.[34]

4.8 STAB RESISTANCE PROTECTIVE ARMOR DESIGN CRITERIA

Armor design reflects an optimization between protection, comfort, and mobility it offers to the wearer. Lighter and thinner armor is a top criterion for most manufacturers of protective armor. The important obstructive aspects of protective armor are its heat retention factor and weight that makes wearer extremely uncomfortable. The heat retention factor elevates with temperature, especially when the vest is being worn in hot or humid conditions.[74]

The protective armor should fit the protected person properly to ensure full cover for the areas to protect all the vital organs and unconstraint movement of the wearer due to armor ergometry. The protective armor is usually manufactured in five sizes and provided with Velcro straps to make it a comfortable fit. It was revealed that components of the force during the impact phase are: axial force, cutting force, lateral force, and torque which is the smallest one.[75,76]

4.8.1 SOFT ARMOR

The design criteria of the protective armor rely on the type of threat and the level of protection for which it is designed (stab, punching, slashing). Some published works suggest that skin penetration typically requires ~ 35–55 N, depending upon the weapon involved,[77] while forces generated by stabbing samples with knives and screwdrivers are completely different. The forces, generated by mild, moderate or severe stabbing are critically much greater than the forces required for skin penetration.[78] The average value depends on the type of weapon and varies between 45 and 250 N. The flat head screw driver shows the highest value, while the steak knife has lowest stab resistance. It was exposed that the maximum average punching energy of the underarm was 63.4 J and the overarm tests showed a maximum of 115 J (the 95% values of underarm and overarm were 54.9 J and 77 J, respectively), that should be considered when designing a protective armor.[79]

According to the standards, the armor should withstand stabbing energy conferring to level of protection at 24–65 J, allowing maximum penetration 7 mm at low energy threats (24–43 J) and 20 mm over test strike (36–65 J).[33] The median organ/artery to skin distances is typically 15 mm or greater, the blade penetration more than this depth could cause serious harm.[80] Additional cutting energy is absorbed by each successive layer of fabric of the armor, till the knife has been stopped. The armor should absorb these levels of stabbing energy.

The design of the armor has several fabric layers that enlarge the area of the fabric involved in the confrontation with the stabbing object.

Fibers recommended for the ant-stab protective fabric should have high values of:

1. cutting resistance,
2. strain energy, and
3. coefficient of friction.

Several types of HPF, such as Kevlar® brand fibers (Kevlar® 29, Kevlar® 49, Kevlar® 129, Kevlar® Protera), Spectra® fiber, Twaron, Dyneema®, Polyethylene (UHMWPE), Zylon®, Technora®, GoldFlex®, are good candidates for flexible stab-resistant fabrics for their low density, high tenacity and modules and high cutting force. Flexible protective armor can be also made from multilayered fiber material or a blend of stapled fibers with a metal core.[81]

The high tightness plain weave fabric design is recommended for more effective stop of stabbing. The number of fabric layers is a basic concept in the design of armor performance with minimum weight and maximum comfort. New high-performance fibers have decreased the weight of the protective armor with improved comfort. The weight of the armor to the protected area was suggested as one of the measures for the protective armor.[82]

4.8.2 IN-CONJUNCTION ARMORS

Many hard plates are designed to be used with a specific soft armor panel to achieve a desired level of protection. The hard plates are inserted in a pocket in the flexible armor in the areas which required extra protection. Both soft armor and hard plate must be in compliance with the protection standard levels separately.

4.8.3 COMBINATION ARMORS

Those are designed to protect against both firearms and edged or stabbing weapons. In these armors, the panels contain layers of materials that are stab resistant as well as layers of materials that are ballistic resistant.[83]

4.8.4 TRAUMA PLATES

They are designed with the main purpose to reduce blunt force trauma injury, nonpenetration injuries,[84] and work as a cushion media. If the back-face signature exceeds a depth of 44 mm, there is a critical chance that a fatal injury could occur. The design should comply with the standards.[85–88] As it was discussed above, the stab resistance failure mode is different when target is hit with knife or spike. The main influencing factors for the cutting and puncture performance of the trauma plate are as follows:

1. fiber cutting force,
2. fiber elastic modulus,
3. yarn structure,
4. yarn tenacity,
5. yarn to yarn friction,
6. weave design,
7. fabric architect,
8. fabric tightness,
9. fabric weight,
10. fabric comfort, and
11. plate structure.

As it was proven, the number of fabric layers is proportional to its quasi-static puncture resistance of the armor, this will help the designer to calculate the number of fabric layers required to get the anticipated level of armor rank. The stabbing energy of the armor must comply with the standard, NIJ Standard–0115.00, and varies for energy level E1 between 24 and 43 J with maximum penetration 7 mm and for energy level E2 from 36 to 65 J with maximum penetration 20 mm.

The energy absorbed (E_{fabric}) by one layer of the fabric under the applied stabbing force (P_{fabric}) for the distance (h), over which the interaction takes place, is:

$$E_{\text{fabric}} = P_{\text{fabric}}\, h. \tag{4.31}$$

For n_{fabric} layers of the fabric

$$P_{\text{n}} = P_{\text{fabric}}\, n_{\text{fabric}} = P_{\text{standard}}. \tag{4.32}$$

$$\text{The number of fabric layers } (n) = P_{\text{standard}} / P_{\text{fabric}} = E1/E_{\text{fabric}}. \tag{4.33}$$

The fabric stabbing energy relies on the fabric structure, for instance for pain weave, twill, satin Kevlar fabrics the punching energy was found to be 1.35 J, 0.9 J, 0.5 J, respectively.[56] Consequently, the number of layers required for the armor of the first class at level E1 is 71, 87, and 192 layers of plain, twill, satin fabric, respectively. The stabbing energy of the single layer of UHMW-PE fabric is 1.82 J when it is punched with a conical probe,[89] then the number of layers to pass NIJ Standard–0115.00 Class 1 will be equal to 14 layers.

The total weight of the armor determines the choice of the fabric specifications that satisfy the two parameters—weight and the type of fiber which have the highest stabbing resistance energy.

The specific fabric stabbing energy will be:

$$\text{weight of fabric } W_{\text{fabric}}\ (\text{GMS}) = [N_{\text{fill}}\ (1 + C_{\text{fill}}\ \%)\ tex_{\text{fill}} + N_{\text{warp}}\ (1 + C_{\text{warp}}\%)\ tex_{\text{warp}}]/1000. \quad (4.34)$$

$$\text{Specific stabbing energy of fabric} = E_{\text{fabric}}/W_{\text{fabric}} \quad (4.35)$$

$$= E_{\text{fabric}} / \{[N_{\text{fill}}\ (1{+}C_{\text{fill}}\ \%)\ tex_{\text{fill}} + N_{\text{warp}}\ (1{+}C_{\text{warp}}\%)\ tex_{\text{warp}}]/1000\}. \quad (4.36)$$

The next consideration is the penetration distance to stop the knife with no more than the stated values in the standards. Fiber energy absorption capacity index (FCI) can give preliminary selection of the type of a fiber the designer is looking for, to optimize the different parameters that determine the armor performance. To design the proper protective armor with a desired level of protection can be quite inspiring. It is a closed loop of innovation, implementation, and testing for all the elements of the protective armor to fulfill the requirements of the compliancy to the standard regulations, such as NIJ 0115.00, VPAM KDIW, or HOSDB.

KEYWORDS

- **comfort**
- **cut resistance**
- **cutting resistance**
- **protective fabrics**
- **soft armor**
- **stab resistance**
- **trauma plates**

REFERENCES

1. Cavallaro, P. Soft Protective Armor: An Overview of Materials, Manufacturing, and Ballistic Impact Dynamics. http://www.dtic.mil/dtic/tr/fulltext/u2/a549097.pdf (accessed Mar 13, 2018).
2. Choi, H.; Hong, T.; Lee, E. et al. In *Stab Resistance of Aramid Fabrics Reinforced with Silica STF*, 18th International Conference on Composite materials, Jeju Island, South Korea, 21–26 August 2011; The Korean Society of Composite Materials.
3. Yang, D. Design, Performance and Fit of Fabric for Female Protective Armor. Ph.D. Thesis, School of Materials, University of Manchester, Manchester, UK, 2010.
4. El Messiry, M. In *Study of Puncture Cut Resistance in Protective Fabrics*, ICCE-21 Tenerife, Spain, 21–27 July 2013. www.academia.edu/ (accessed Sep 3, 2014).
5. Egres, Jr.; Decker, M.; Halbach, C. et al. In *Stab Resistance of Shear Thickening Fluid (STF)–Kevlar Composites for Protective Armor Applications*, Proceedings of the 24th Army Science Conference, Orlando, FL, 29 November–2 December 2004.
6. Thang, C.; Vu-Khanh, T. A Study on the Cutting Resistance Characterization of Protective Clothing Materials. https://www.irsst.qc.ca/media/documents/PubScientifique/Affiches-posters/Nguyen-puncture-protective-clothing.pdf (accessed Mar 9, 2014).
7. Li, T.; Wang, R.; Wen, C. et al. Evaluation of High-modulus, Puncture-resistance Composite Nonwoven Fabrics by Response Surface Methodology. *J. Ind. Text* **2013,** *43* (2), 247–263.
8. Hassim, N.; Ahmad, M.; Ahmad, W. et al. Puncture Resistance of Natural Rubber Latex Unidirectional Coated Fabrics. *J. Ind. Text* **2012,** *42*, 118–131.
9. Militky J. *Cut Resistance of Textile Fabrics, Selected Topics of Textile and Material Science*; Liberec: TUL, CZ, 2011.
10. Ulbricht, V.; Franeck, J.; Schirmacher, F. et al. Numerical Investigations for Cutting of Wires and Thread. *AUTEX Res. J.* **2003,** *3*, 9–15.
11. Lara, J. In *The Effect of the Glove Material Stretch Deformation on Cut Resistance*. 3rd European Conference on Protective Clothing (ECPC) and NOKOBETEF 8 (3rd), 10–12 May 2006; Gdynia, Poland.
12. Scardino, F. L.; Ko, F. Triaxial Woven Fabrics. *Text Res. J.* **1981,** *51*, 80–89.
13. El Messiry, M. *Natural Fiber Polymer Composites Engineering*; Apple Academic Press Inc: USA, 2017.
14. Miller, S.; Jonesb, M. Kinematics of Four Methods of Stabbing: A Preliminary Study. *Forensic Sci. Int.* **1996,** *82*, 183–190.
15. Smith, S.; Dyson, R.; Hale, T.; Janaway, L. Development of a Boxing Dynamometer and its Punch Force Discrimination Efficacy. *J. Sports Sci.* **2000,** *18* (6), 445–450.
16. Neckář, B.; Ibrahim, S. Theoretical Approach for Determining Pore Characteristics Between Fibers. *Text. Res. J.* **2003,** *73* (7), 611–619.
17. Neckář, B. Compression and Packing Density of Fibrous Assemblies. *Text. Res. J.* **1997,** *67* (2), 123–130.
18. Pan, N.; Shang, X. Shear Strength of Fibrous Sheets: An Experimental Investigation. *Text. Res. J.* **1997,** *67* (8), 593–600.
19. El Messiry, M. Investigation of Puncture Behavior of Flexible Silk Fabric Composites for Soft Protective Armor. *Fibres Text. East. Eur.* **2014,** *22*, *5* (107), 71–76.

20. El Messiry, M.; Eltahan. E. Stab Resistance of Triaxial Woven Fabrics for Soft Protective Armor. *J. Ind. Text.* **2016,** *45* (5), 1062–1082.
21. Jing, L.; Wanga, Z.; Ningb, J.; Zhaoa, L. The Mechanical Response of Metallic Sandwich Beams Under Foam Projectile Impact Loading. *Lat. Am. J. Solids Struct.* **2011,** *8*, 107–120.
22. BLAD. https://en.wikipedia.org/wiki/Blade (accessed June 29, 2017).
23. Jeremie, Plane. 5 Ways to Choose a Quality Knife. https://kotaikitchen.com/blogs/news/5-ways-to-choose-a-quality-knife (accessed June 29, 2017).
24. CASEY, M. Sharpening knives. https://www.caseyspm.com/knives.html (accessed Feb 28, 2018).
25. Needham, C.; Alexandrou, M. How to Choose and Use a Survival Knife. https://infolific.com/leisure/wilderness-survival/survival-knives/ (accessed Feb 28, 2018).
26. Knives: Sharpening Angles. http://woodlooking.blogspot.com/2014/10/knives-sharpening-angles.html (accessed Feb 28, 2018).
27. McCarthy, C.; Hussey, M.; Gilchrist, M. On the Sharpness of Straight Edge Blades in Cutting Soft Solids: Part I - Indentation Experiments. *Eng. Frac. Mech.* **2007,** *74*, 2205–2224.
28. Meehan, R.; Kumar, J.; Earl, M.; Svenson, E.; Burns, J. The Role of Blade Sharpness in Cutting Instabilities of Polyethylene Terephthalate. *J. Mater. Sci.* **1999,** *18*, 93–95.
29. Marsota, J; Claudonb, L.; Jacqmina, M. Assessment of Knife Sharpness by Means of a Cutting Force Measuring System. *Appl. Ergon* **2007,** *38*, 83–89.
30. Mulder, J.; Scot, J. The Measurement of Knife Sharpness and the Impact of Sharpening Technique on Edge Durability. https://core.ac.uk/download/pdf/44290183.pdf (accessed Feb 26, 2018).
31. Singha, V.; Dasa, M.; Kumar, S. Effects of Knife Edge Angle and Speed on Peak Force and Specific Energy When Cutting Vegetables of Diverse Texture. *IJFS* **2016,** *5*, 22–38.
32. Ankersen, J. Quantifying the Forces in Stabbing Incidents. Ph.D. Thesis, Department of Mechanical Engineering University of Glasgow, Scotland, 1999.
33. Stab Resistance of Personal Protective Armor NIJ Standard–0115.00. https://www.ncjrs.gov/pdffiles1/nij/183652.pdf (accessed June 29, 2017).
34. Nayak, R.; Ian Crouch, I; Kanesalingam, S.; Ding, J.; Tan, P.; Lee, B.; Miao, M.; Ganga, D.; Wang, L. Protective Armor for Stab and Spike Protection, Part 1: Scientific Literature Review. *Text. Res. J.* **2018,** *88* (7), 812–832.
35. Szymanek, M. Analysis of Cutting Process of Plant Material. *TEKA Kom. Mot. Ener. Roln. OL PAN* **2007,** *7A*, 107–113.
36. Kothari, V. Cut Resistance of Textile Fibers—Theoretical and Experimental Approach. *IJFT Res.* **2007,** 306–311.
37. Heuse, O. Damage to Clothing Caused by Stabbing Tools. *Archiv fuer Kriminologie* **1982,** *170*, 129–45.
38. Johnson, N. Physical Damage to Textiles. https://aic.gov.au/sites/default/files/publications/proceedings/downloads/18-johnson.pdf (accessed Feb 28, 2018).
39. Mayo, Jr.; Wetzel, E. Cut Resistance and Failure of High-performance Single Fibers. *Text. Res. J.* **2014,** *84* (12), 1233–1246.
40. Ding, H.; Zhan, Y.; He, Z. Fracture Failure Mechanisms of Long Single PA6 Fibers. *Polymers* **2017,** *9*, 243, 1–11 (accessed Mar 5, 2018).

41. Wang, D.; Jialu, L.; Yanan, Y. Stab Resistance of Thermoset-impregnated UHMWPE Fabrics. *Adv. Mater. Res.* **2010,** 134–136.
42. Kusińska, E.; Starek, A. Effect of Knife Wedge Angle on the Force and Work of Cutting Peppers. TEKA. *Comm. Motor. Ener. Agric.* **2012,** *12* (1), 127–130.
43. El Messiry, M., Sheet Forming of Woven Fabric Composite by Combined Cyclic Stretch and Deep Drawing. *J. Text. I* **2018,** *47* (7), 1681–1701.
44. El Messiry, M.; Mohamed, A. Analysis of Cyclic Load Die Forming for Woven Jute Fabric 3-d Reinforcement Polymeric Composites. *J. Ind. Text.* **2018,** *47* (7), 1681–1701.
45. Sard, S.; Bailey, A.; Briscoe, B. Extensions, Displacements and Forces Associated with Pulling a Single Yarn From a Fabric. *J. Appl. Phys.* **1987,** *20*, 130–139.
46. Ning, Pan. Analysis of Woven Fabric Strengths: Prediction of Fabric Strength Under Uniaxial and Biaxial Extensions. *Comp. Sci. Technol.* **1996,** *56*, 311–327.
47. Mahdi, S.; Kadivar, N.; Sajjadi, A. Analytical Assessment of Woven Fabrics Under Vertical Stabbing – The Role of Protective Clothing. *For. Sci. Int.* **2016,** *259*, 224–233.
48. Tables. http://www.roymech.co.uk/Useful_Tables/Mechanics/Plates.html_(accessed Mar 14, 2018).
49. El Messiry M.; Aloufy, A.; Abd-Ellatif, S.; Elmor, M. In *Enhancement of Quasi-static Puncture Resistance Behaviors of Triaxle Fabric/Polyester Hybrid Composites*. 7th International Conference on Recent Trends in Science, Engineering and Technology (RTSET—2017) London (UK), 29–30 June 2017.
50. Shin. H.; Erlich, D.; Simons, J.; Shocky, D. Cut Resistance of High-strength Yarns. *Text. Res. J.* **2006,** *76* (8), 607–613.
51. Technology of Short Staple-spinning/Yarn-formation. http://www.rieter.com/en/rikipedia/articles/technology-ofshort-staple-spinning/yarn-formation/assembly-of-fibers-to-make-up-a-yarn/yarn-structure/ (accessed Mar 14, 2018).
52. Shin, H.; Erlich, C.; Simons, J.; Shocke, D. Cut Resistance of High-strength Yarns. *Text. Res. J.* **2006,** *76* (8), 607–613.
53. Vectran. http://imattec.com/linked/vectran%20-%20technical%20data.pdf (accessed Mar 5, 2018).
54. Kothari, V.; Das, A.; Sreedvi, R. Cut Resistance of Textile Fabrics- A Theoretical and An Experimental Approach. *IJFT Res.* **2007,** *1*, 306–310.
55. Mahbub, R. Comfort and Stab-Resistant Performance of Protective Armor Fabrics and Female Vests. Ph.D. Thesis, School of Fashion and Textiles, RMIT University, Melbourne, Australia 2015.
56. Wang, Q.; Sun, R; Tian, X.; Yao, M.; Yan Feng, Y. Quasi-static Puncture Resistance Behaviors of High-strength Polyester Fabric for Soft Protective Armor. *Resul. Phys.* **2016,** *6*, 554–560.
57. Roylance, D. Ballistic of Transversely Impacted Fibers. *Text. Res. J.* **1977,** *47*, 679–684.
58. Sun, D.; Chen, X. Plasma Modification of Kevlar Fabrics for Ballistic Applications. *Text. Res. J.* **2012,** *82* (18), 1928–1934.
59. Křemenáková, D.; Militký, J. In *Acoustic Dynamic Modulus of Staple Yarns*. The 4th RMUTP International Conference: Textiles & Fashion 2012, July 3–4, 2012, Bangkok, Thaila.

60. Křemenáková, D.; Militký, J.; Pivonková, D. In *Structure and Mechanical Behavior of Polypropylene Yarns*. 7th International Conference—TEXSCI 2010, September 6–8, Liberec, Czech Republic.
61. Yan, A.; Postle, R. Application of Sonic Wave Theory to the Measurement of the Dynamic Elastic Moduli of Woven and Knitted Fabrics. *Text. Res. J.* **1981,** *51*, 732–740.
62. Blyth, G.; Postle, R. Measurement and Interpretation of Fabric Dynamic Modulus. *Text. Res. J.* **1979,** *49*, 601–608.
63. El Messiry, M.; Ibrahim, S. Investigation of Sonic Pulse Velocity in Evaluation of Knitted Fabrics. *J. Ind. Text.* **2016,** *46* (2), 455–472.
64. Bilisik, K. Multiaxis Three-dimensional Weaving for Composites: A Review. *Text. Res. J.* **2012,** *82* (7), 725–743.
65. Lambert, V. Enhancement of Spike and Stab Resistance of Flexible Armor Using Nanoparticles and A Cross-Linking Fixative. MSc. Thesis, Florida Atlantic University, Boca Raton, Florida, April 2009.
66. Gong, X.; Xu, Y.; Zhu, W.; Xuan, S.; Jiang, W.; Jiang, W. Study of the Knife Stab and Puncture-resistant Performance for Shear Thickening Fluid Enhanced Fabric. *J. Comp. Mat.* **2014,** *48* (6), 641–657.
67. Bao, L.; Wang, Y.; Baba, T.; Fukuda, Y.; Wakatsuki, K.; Morikawa, H. Development of a High-density Nonwoven Structure to Improve the Stab Resistance of Protective Clothing Material. *Ind. Health* **2017,** *55* (6), 513–520.
68. El Mor, M. Investigation of the Factors Affecting the Puncture Force of Hybrid Fibrous Structures. Ph.D. Thesis, Alexandria University, Egypt, 2018.
69. Hassan, T.; Rangari, V.; Jeelani, S. Synthesis, Processing and Characterization of Shear Thickening Fluid (STF) Impregnated Fabric Composites. *Mater. Sci. Eng.* **2010,** *527* (12, 15), 2892–2899.
70. El Messiry, M.; El-Tarfawy, S. In *Performance of Weave Structure Multilayer Bullet Proof Flexible Armor*. 3rd Conference National Campaign of Textile Industry, 9–10 March 2015, NRC, Cairo, Egypt.
71. Deckera, M.; Halbacha, C.; Nama, C.; Wagnera, N.; Wetzel, E. Stab Resistance of Shear Thickening Fluid (STF)-Treated Fabrics. *Compos. Sci. Technol.* **2007,** *67* (3–4), 565–578.
72. Egres, R.; Halbach, C.; Decker, M.; Wetzel, E.; Wagner, N. In *Stab Performance of Shear Thickening Fluid (STF)–Fabric Composites for Protective Armor Applications*. Proceedings of SAMPE 2005: New Horizons for Materials and Processing Technologies. Long Beach, CA, 1–5 May 2005.
73. Yu, K.; Cao, H.; Qian, K.; Jiang, L.; Li, H. Synthesis and Stab Resistance of Shear Thickening Fluid (STF) Impregnated Glass Fabric Composites. *Fibr. Text. East. Eur.* **2012,** *20, 6A* (95), 126–128.
74. SafeGuard Armor. https://www.safeguardarmor.com/support/research/ (accessed Mar 10, 2018)
75. Chadwick, E.; Nicol, A.; Lane, J.; Gray, T. Biomechanics of Knife Stab Attacks. *Forensic Sci. Int.* **1999,** *105* (1), 35–44.
76. Chardwick, E.; Lane, J; Nicol, A.; Gray, T. Investigation of Knife Stab Characteristics. *J. Biomech.* **1998,** *31*, 45–51.

77. Gilchrist, M.; Keenan, S.; Curtis M.; Cassidy M.; Byrne, G.; Destrade, M. Measuring Knife Stab Penetration into Skin Simulant Using A Novel Biaxial Tension Device. *For. Sci. Int.* **2008,** *177* (1), 52–65.
78. Nolan, G.; Hainsworth, S.; Rutty, G. Forces Generated in Stabbing Attacks: An Evaluation of the Utility of the Mild, Moderate and Severe Scale. *Int. J. Legal Med.* **2018,** *132* (1), 229–236.
79. Horsfall, I.; Prosser, P.; Watson, C.; Champion, S. An Assessment of Human Performance in Stabbing. *For. Sci. Int.* **1999,** *102* (2–3), 79–89.
80. Venara, A.; Gaudin, A.; Lebigot, J.; Airagnes, G.; Hamel, J. F.; Jousset, N.; Ridereau-Zins, C.; Mauillon, D.; Rouge-Maillart, C. Tomodensitometric Survey of the Distance Between Thoracic and Abdominal Vital Organs and the Wall According to BMI, Abdominal Diameter and Gender: Proposition of an Indicative Chart for the Forensic Activities. *For. Sci. Int.* **2013,** *229* (1–3), 167–171.
81. Kolmes, N.; Pritchard, C.; Mussinelli, M. High Performance Fiber Blend and Products Made Therefrom. Patent no. US 5,628,172, 2007.
82. National Research Council. Opportunities in Protection Materials Science and Technology for Future Army Applications; the National Academies Press: Washington, DC. https://www.nap.edu/catalog/13157/opportunities-in-protection-materials-science-and-technology-for-future-army-applications (accessed July 26, 2018).
83. National Institute of Justice. Selection and Application Guide to Ballistic-Resistant Protective Armor For Law Enforcement, Corrections and Public Safety. https://www.ncjrs.gov/pdffiles1/nij/247281.pdf (accessed Mar 10, 2018).
84. Johnson, A. Establishing Design Characteristics for the Development of Stab Resistant Laser Sintered Protective Armor. Ph.D. Thesis, Loughborough University's Institutional Repository, 2014.
85. Sinnappoo, K.; Arnold, L.; Padhye, R. Application of Wool in High-velocity Ballistic Protective Fabrics. *Text. Res. J.* **2010,** *80* (11), 1084–1092.
86. Cavallaro, P. Soft Protective Armor: An Overview of Materials, Manufacturing, Testing, and Ballistic Impact Dynamics. Naval Undersea Warfare Center Division, 2011. Report No: NUWCD-NPT-TR-12-057. http://www.dtic.mil/dtic/tr/fulltext/u2/a549097.pdf (accessed Mar 10, 2018).
87. Tien, T.; Kim, S.; Huh, Y. Stab-resistant Property of the Fabrics Woven with the Aramid/Cotton Core-spun Yarns. *Fibers Polym.* **2010,** *11* (3), 500–506.
88. Shin, H.-S.; Erlich, C.; Shockey, A. Test for Measuring Cut Resistance of Yarns. *J. Mat. Sci.* **2003,** *38* (17), 3603–3610.
89. Fangueiro, R.; Carvalho, R.; Silveira, D.; Ferreira, N.; Ferreira, C. et al. Development of High-performance Single Layer Fill Knitted Structures for Cut and Puncture Protection. *J. Text. Sci. Eng.* **2015,** *5*, 1–6.

FIGURE 1.15 Modern bulletproof vest.

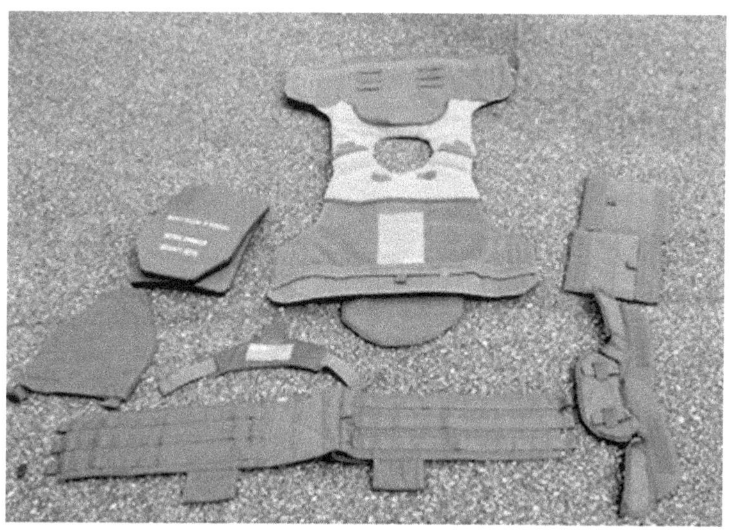

Bulletproof vest parts

FIGURE 3.12 Protective armor.

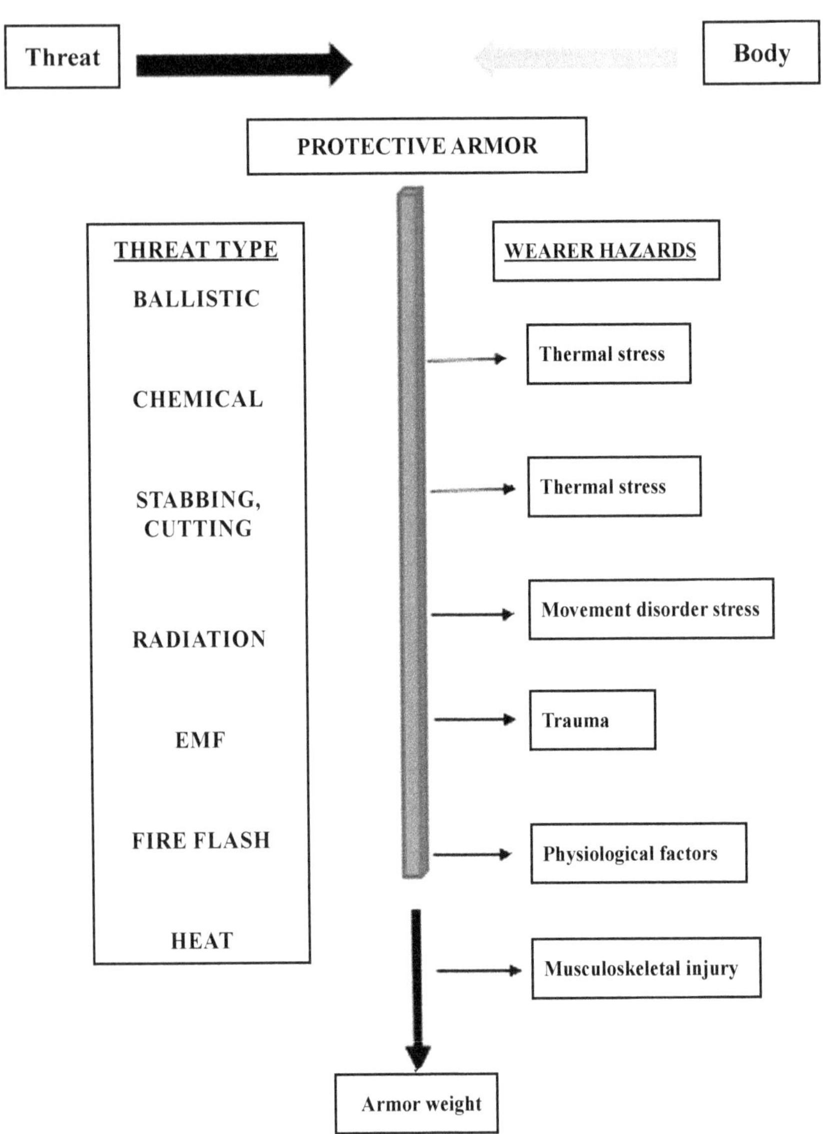

FIGURE 4.1 The interaction between body and threat.

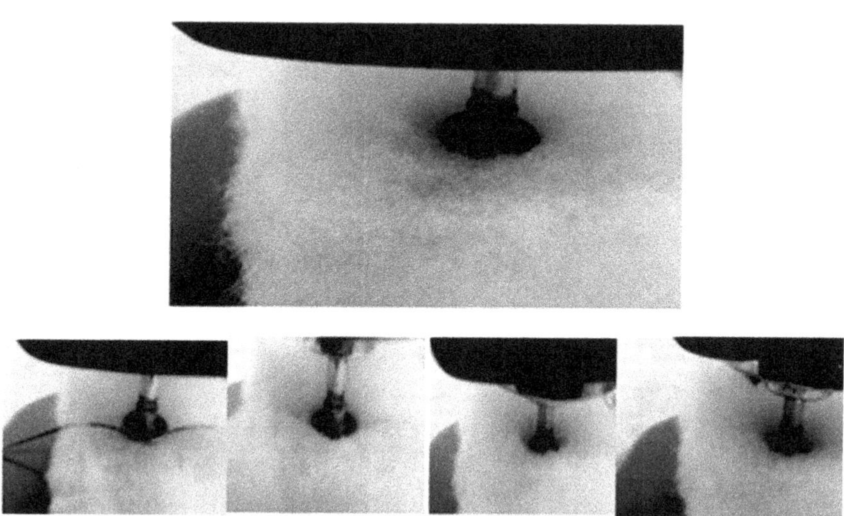

FIGURE 4.5 Compactness of fibrous structure.

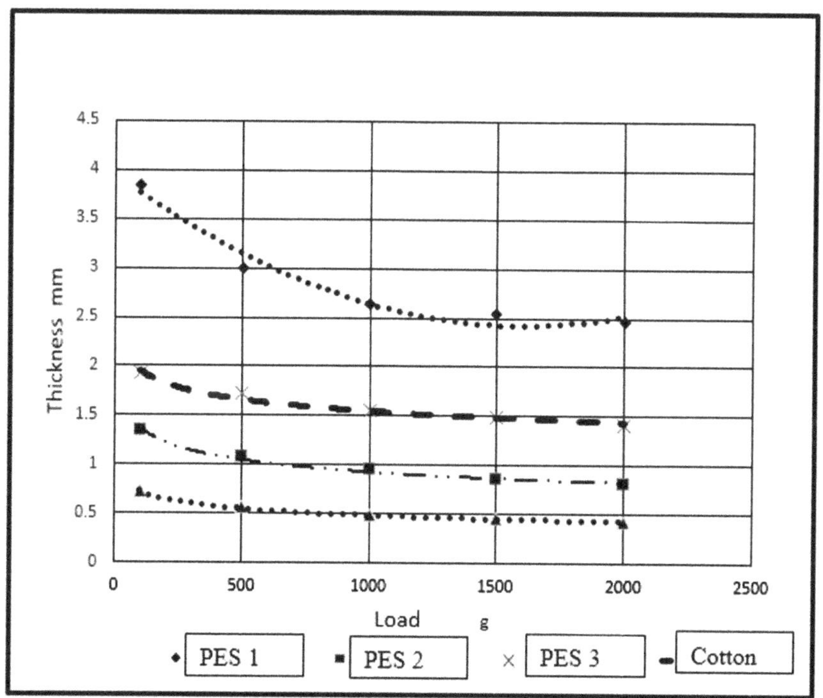

FIGURE 4.6 Thickness–load curve of some types of fibrous structures.

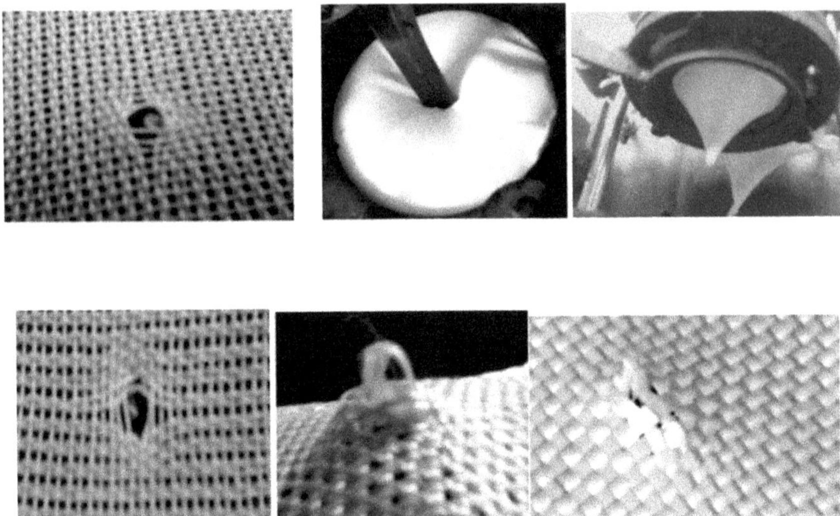

FIGURE 4.31 Different examples of stabbing cut shapes.

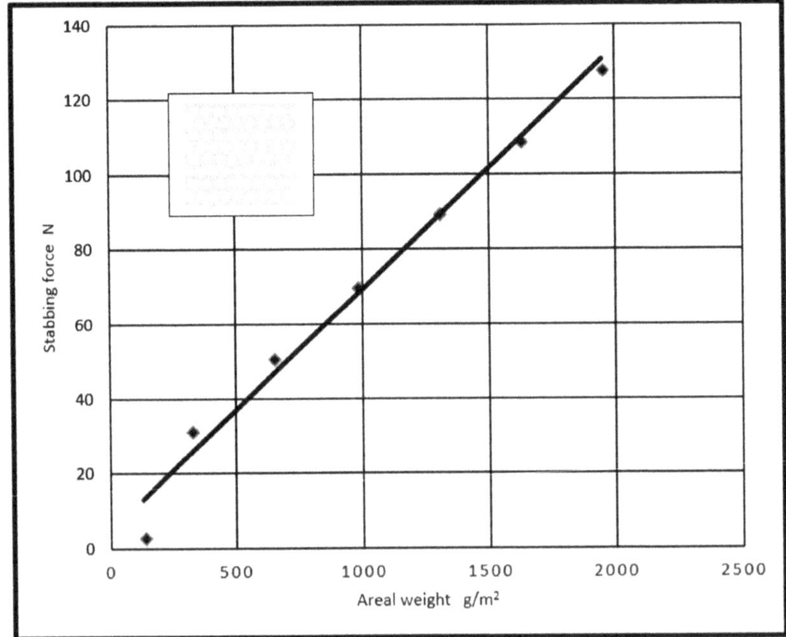

FIGURE 4.38 The stabbing force of the composite structure of silk fabric supported by layer of Kevlar fabric.

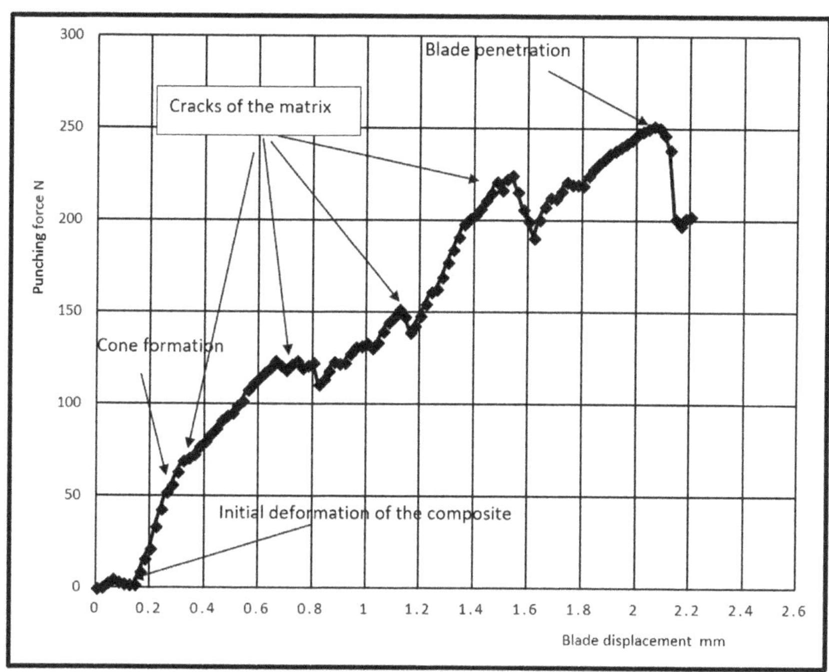

FIGURE 4.43 Punching resistance of the triaxial fabrics reinforced polymer composite.

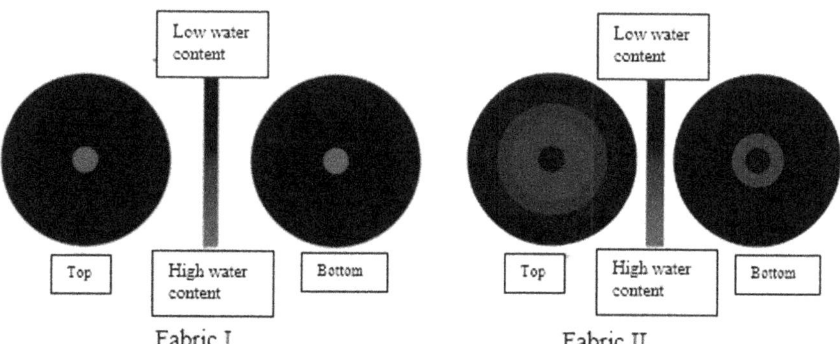

FIGURE 5.2 The water spreading on the top and bottom surface of the fabric.

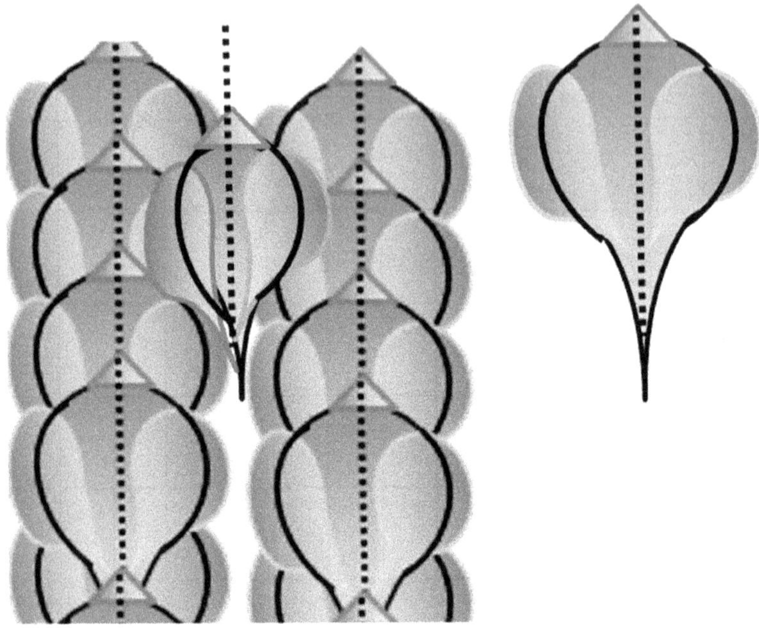

FIGURE 5.18 Placoid scales.[21]

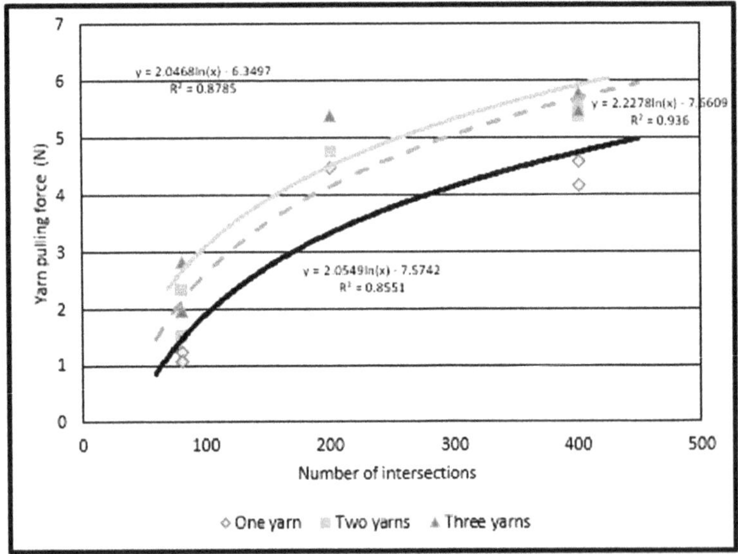

FIGURE 5.22 Yarn pulling—off force versus the number of intersections.

Ceramic components

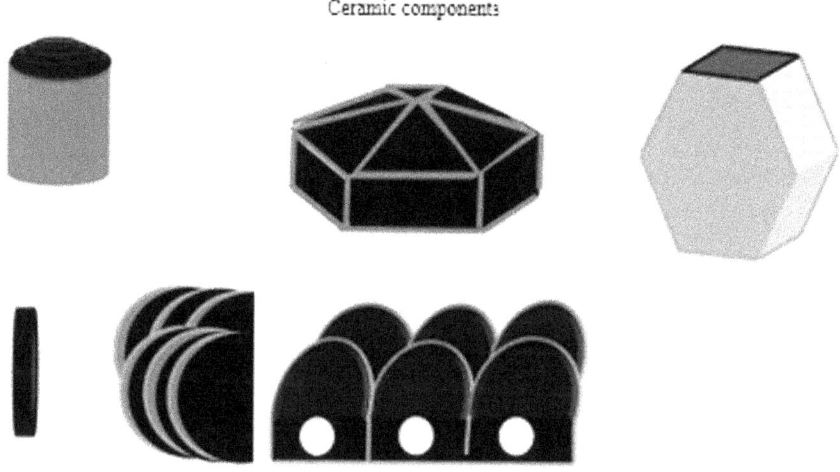

Ceramic protective insert

(a) Ceramic plate's elements

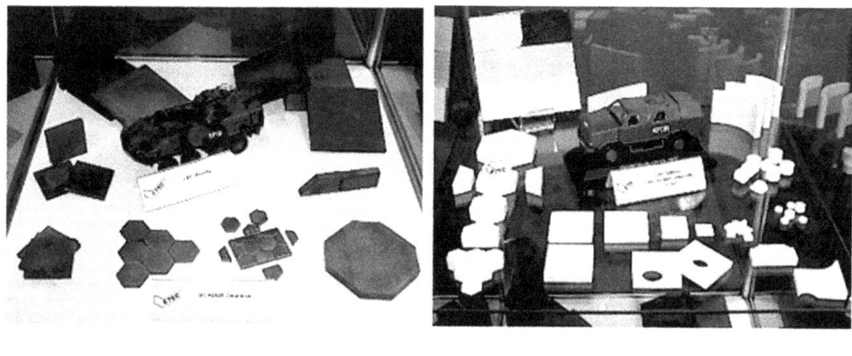

(b)

FIGURE 5.25 Some elements of the ceramic plates.[101]

CHAPTER 5

Bulletproof Flexible Protective Armor

ABSTRACT

During the last decades, intensive scientific and industrial researches were dedicated to the development of lightweight bulletproof body armor systems to satisfy the standards such as Ballistic Resistance of Body Armor NIJ Standard-0101.06 or UK Body Armor Standard (2017). The main requirements (level of protection, comfort, fabric hand, moisture management, vapor resistance, thermal comfort) for the different designs of modern protective armor are assessed. The designs of armor have been largely inspired by the mother nature; biomimicry applications for armor design are explained.Factors involved in the body armor effectiveness are defined.

5.1 INTRODUCTION

Threat and protection have been the main challenges to the humanity since the early ages. The science of designing the protective armor proceeds to protect the wearer from the possible injury by the projectile of increase in mass and velocity. The modern protective armor (PA) starts at the early of the 19th century with the appearing of new synthetic fibers, accomplished during the World War I and World War II and really motivated by the progress in the textile fiber science. The armor materials began with leather, steel plates, and natural fibers and proceed to high-performance fibers to protect soldiers, animals, and finally the war vehicles. To meet the high speeding-up of the developed weapons, the design of PA is open to provide all the required functionality to it.[1,2]

During the last decades, intensive researches were devoted to the development of lightweight armor systems for personnel and their ballistic performance. To visualize the energy required to stop the bullet, let us consider the protection levels given by NIJ in Table 5.1. For the protection level IV, the armor should stop a bullet as in the case to instantaneously breaking a car of 500 kg weight moving at a speed of 15 km/h.

5.2 CLASSIFICATION OF ARMOR

The main target of the PA is to provide bullet multihit protection with reduced shock. Different types of PA exist: concealable, which is a concealable carrier and a flexible armor provided with soft armor ballistic inserts; low visibility armor; tactical armor; and full spectrum (full-body armor). Ballistic vests can be concealable, semirigid, or rigid. Concealable ballistic vests (Levels I, IIA, II, and IIIA) usually provide full protection of the torso from handguns or long rifles.[4-6] Body armor is constructed from the multiple layers of fabric of high-performance fibers, such as high-performance polyethylene, polyphehylenebenzobisoxazole, Kevlar®, Spectra®, Dyneema®, Zylon®, Kevlar®, Protera, and GoldFlex®.

The armor plates, hard or flexible, which are a separate part, improve the protective armor level of protection against high-velocity bullet threats, they may be permanently sewn into the flexible vest or may be removable. The dimensions of panel measures are 250 × 300 mm and weigh 2 kg. The semi-rigid (SRBV) or rigid ballistic vests (RBV) are recommended for higher threat levels (Levels III and IV). SRBV uses small plates of ballistic material, consisting of a less flexible material, with the impregnated ballistic fabrics. RBV consists of molded ballistic material to cover a specific part of the torso. Combination armors are designed to protect against both firearm and stabbing. The design of the plates should contain the layers of fabrics to resist the stabbing and defeat projectiles. The trauma plates are designed with the main purpose to prevent the trauma injury. The next generation of the protection gear is "soldier protection system" which would equip troops with lighter body armor, along with upgraded equipment including health sensors and new protective eyewear.[7] This system will consist of protection against ballistic/nonballistic impact, stab, and chemical, antimicrobial, and gas hazards. That means complete capsulation of the wearer to protect against all the expected threats all over whole body area, securing his comfort requirements and freedom in movement.

TABLE 5.1 The Level of Kinetic Energy for Different Levels.[3]

Level	Projectile type	Weight (g)	Velocity (m/s)	Kinetic energy (J)
Type IIA	Full-metal-jacketed round nose (FMJ RN) of 9 mm and 40 S&W FMJ	8.0	373 ± 9.1	556.5
		11.7	352±9.1	724.8
Type II	FMJ RN of 9 mm and .357 magnum jacketed soft point (JSP)	8.0	398 ± 9.1	633.6
		10.2	436 ± 9.1	969.5
Type IIIA	.357 SIG FMJ flat nose (FN), .44 magnum semijacketed hollow point (SJHP)	8.1	448 ± 9.1	812.9
		15.6	436 ± 9.1	1482.7
Type III (rifles)	FMJ (7.62 mm), steel-jacketed bullets (US military designation M80)	9.6	847 ± 9.1	3443.6
Type IV (armor-piercing rifle)	.30 caliber armor-piercing (AP) bullets (US military designation M2 AP)	10.8	878 ± 9.1	4162.8

5.3 COMFORT OF BODY ARMOR

5.3.1 INTRODUCTION

The objective of the body armor is to protect the wearer from the surrounding threats, from the stabbing to other different ballistic threats. The design of the flexible armor consists of multilayers weave of high-performance fibers. The number of layers depends on the level of the energy it designed to protect against, it may be more than 25. The areal density of the final layered fabric per square meter may reach 500 g/m^2. Some designs of flexible armor are provided with a flexible or hard panel to enhance the protection level. Hard body plate insertions are designed from composites of ceramic, metal or plastic to guarantee protection against a high level of threat. PV design aimed to increase the wearer safety and ensures a suitable degree of comfort but not affects negatively on thermo-physical stress, fatigue, limitation of movement, and reduction of the blunt trauma associated with bullet impacts. Consequently, the increase of the vest comfort with maximizing the wearer's safety is the challenges facing the armor vest designer. The comfort has no concrete definition since it depends on the interaction between the physiological, sociopsychological and physical balance among a person, clothing design, and surrounding environment to fulfill a certain level of protection. The comfort is one of the engineering problems that should be considered hence the wearer may be enforced to wear protective ballistic armor for a long time in unpleasant surrounding conditions like high moisture, elevated temperature. Combat efficiency of the troops hinges on the comfort of a protective armor.[8] The performance of a protective vest and its interaction with the wearer's body should be intensively studied while designing body armor.[9] The satisfactory thermal equilibrium, the interaction of physiological and mobility limiting factors, the environmental conditions as temperature, humidity, air movement, and radiant heat may identify the criteria for the PA comfort. Consequently, clothing comfort is a result of direct interactions of various physical and nonphysical stimuli of a person wearing a given ensemble under specified environmental conditions.[10]

5.3.2 COMFORT MODELS

There are several models to consider the cloth comfort by explaining the relations between a person, clothing, and environment. Figure 5.1 shows the interaction between the different elements of the cloth comfort.

The next important factor is to analyze the parameters that fabric comfort depends on, such as:

a. Clothing Attributes

1. Fabric geometry,
2. Moisture sorption properties,
3. Air permeability,
4. Pore volume,
5. Fabric thermal resistance,
6. Vapor permeability,
7. Fabric pilling resistance,
8. Fabric drape,
9. Fabric crease recovery,
10. Fabric softness,
11. Weight of the fabric per unit area,
12. Size and fit,
13. Compressibility,
14. Thickness,
15. Fullness per unit volume,
16. Drape,
17. Hand friction,
18. Fabric resilience,
19. Cloth retain heat,
20. Wind resistance,
21. Thermal resistance,
22. Evaporative resistance, and
23. Moisture management.

FIGURE 5.1 Comfort elements.

Most of the above factors influence PA comfort. It is obvious that comfort involves a complex combination of properties and moreover, some of these properties behave opposite to each other while the other parameters significantly interact.[18]

b. Person Attributes

1. Sex,
2. Age,
3. Race,
4. Weight,
5. Height,
6. Physical condition,
7. Activity,
8. Exposed area, and
9. Social-psychological.

c. Environmental Attributes

1. Temperature,
2. Relative humidity,
3. Pressure,
4. Wind speed, and
5. Oxygen level.

The comfort of the vest wearer is evaluated by Sztandera[11]:

- Subjective comfort,
- Quantitative comfort, and
- Combination of the two above models.

Subjective sensorial comfort can be measured subjectively by wearer's acceptability/satisfaction questionnaires. On the contrary, the objective sensorial comfort can be measured using different testing methods associated with comfort: physical, mechanical, thermal, etc. Many researches try to find out the algorithm for the comfort evaluation assuming a smaller number of variables. Some results indicate that the measuring of three main sensory factors (thermal–moisture comfort, tactile comfort, and pressure comfort) explains 81.6% of the total in the perception of clothing comfort.[12] Several studies combined oral temperature, heart rate, and sweat loss measures to compare the physiological implications to the other mechanical properties in evaluating the comfort of the cloth.[13] While the others considered the fabric hand as a quantitative measure of the fabric objective sensorial comfort.[11]

5.3.3 FABRIC HAND

There are several methods for assessing fabric hand: the direct sensory evaluation,[14,15] the assessment of some comfort parameters that corresponds to the human psyche, the measurement of some mechanical physical fabric properties by KES System, FAST System, FTT (Fabric Touch Tester) System, BM Technique, etc. Table 5.2 gives fabric properties defining fabric hand, conferring to the system applying it.

The aim of the objective evaluation of fabric hand is to replace the subjective one, which is often time-consuming. All these systems are based on the objective measurement of properties having a relation with

TABLE 5.2 Fabric Hand Systems and the Fabric Properties.

Properties	Parameters	System
Tensile	Work of rupture, tensile resilience, and initial tensile modulus	*KES, BM*
Extension	Extension at 5 cN/cm, extension at 20 cN/cm, and extension at 100 cN/cm	*FAST*
Bending	Bending length, bending rigidity, hysteresis of bending moment, bending average rigidity, and bending work	*KES, FAST, FTT*
Shear	Hysteresis of shear force at angle $\varphi = 0.5°$ of shear angle, hysteresis of shear force at angle $\varphi = 5°$ of shear angle, and shear rigidity	*KES, FAST*
Compression	Compressibility, compression work, compression recovery work, and compression average rigidity	*KES, FAST, FTT*
Surface friction	Surface friction coefficient, surface roughness amplitude, and surface roughness wavelength	*KES, FAST, FTT*
Construction	Thickness	*KES, FAST, BM*
Formability	Formability	*FAST*
Area weight	Area weight	*KES, FAST, FTT, BM*
Thermal	Thermal absorptivity, thermal conductivity when compression, thermal conductivity when recovery, and thermal maximum flux	*FTT, BM*

the hand of textiles.[16] Comfort is not a sensory dimension, because it is not associated directly with any single human sense organ, thus there is no underlying physical dimension of the stimulus that varies continuously and is monotonic with the perception of comfort.[17] Material selection is critical to ensure the least interference with the soldiers' ability to perform physical or cognitive tasks.

Ample approaches were introduced to evaluate fabric comfort.

- *Multiple regression approach.* Four parameters are chosen to study their effect as well as their interactions: fiber content, stiffness, roughness, and thickness.[18] The interaction of these four parameters may be or may not significantly influence fabric comfort. Stiffness, thickness, roughness, and the interaction between fiber content were found to intensely affect the fabric hand.
- *Physiological parameters approach.* Certain physiological factors, such as skin temperature, degree of skin wetness, rate of sweating, total sweat loss, and heart rate were measured as presently accepted criteria of human comfort. Physiological responses indicated that comfort largely depends upon the geometry of fabric construction and the manner in which the fabric is worn on the body.[19]
- *The pressure–comfort approach.* It is suggested as an important property that determines a fabric comfort. Pressure–relief property of a fabric is an important characteristic for the ergonomic applications. There is a relation between the pressure–relief property and the pressure–comfort property, especially for the spacer fabrics. The compression indexes, four indexes, namely compression work, recovery work (RW), hysteresis work (HW), and maximum compression force (CF) are featured from spherical compression force-displacement curves and structure parameters (diameter of the spacer filament D_f and fabric thickness T).[20] The fabric pressure–comfort performance evaluation score:

$$E_s = 5.851 + 0.020RW - 0.055HW + 0.025CF - 0.137T - 0.011D_f \quad (5.1)$$

For convenience, the symbols used throughout are listed as follows in Table 5.3.

TABLE 5.3 List of Symbols and Abbreviations.

PA	Protective armor
PF	Protective fabric
SRBV	Semi-rigid ballistic vest
RBV	Rigid ballistic vest
S1	Vapor resistance between the skin and the fabric
S2	Vapor resistance between the fabric and the ambient environment
S'1	Vapor resistance between the skin and the fabric in a lower position
S'2	Vapor resistance between the skin and the ambient environment in a lower position
S"1	Vapor resistance between the skin and the fabric in the upper position
S"2	Vapor resistance between the skin and the ambient environment in the upper position
M_{skin}	Moisture transport from the skin
C_{skin}	Water vapor concentration of the air immediately adjoining the skin,
$C_{atmosphere}$	Water vapor concentration in the ambient atmosphere
R1	S'1 + S'2
R2	S"1 + S"2
OMMC	An index to indicate the overall ability of the fabric to Manage the transport of liquid moisture
MWR (*t or b*)	Maximum wetted radii (top or bottom)
OWTC	Accumulative one-way transport capacity (is the difference in the cumulative moisture content between the two surfaces of the fabric in the unit testing time period
SS	Spreading Speed (mm/sec) (top/bottom) = MER*t/twrt* or MER*b/twrb*
t_{wrt} and t_{wrb}	time to reach the maximum wetting rings on the top and bottom surfaces
C1, C2, C3	Weights of the indexes of the absorption rate (MARb), equals to 0.25, 0.5, and 0.25 respectively
t_{wet} and t_{wrb}	Time to reach the maximum wetting rings on the top and bottom surfaces
$SS_{b'}$, SS_t	Spreading/drying rate on the bottom, top fabric surface
MH	Metabolic heat
MW	Heat converted into mechanical power
HC	Heat transfer by convection to the environment
HR	Heat transferred by radiation to the environment
HE1	Heat transferred to the environment by evaporation of sweat from the skin surface
HC1	Heat transferred to the environment by convection through breathing

TABLE 5.3 *(Continued)*

HE2	Heat transferred to the environment by evaporation through breathing
HS1	Heat accumulated in the skin
HS2	Heat accumulated by the body
S	Shear modulus
N	Number of intersections
T	Fabric tightness
HV	Vickers hardness
E	Young's modulus
K	Fracture toughness
C	Material sonic velocity
ρ_{bac}	The density of the backing material
ρ_{cer}	The density of the ceramic
P_{bac}	Penetration depth into the unprotected backing material
$Tcer$	The thickness of the ceramic
DOP	Depth of penetration into backing after striking the ceramic target,
N	number of layers
ρ_i	Density thickness of the *i*th layer
ti	The thickness of the *i*th layer
I	1,2,3,…,n
THV	Total hand value
KES	Kawabata evaluation system,
TCS	Tactile comfort score
EMT	Extensibility
BR	Bending rigidity
LT	The linearity of the load/extension curve
HG	Hysteresis of shear force
COM RR	Compressional Resilience Rate
KW	Fabric weight/unit area in mg/cm^2 (in Kawabata system),
Grainy	Amount of small, round particles in the surface of the sample
Gritty	Amount of small, abrasive, picky particles in the surface of the sample
Wt	Fabric weight /unit area oz/sq.yd,
MMT	Moisture management tester
$M1$	Moment acting on the scales during its rotation
$M2$	Moment acting on the tip of scale due to the rotation of the supporting scale

TABLE 5.3 *(Continued)*

H	Degree of overlap
l_1	Scale's length,
T_c	Distance between scales
P	Scale's pitch
B	Scale's width
Θ	Scale orientation angle relative to tissue
γ_S	Scales volume fraction
T_s	Scale's thickness
Tc	Distance between scales
I	Number of intersections
TI	Fabric tightness
PF	Yarn pulling force
E	Young's modulus
HV	Vickers hardness
K	Fracture toughness
B	Brittleness
E_m	Mass efficiency
η_b	Ballistic efficiency
t_{crit}	Critical thickness
ρ_{bac}	The density of the backing material
ρ_{cer}	The density of the ceramic
P_{bac}	Penetration depth into the unprotected backing material
t_{cer}	The thickness of the ceramic
DOP	Depth of penetration into backing after striking the ceramic target
V_{50}	Armor ballistic limit

- *Regression approach.* Several regression equations were proposed for objective evaluation of fabric hand[13, 21-22] as a comfort index (y) to estimate the total hand value.

x_i is the measured nth mechanical properties, $i = 1,2\ldots n$ are given in Table 5.2.

$$y = b_0 + \sum b_i x_i \tag{5.2}$$

$$y = b_0 + \sum b_i \log x_i \tag{5.3}$$

$$y = b_0 + \sum b_i x_i + \sum b_i \log x_i \tag{5.4}$$

$$\log y = b_0 + \sum b_i \log x_i \tag{5.5}$$

$$\log y = \log b_0 + \sum b_i \log x_i \tag{5.6}$$

$$\log y = b_0 + \sum b_i x_i \tag{5.7}$$

The different systems for fabric hand measuring result in different hand assessment for the same fabric. However, there exists a good correlation between the hand property of the silk fabrics measured on the FAST technique and KES system.[24,25] Equation (5.4) is one mostly used for the estimation of fabric hand when taken into consideration $n = 16$ properties.[26]

Applying Kawabata component score to calculate the Military Clothing Fabrics, the following regression equation was concluded[16]:

$$\text{Comfort} = 11.8\,(\text{shear}) - 3.1\,(\text{bending}) - 0.3\,(\text{compression/friction}) - 11.9\,(\text{tensile}) + 0.4\,(\text{surface roughness}) + 27.5. \tag{5.8}$$

Tactile comfort score (TCS) can be expressed by regression equation[27]:

$$= 48.63 + 1.29\,\text{EMT} - 43.47\,\text{BR} - 52.35\,\text{LT} + 0.992\,\text{HG}. \tag{5.9}$$

These two equations express the comfort of the same fabric through the choice of different mechanical prosperities. Despite expanding the literature on the comfort evaluation topic and studies of human responses to clothing materials, it suffers from a lack of theoretical models due to the large number of variables that controlling the response of cloth, environmental parameters and the physiological and sensory responses of the wearer. Moreover, the two criteria that affect the function of the comfort–time relation of the cloth are its moisture management and thermal comfort properties.

5.3.4 FABRIC COMFORT EVALUATION PARAMETERS

- *Subjective and objective approach.* Several systems for the evaluation of fabric comfort are a combination of subjective and objective tests. It was found that out of the 40 fabric parameters, that influence the fabric comfort, five (three hand feel, one mechanical, and one constructional) were included in the final equation based on their contribution to calculate the fabric comfort.[27]

Tactile Comfort Score (TCS)

TCS = 136.130 − 5.252 Grainy − 4.555 Gritty − 9.434 COMRR − 162.129 K MMD + 1.987 K W − 4:248 Wt (5.10)

Where COMRR is Compression Resilience Rate and the variables starting with K_ indicate Kawabata (mechanical) properties, W is fabric weight/unit area in mg/cm^2 (in Kawabata system) and Wt is also weight but measured as a construction variable measured in oz/sq.yd.). MMD Mean deviation of coefficient of friction (frictional roughness of the surface), Grainy – amount of small, round particles in the surface of the sample, Gritty – amount of small, abrasive, picky particles in the surface of the sample.

- *Subjective approach.*

1. The testing of the of hand feel ranking conferring to the judge of panelists, they rated the perceived comfort on the CALM scale[26, 16] that classifies the comfort of the cloth between the following categories: Greatest imaginable discomfort, extremely uncomforted, very uncomforted, moderately uncomforted, slightly uncomforted, neither comfortable nor uncomfortable, slightly comfort, moderately comfort, and very comfort, extremely comfort.
2. Several researches have been conducted to measure the cloth comfort for various types of fabric. Through the suggestions of another uncomfortable scale of seven classes (1 = "never appropriate"; 7 = "always appropriate"), the subjective evaluation of the fabrics was classified conferring to the end use. It is interesting to compare the comfort of the fabric from the different fibers applying the above classification method as given in Table 5.4,[28] which indicates the advance of the natural fibers like silk, cotton, and their blends over the synthetic fibers. The judgment on the fabric conferring to the suitability for a certain end-use will change the ranking order of the different fiber. For example, silk is a bad choice as a water resistance fabric due to its high hygroscopic nature and has low appropriateness for desert and jungle applications. The same principle can be used to rank the

fabrics conferring to different factors: Comfortable (breathable, cool, wear a long time, and good in the desert) by calculating the percentage of appropriateness ratings for the five highest weighted fabrics on each fabric factor and for the highest weighted uses on the seven use factors.

TABLE 5.4 Comfortability Ranking of Some Fibers.

Type of fibers	Comfortability
Kevlar	III
Nomex	V
Nylon	VI
Polypropylene	VI
Polyester	III
Silk	VI
Cotton	V
Wool	III
Ideal fabric	VII

5.3.5 MOISTURE MANAGEMENT

The evaporation of sweat to remain comfortable and prevent overheating in a hot environment and during movement is another approach for comfort evaluation. The discomfort associated with the rate of evaporation is less than the rate of sweat discharge. The fabric moisture management is rather complicated since the fabric during the movement is not continuously in the interaction with the skin. However, each part of the fabric has its own 3D movement. Mechanism of water vapor transferred through fibrous materials[29–31] can occur by several ways: (1) diffusion of the water vapor through the fibrous layers; (2) absorption, transmission, and desorption of water vapor by the fibers; and (3) transmission of water vapor by forced convection. Diffusion of water vapor molecules through the air spaces in fabrics is a major contributor to moisture vapor transport. The other transfer processes involve smaller amounts of moisture vapor. The water vapor diffusion is reliant on the air permeability and porosity of the fabric.[32]

The tester MMT was developed to measure dynamics of water transport properties in the three dimensions through and in-plain of the fabrics by the water spreading rate versus time for the top and bottom fabric's surfaces (Ut and Ub).[33] The following parameters can be calculated to indicate the water–fabric relation: *Absorption rate*, moisture absorbing time of the fabric's inner and outer surfaces; *one-way transportation capability*, one-way transfer from fabric's inner surface to outer surface (Fig. 5.2); *spreading/drying rate*, speed of liquid moisture spreading on fabric's inner and outer surfaces.

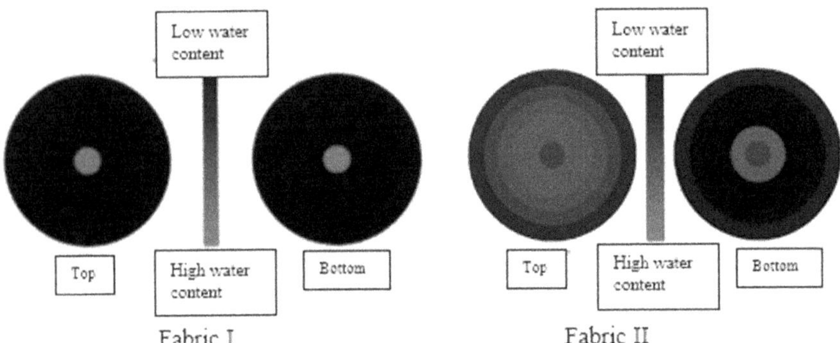

FIGURE 5.2 **(See color insert.)** The water spreading on the top and bottom surface of the fabric.

Overall moisture management capacity (OMMC) is an index to indicate the overall ability of the fabric to manage the transport of liquid moisture:

$$\text{OMMC} = C1\ \text{MAR}_b + C2\ \text{OWTC} + C3\ \text{SS}_b \tag{5.11}$$

Where: MAR_b is the maximum moisture on bottom surfaces, OWTC is the one-way transport capacity, and SS_b is the spreading/drying rate at the bottom surface.

The values of $C1$, $C2$, and $C3$ are adjustable in practice to suit the end-use purposes.[33] The fabric can be evaluated by AATCC 195–2012 standard, the fabrics are graded into five-grade scale, Table 5.5, based on their OMMC values. The higher the OMMC value, the better will be the moisture management ability of the fabric.

TABLE 5.5 Grading of Moisture Management, AATCC 195–2012.

Grade AATCC 195–2012	1	2	3	4	5
OMMC	0–0.2	0.2–0.4	0.4–0.6	0.6–0.8	>0.8
	Poor	Fair	Good	Very good	Excellent

Table 5.6 gives the Evaluation of Fabric Water Management Parametersfor different levels of WT_T and WT_B, AR_T or AR_B, MWR_T and MWR_B, and SS_t/SS_b. Where: Wetting time WT_T (top surface) and WT_B (bottom surface) (seconds), Maximum absorption rates MAR_T and MAR_B (%/second), t_{wrT} and t_{wrB} are the times to reach the maximum wetted rings on the top and bottom surfaces, respectively, $SS_T = (MWR_T/ t_{wrT})$, $SS_B = (MWR_B/ t_{wrT})$.

5.3.6 *FABRIC VAPOR RESISTANCE*

The protective vest is mostly made of a fabric that has no effect on the diffusion between the skin and atmosphere because it is assumed to have no vapor resistance or moisture capacity. It was found that the air permeability increases as the porosity of the fabric increases, which also results in a higher moisture transmission through the air spaces within the fabric. The diffusion of vapor across a region without any moisture capacity is equal to the concentration difference divided by the vapor resistance.[34] The sweat can quickly transfer from the surface of the skin when the fabric has a good OMMC and high OWTC.[35] The mechanism of vapor resistance of the fabric controls the amount of the perspiration evaporated from the surface of the skin through the fabric to the ambient environment. Figure 5.3 shows a simple model of this mechanism. The rate of transportation of the moisture through the fabric will be a function of skin temperature, fabric structure, fiber material, and the ambient environment relative humidity and temperature. Simply, the mechanism of moisture transport from the skin (M_{skin}) is given by the following equation[34]:

$$M_{skin} = (C_{skin} - C_{atmosphere}) / (S_1 + S_2). \tag{5.12}$$

TABLE 5.6 Evaluation of Fabric Water Management Parameters.

Grade AATCC 195–2012	1	2	3	4	5
WT_T or WT_B	>120	20–119	5–19	2–5	<3
AR_T[a] or AR_B[a]	>10–10	10–30	30–50	50–100	>100
MWR_T or MWR_B	0–7	7–12	12–17	17–22	>22
SS_T or SS_B	0–1	1–2	2–3	3–4	>4
OWTC	>–50	50 to 100	100–200	200–400	>400

[a]Absorption rate (AR_T—top surface and AR_B—bottom surface).

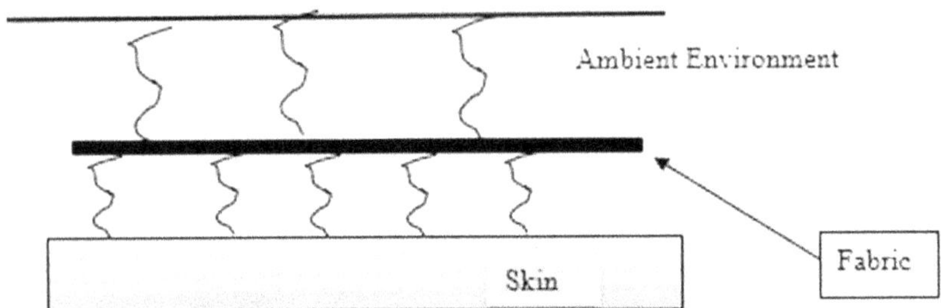

FIGURE 5.3 Model of moisture transport through textiles.

The vapor resistance between the skin and the fabric is represented by S_1, and between the fabric and the ambient environment, it is S_2. Assume $S_1 = S_2$, then:

$$M_{skin} = (C_{skin} - C_{atmosphere}) / 2S_1. \tag{5.13}$$

5.3.6.1 *FACTORS AFFECTING THE FABRIC VAPOR RESISTANCE*

In the case of humid condition, the moisture is transported through the fabric as liquid and vapor,[36] the liquid will flow in both direction; through the fabric thickness, liquid water is drawn from the capillaries to the upper surface of the fabric, liquid diffusion, and spread over the fabric area from wet to the dry area. The spreading speed (SS_i is the accumulated rate of surface wetting from the center of the specimen, where the test solution is dropped to the maximum wetted radius) depends on the type of fiber and the fabric structure. In sweating conditions, in clothing with high wicking properties, moisture coming from the skin is spread throughout the fabric offering a dry feeling,[30] moreover the spreading of the liquid enables moisture to evaporate easily. The fiber type, fabric air permeability, and porosity play an important role in the mechanism of water vapor diffusion, as it is influenced by both the collection and the passage of moisture along the fabric.[37] Consequently, the fabric design and specifications of the protective vest should be selected to ensure the best comfort with a suitable rate of perspiration transfer, taking into consideration the environmental conditions: environment temperature, relative humidity, wind speed.[38] It was revealed that the contribution of cold sensation to the perception of fabric moisture was observed in response to greater reduction in skin temperature, which in return, was affected by fabric water content.[39] In the condition of movement, an interaction between the body and the clothing will change the mechanism of the moisture diffusion through the fabric due to the change in the values of the S_1 and S_2. Through the change of the air film thickness between the cloth and body during the motion, the air flow through the fabric changes, too.[34, 39, 40] In this case, the degree of fabric contacts with the skin, fabric thermal resistance and the type of fiber will influence the comfort rating. Consequently, the movements of the wearer, that induce high sweat production, metabolic heat rate (thermo-behavioral responses[41]), reduce the air gaps between the cloth and the body and result in a very uncomfortable situation. The moisture transport from the skin depends on the change of the values of S_1 and S_2, hence the gap between the fabric and skin is changed within the minimum and maximum values. The average moisture transport M_{skin}[42] is given by the following equation:

$$M_{skin} = 0.5\,(C_{skin} - C_{atmosphere}) / [(1/R_1) + (1/R_2)]. \tag{5.14}$$

For protective armor cloth, it is recommended to have a high value of OMMC.

5.3.7 THERMAL COMFORT PROPERTIES

Several researches focused on the thermal comfort of the cloth because thermal stress is a major factor contributing to human performance degradation. The heat will flow through the armor due to the difference between body temperature and the environment temperature.[16] The heat and perspiration continuously emitted from the wearer, especially during the movement, and because of the failure of heat transfer to the surrounding environment, cause a low degree of comfort. Thermal comfort of cloth, as shown in Figure 5.4, depends on the interrelation between several parameters related to: wearer physiology and psychology, as well as surface temperature and fabric specifications, such as vapor resistance, fabric porosity, fabric thickness, fabric air permeability, fabric weight, cloth surface area, which are the main players to define the thermal comfort.

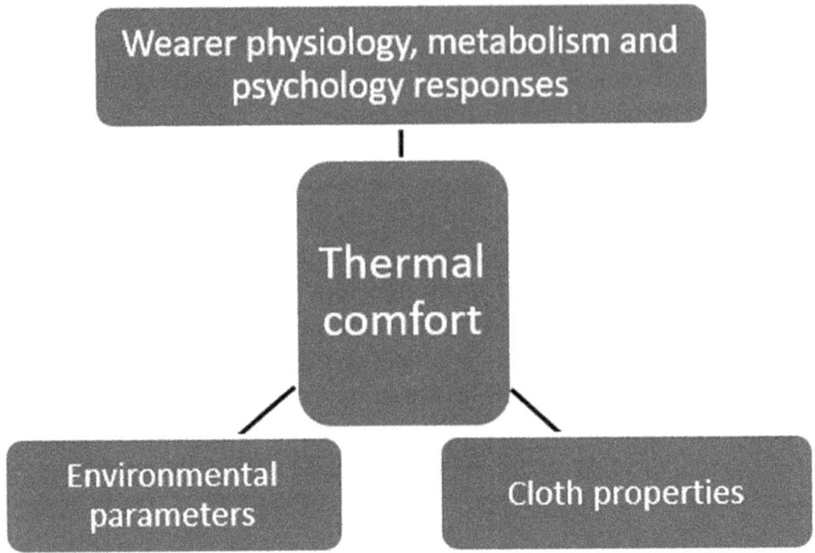

FIGURE 5.4 Parameters affecting the armor thermal comfort.

The physical process of heat exchange between the human body and the environment is[43]:

$$\begin{aligned} MH = {} & MW + (HC + HR + HE1) \\ & + (HC1 + HE2) + (HS1 + HS2). \end{aligned} \tag{5.15}$$

The value of HS2 is responsible for the thermal stress of the armor wearer, which depends on the capability of the armor to transfer more heat components to the environment, that is, higher values of HC, HR, HE1, and low HS1. The heat isolation by the armor causes an accumulation of body heat and thus leading to hyperthermia.

5.3.8 WETNESS PERCEPTION

Skin wetness, due to increases in metabolic heat production or exposure to hot environments, results in reduction of the body temperature, core overheating is prevented, and heat balance is maintained.[44] Skin wetness is usually expressed as a percentage of an upper limit for a full wet skin.[45] In the context of thermal comfort, the level of sweat-induced skin wetness had been shown to positively correlate with thermal and sensorial discomfort[39, 46] and negatively correlate to fabric thickness, to drop in skin temperature induced by the wet fabric and increased with the application of pressure. In practice, factors like the wet weight of the fabric and resultant local skin temperature drop should be considered when designing a garment with reduced wetness perception and related discomfort features.[38, 39] Recently, the scales for measuring wetness perception, thermal sensation and thermal comfort were developed to facilitate the judgment of comfort.[38]

- Wetness perception scale is equally divided from 30 to 0 to describe wet conditions: extremely wet, very wet, wet, slightly wet, barely wet, dry, and extremely dry.
- Thermal sensation scale is equally divided from +10 to −15 to describe thermal sensation: warm, slightly warm, neutral, slightly cool, cool, and cold.
- Accordingly, the two scales define the comfort scale which is divided into four statuses: (7) very uncomforted, (5) uncomforted, (3) slightly uncomforted, and (1) comfortable.

The use of body armor was associated with the increase of the thermal stress, with decline in cardiovascular, balance, strength, and functional field test performance.[47-49] It was revealed that the deep body temperature increases as a function of time up to 39°C after wearing battle armor for 2 h of walking 90 min at 4 km/h and 30 min at 6 km/h, while the heart rate reaches 180 beats/min.[50]

Ballistic vests can be concealable, semi-rigid, or rigid. When choosing between these styles, the wearer considers the end use, protection level, and comfort level of the armor system.[47] Unfortunately, the successful performance of military tasks and the protection against hazards often have conflicting requirements with the wearer's comfort.

5.4 BIOMIMICRY APPLICATIONS FOR ARMOR DESIGN

5.4.1 INTRODUCTION

The Mother Nature has been inspiring the designer of the armor since the early ages, and the human been always looking for the solution of protection and his safety from the different threats, deriving the ideas from natural designs in the area of materials as well as the structural designs of the protective armors. Increasingly, researchers accomplish to create the materials that mimic structures found in nature, a strategy known as biomimetics.[51] Always nature gives a lesson in armor design, in the context of knowledge and needs.[52]

The ancient armor designers were inspired by the fish scales to design a protective armor. Leonardo da Vinci, 1490, imitated the bird wings structure for flying machine invention, as shown in Figure 5.5.

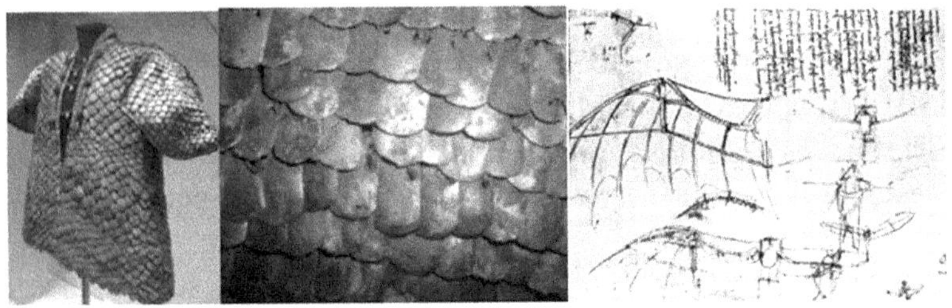

Ancient scales armour   Leonardo da Vinci sketch of flying machine

FIGURE 5.5 Ancient inspirations.

The nature produces a high-performance natural composite made of relatively weak components (brittle minerals and soft proteins) arranged in intricate ways to achieve specific combinations of stiffness, strength, and toughness.[53] Materials like bone, teeth, or mollusc shells are several orders of magnitude tougher than the minerals they contain, mostly made from calcium carbonate with up to 5% organic materials. The biomimicry science starts to be educational subject in the research institutions.

There are several natural armor systems that were imitated for protection from different types of threats. Understanding the fundamental design principles of natural armor systems, like shells, scales, etc., may help engineers design improved body armor systems.[54] The mollusc family build up their protective shells to defend their very soft body. A manmade replica of this structure can be used in stab resistance armor, bulletproof vests, helmets, gloves, and other protective armors. It should be mentioned that it is not an easy task since it is required to adopt the basic concept to the human body, taken into consideration the difference in the applicant's parameters and needs. The innovative design in the context of solving the protective armor problems is greatly anticipated on the market.

5.4.2 SEASHELL ARMOR

The objective of the designer of the protective armor is to achieve higher protection from the threat, but also comfort and lightweight. Figure 5.6 shows several types of seashells which have the thousand species of different structure, material, shape to fulfill their essential protection level. The MIT revealed that "windowpane" oyster's shell uses long diamond-shaped crystals that can dissipate energy from penetrating forces nearly ten times better than calcite-based rocks such as limestone and chalk.[55] Several seashell armor structures are existing, for instance:

1. The mollusc shell deformed through a process called "twinning," a process assisted in localizing damage and contained cracks from spreading throughout the shell.[55] Figure 5.7 shows one of the twinned crystal arrangements.[56] Under a load, the twins deformation most likely to form when strained, because they rarely have a sufficient number of slip systems for an arbitrary shape change.

FIGURE 5.6 Different types of seashell armor.

FIGURE 5.7 Twinned crystal group.[56]

2. *Seashell armor:* windowpane oyster armor is made of layers of long diamond-shaped crystals of calcite joined together by organic material. When dented, the shell deforms via twinning, then mirror images of calcite crystals formed around the penetration zone. Such twinning helped dissipate energy and localize damage.[56]
3. *Mother-of-pearl:* structure, where hexagonal calcium carbonate platelets are bound by an organic matrix, as shown in Figure 5.8.

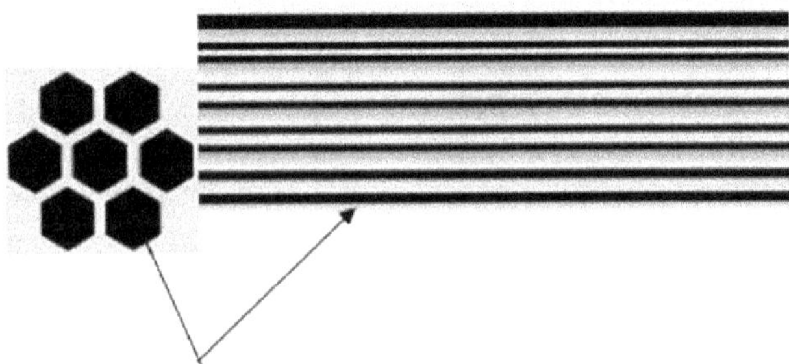

FIGURE 5.8 Mother-of-pearl armor structure.

4. *Deep-sea snail shell:* The deep-see snail shell has a variety of shape in helical thin curved shell design in order to withstand the pressure. The shell consists of a 250-μm-thick inner layer of aragonite, a common shell material, sheathed in a 150-μm-thick layer of squishy organic materials. The organic layer is encased in a thin, stiff outer layer (about 30 μm thick) made of hard iron

sulfide–based scales. The gastropod wears larger versions of the scales on its exposed foot.[57] The mechanism of failure of the shell under the pressure indicates that the crack in one layer is branched and more impact energy will be absorbed due to the presence of a thick layer of organic material between layers.

5. *Seashells:* The seashell armor is formed from diamonded-shape crystals, is a lamellar bioceramic material composed of brittle aragonite platelets (hard phase) closely staggered and bonded together by less tough proteins (soft phase).[58,59]
6. *Queen conch shell*: The structure of this shell material is different, which features crisscrossed calcium carbonate layers laid out in different orientations and separated by softer proteins.[63] This arrangement gives the great resistance to bunching force since it resolves the force into several directions, thus increases the area participating in the energy absorption.

5.4.3 MECHANISM OF FAILURE OF SHELL ARMOR

The structure of the shell armor consists of two layers: the hard surface one (large crystals of calcite) and a softer inner one (crystallographic forms of calcium carbonate polysaccharides).[60] The hard layer is formed from small polygon aragonite shapes tablets meshed to each other to form a convoluted surface with surface waviness. The interface between tiles reaches 30 microns of organic material. Figure 5.9 shows the sketch of the shell structure.[61]

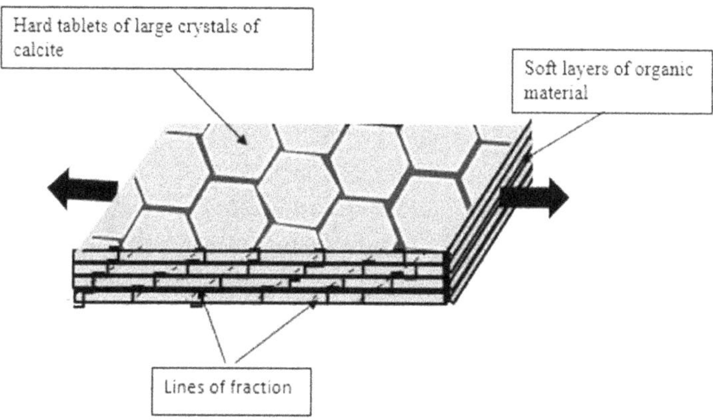

FIGURE 5.9 Mother of shell armor structure.

The mechanism of failure of such material under tension is accomplished by the sliding of the small tablets over each other, and the strain at failure may reach 1.5% under stress 60 MPa.[62] During sliding of the tablets, the soft layer will provide resistance to their sliding, distributing the mechanism of tablets sliding over a large area, and higher resistance to failure is granted. The more crack is extended, the more energy dissipated. The length of the crack path will depend on the aspect ratio of the tablets; the higher the aspect ratio, the longer will be the crack path and consequently, more the dissipated energy and the area participating in the resistance to the punching force.

The shell is 3D twisted curved construction as shown in Figure 5.6. The shape, dimensions and size differ from one species to the other. Figure 5.10 presents a sketch of the shell which, as mentioned before, consists of hard and soft layers.

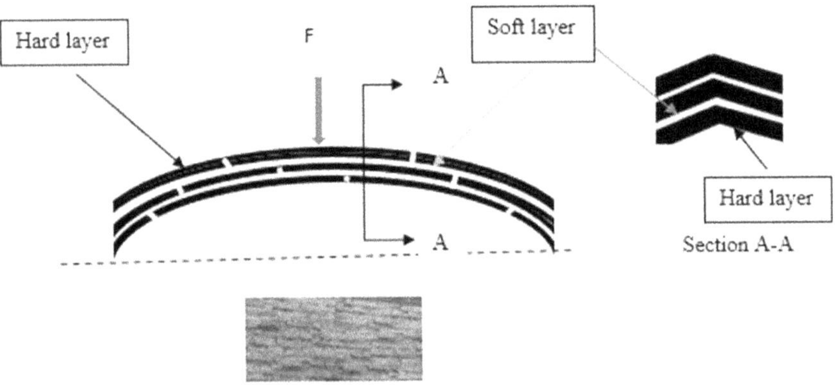

FIGURE 5.10 Schematic of the curved shells structure.

The arch is a multilayer curved beam of the hard plate over a soft material. The crack path between the tablets takes a long path, represented in Figure 5.11. The interfacial crack of the soft layer depends on its shear strength and the path length (a-b-c-d-e-f-g-h). Also, it relies on the aspect ratio and thickness of the soft layer. The fraction energy will depend on the plates bending stiffness, plate aspect ratio and the volume fraction.

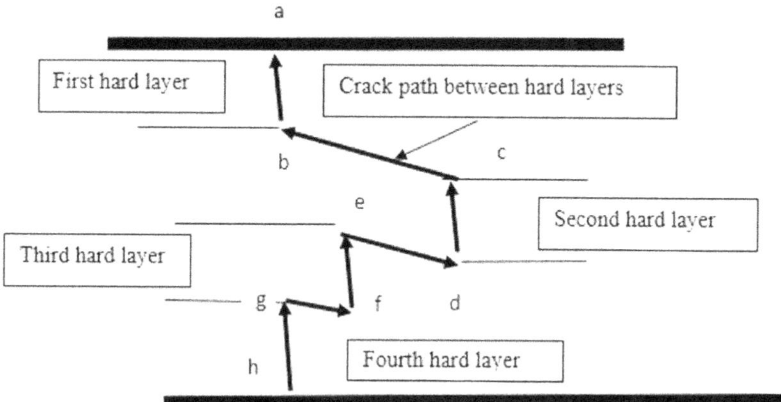

FIGURE 5.11 Crack propagation between layers.

5.4.4 FISH SCALES

Most of the fish species have scales on their body to provide protection from the external threat. Based on the shape, size, and arrangement of the scales, which are a hard material fixed to the fish skin, there is a wide variety of fish scales structures.[64] They may be classified into Placoid, Leptoid, Cycloid, Ctenoid, Ganoid, Elasmoid, Cosmoid, and Thelodont scales. In its initial configuration, the scale structure is represented as a two-dimensional arrangement of initially straight scales, lying on a straight support. Scale deformation is then restricted by its attachment to the support (on its left end) and a rotational spring resisting scale rotation at its attachment point.[64] To analyze the mechanism of working of the scales, Figure 5.12a shows the overlapping of the scales in the initial position, while Figure 5.12b indicates the position of the scales when applying on them a force F.

The force F will cause compression in the under-skin layer. This will lead to the rotation of the scales with different angles θ_i, and the contact points of the overlapped scales will press the successive scales, generating a resistive force to create a moment in the direction opposed to that initiated by the force F. The deformation of the skin layers in the area will resist the applied force F. The shape of the scales, their dimension and the type of the overlaying are the determinant factors of the number of the scales in contact with each other. The resisting energy, required to cause the failure of the scale under the force F, depends on its stiffness. The point of contact with

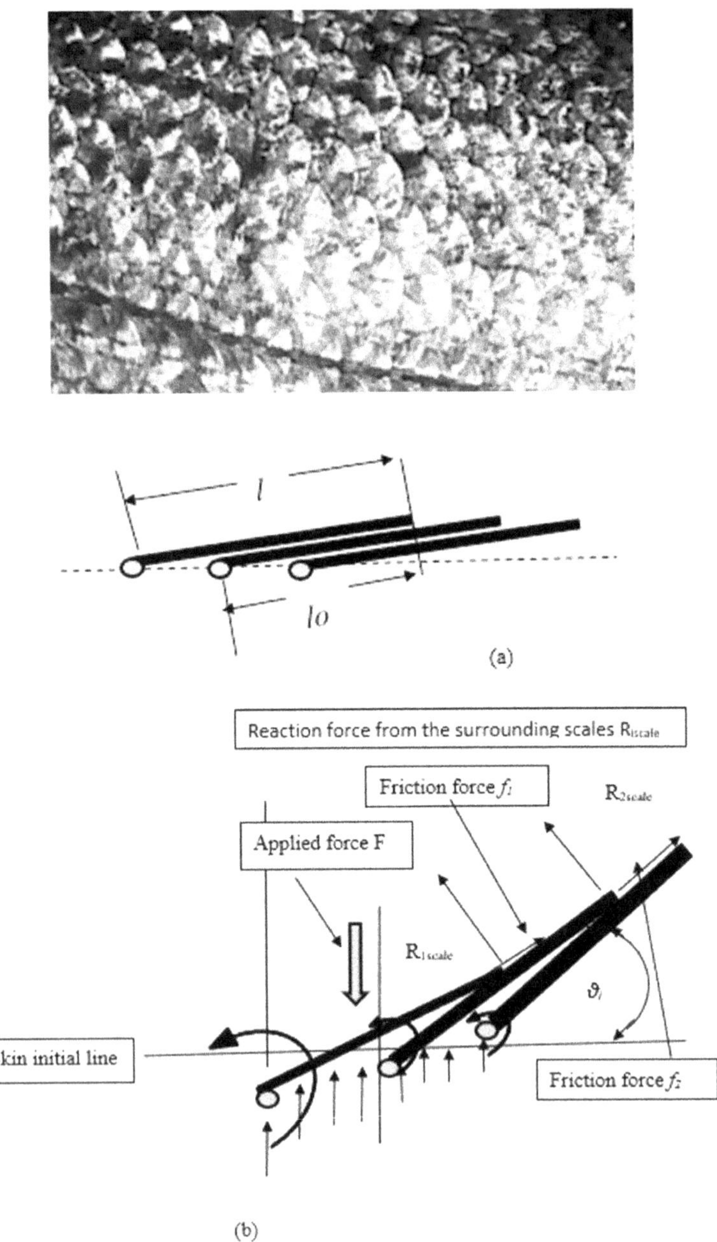

FIGURE 5.12 Model of the scales arrangement under load; (a) illustrates the overlapping of the scales in the initial position; (b) indicates the position of the scales when applying on them force F.

the scale under a load with the other scales is defined by its deformation under the load. The elastic deformation of the skin is the main initiator of the mechanism of scales resistance to the applied external force that preventing the localizing effect of the applied force. Flexibility and anchoring of the scale in the skin are insured at the fixation points. The length of the scale and the scale density are the main factors that determine the penetration through the skin by punching force. It was revealed that the contact force between scales increases exponentially with a measure of scale density.[64] The scales overlap length "l_0" influences the value of the stress on the scales, which may reach 0.46 of the scale length "l." The dimensions of scale vary conferring to the fish species.[65] From the analysis of the scale mechanism, it can be concluded that to prevent the stress concentration, the skin should be flexible, soft in compression, the scales are stiff, but can rotate at the point of support with the skin.[66] The mechanism of resistance of the punching force to the scales ensures the participating of a large number of the surrounding scales as well as the resistance force of the flexible body and the muscles at the point of scale fixations. Figure 5.13 shows the forces acting on the scales under punching force "F."

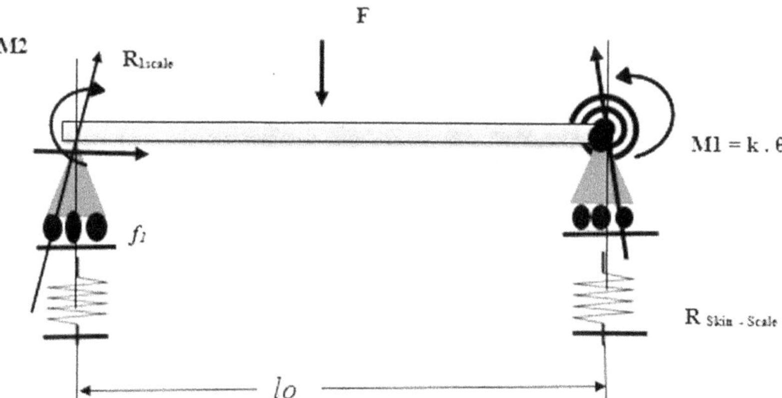

FIGURE 5.13 The scale under punching force.

The fish scales structure plays a great roll in their deformation under load. In the example of the Cycloid scale, the outer layer of the scale is significantly more mineralized and often referred to as "bony layer," whereas the inner layer (collagen layer) is mineralized mostly near the bony layer, but with mineralization, pockets proceeding well into the collagen fibril layer.[67] The scales are a composite laminated structure; bony layers are mineralized in collagen layers, which are soft cross-ply layers formed with Nanoscale fibrils, laid as laminated 90°/0°/90° (Fig. 5.14). Collagen fibrils type-I are partially mineralized with hydroxyapatite which represents 46%.[67]

In some cases, the scales are imbedded in the skin, forming a thin hard plate on elastic foundation (Fig. 5.15).

The maximum deflection under the force of the scales δ_{max} in the case of flat bony plates under the punching force P_p, can be simply expressed as a function of:

$$\delta_{max} = f^{n}(P_p, \gamma_1, b, E_S, J), \tag{5.16}$$

where γ_1 is a coefficient of the stiffness of foundation per unit area, E_S is the modulus of elasticity, J is the area second moment of inertia, and b is the width of the scale

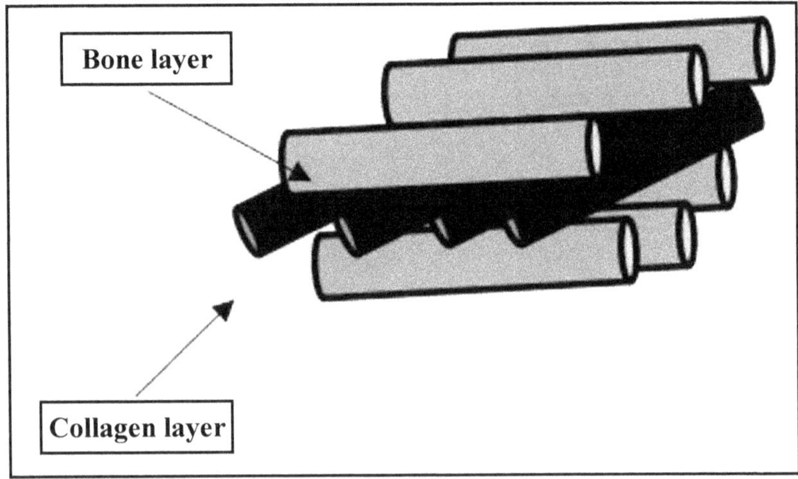

FIGURE 5.14 Fish scales structure.

The Young's modulus, dimensions of the plate and the stiffness of the foundation determine the scales absorption energy. The lower stiffness of the collagen layer increases the absorption energy, while high Young's modulus increases the plate failure force.[68]

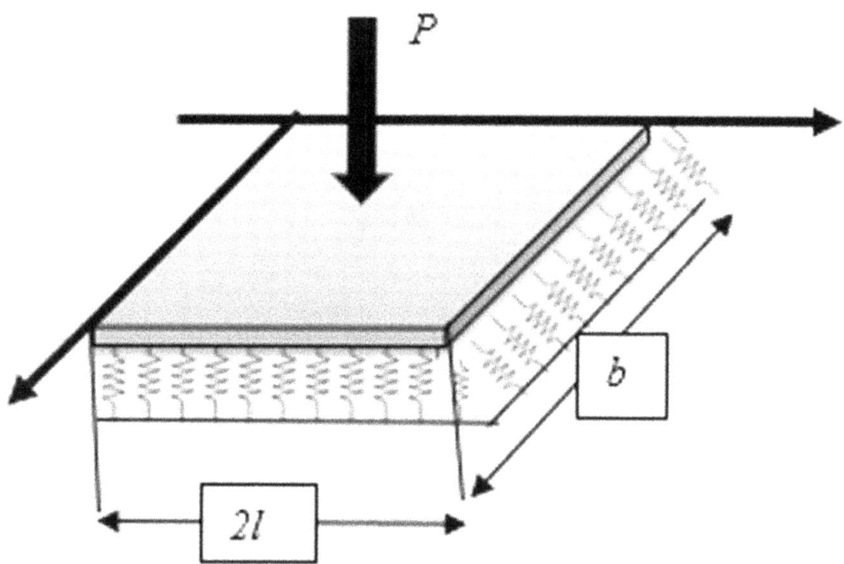

FIGURE 5.15 Scale embedded over flesh layer.

5.4.4.1 ANALYSIS OF THE SCALES STRUCTURE

The objective of these analyses is to explore how a limited palette of constituents with low mechanical properties has a high combination of strength, toughness, and energy absorption.[69] The scales are arranged in such a way that allows geometric flexibility.[70] This will inspire in the design of a protective vest.

The structure of the collagen layer consists of collagen nanofibrils arranged in 3D laminates oriented in different inclinations to each other (Fig. 5.16). The direction of the collagen fibrils in each laminate allows redistributing the stress of applied punching force in multi 3D axis (Fig. 5.16), in such a way that the punching force will be resolved into several components supported by the different laminates, as in the case of simple space truss. Hard and stiff outer bone layers are anchored in the collagen layer.

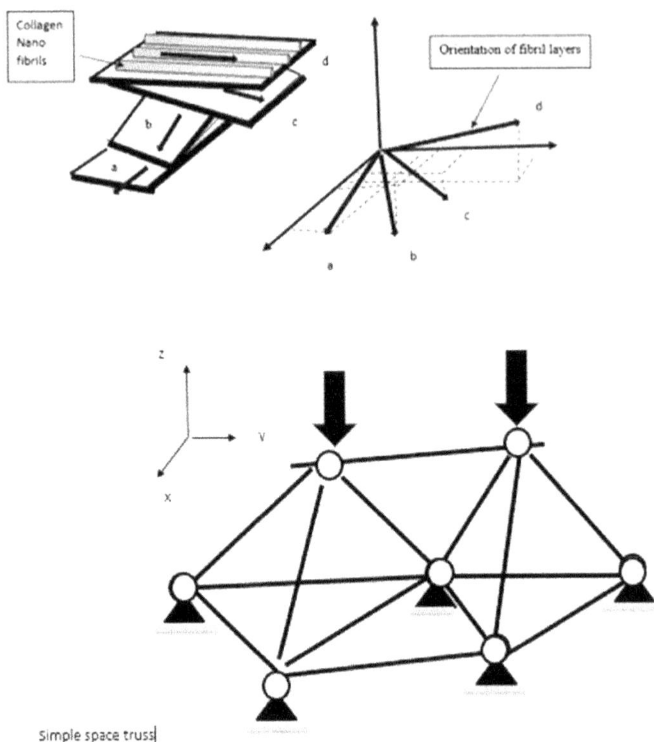

FIGURE 5.16 Multilaminate structure of collagen fibril layers.

I. *Elasmoid scales*

This structure is also consisting of collagen fibril laminates in different arrangements (Fig. 5.17). Two sets of laminates are in the perpendicular arrangement to each other so that the fibril arrangement in *x* direction is 90° to the set of laminates in *y* direction.[71] The angle of the fibril orientation in the alternate layers of set one is 0, θ_1, θ_2, θ_3, …0 (Fig. 5.19), and the fibril orientation in alternate layers of set two is 90, 90 + θ_1, 90 + θ_2, 90 + θ_3, … 90.

This arrangement gives high tensile strength and modulus in both, longitudinal and transverse directions. The bridging of the laminate of collagen helps in preventing the spread of the crack under the punching force. It was proved that the fracture of the laminates, the stretching, rotation, and delamination of the collagen fibrils dissipate a significant amount of energy prior to failure.[72]

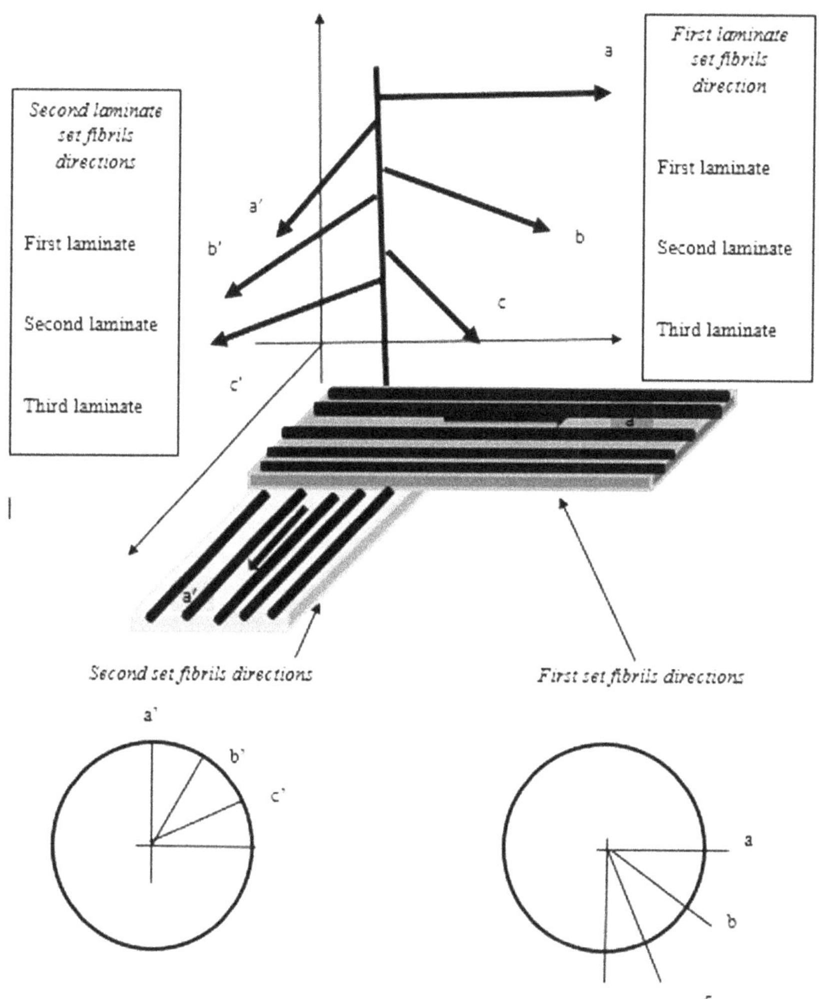

FIGURE 5.17 Multilaminate structure of collagen fibril layers in the perpendicular arrangement.

II. Alligator gar scales

The scales have an outer layer of mineral with the arrangement of crystals which increase the toughness of the scales. Besides, its top has an irregular surface that makes the punching force inclined to the scale surface and more energy is required to punch the scales. The bone of the scale consists of mineralized collagen fibrils normal to the scale surface, embedded in a mineral matrix.[70]

III. Placoid scales

Placoid scale is flat corrugated multi tip, the scales are overlapping (Fig. 5.18). The base of each scale is imbedded in the skin.

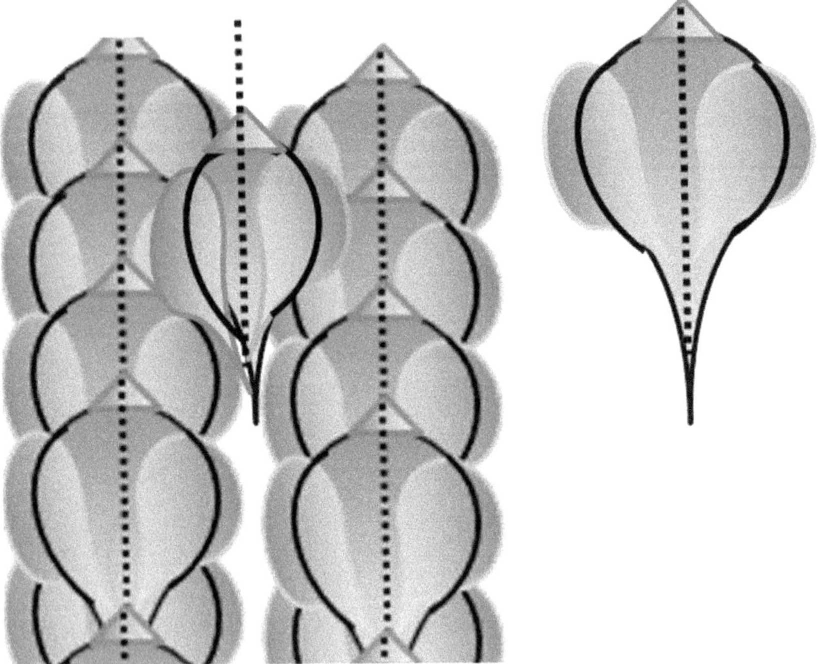

FIGURE 5.18 (See color insert.) Placoid scales.[21]

IV. Ganoid scales

The scale has a rhombus shape with up to 4 cm length and about 5 mm thickness (Fig. 5.19); it consists of a bony structure that resists the cracking propagation and is embedded in the mineralized ganoine material which is responsible for giving the required flexibility. The scales are of serrated edges to allow the deformation under loading.

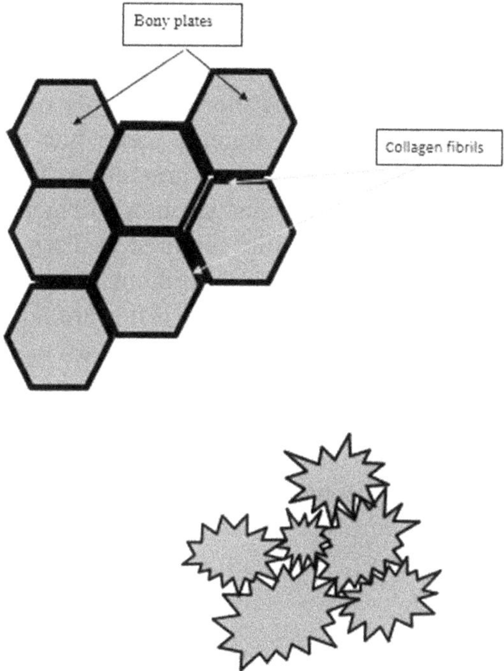

FIGURE 5.19 Scales shapes.

This bony structure will deform under the applied load (Fig. 5.20); the mineralized ganoine layers will deform to resist the applied force and dissipate the punching energy into a larger area.

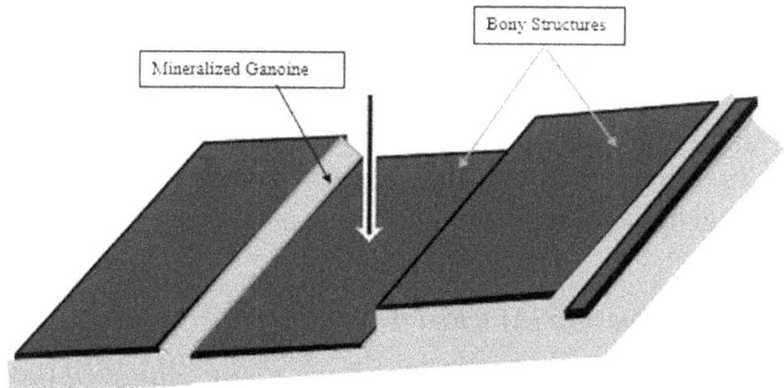

FIGURE 5.20 Displacements of bony plates under load.

5.4.5 ANIMALS FLEXIBLE ARMOR

Mammal armors are different than the fish by using osteoderm armored tiles, bony plates. The shape of the tile varies conferring to the location, hexagonal for the top part and the triangle shape over the other parts, as in the case of turtles, snakes, lizards, and armadillos. The hexagonal hard plates are connected with each other through collagen fibrils that give the armor the strength and flexibility required to dissipate the energy of punching force to ensure maximum absorption of the punching energy. In the other structure, the shapes of bony plates are not regular, and the neighbor plates are meshed in each other to give more stiffness to the structure.[73] The designers have learned to not simply copy the design of nature but also the qualities of structures and have managed to reverse-engineer solutions to problems. Geometrical analysis of the scale's design is a necessity for an armor construction.

5.5 SCALE DESIGN

The energy absorption will be a function of the scale volume fraction. The design of the scale armor, Figure 5.21, can be identified by the following parameters:

I. Scale aspect ratio:

$$\alpha = (\text{Scale length / Scale thickness}), \quad (5.17)$$

or

$$\beta = (\text{Scale length / Scale width}). \quad (5.18)$$

Both parameters are varied conferring to the age, size of the fish or the mammal. It was revealed that the value of α varied between 0.4 and 3, while β varied between 0.01 and 0.5.[74]

II. The degree of overlap $(\eta) = lo/l$ (5.19)

III. The scale volume = [volume of the scale / (the volume of the scale + volume of collagen between the neighbor scale)] (5.20)

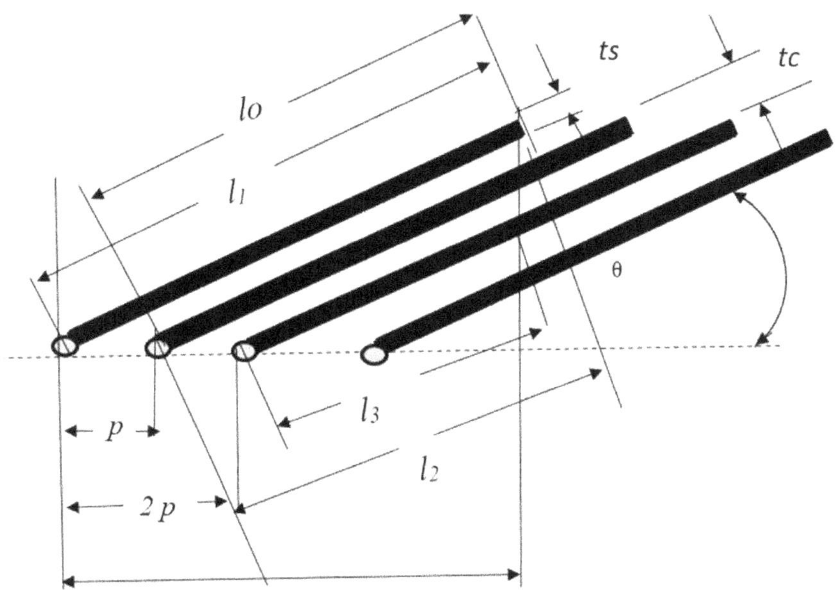

FIGURE 5.21 Analysis of the scales design.

The volume of scales

$$V_{scales} = (l_1 + l_2 + l_3 + \ldots + l_{n)}\, t_s\, b \tag{5.21}$$

Volume of collagen

$$V_{collagen} = 0.5b.\ (n\, p)\ (l_1 \sin\theta) - V_{scales} \tag{5.22}$$

$$\text{Scales volume fraction } (\gamma_S) = V_{scales}/(V_{collagen} + V_{scales}) \tag{5.23}$$

$$= 2\ (l_1 + l_2 + l_3 + \ldots + l_n)\ t_s/(n\,(t_s + t_c) \cos(\theta)) \tag{5.24}$$

$$= 2\ l_1\ (\textstyle\sum^{n}_{i=2} (1+\eta^{(n-1)}))\ t_s/(n\,(t_s + t_c) \cos(\theta)) \tag{5.25}$$

$i = 2, 3, 4 \ldots\ldots\ldots n,\ n = \text{integer}((l_1 \cos(\theta))/p) - 0.5)$

$$\gamma_S = k_1\, k_2\, l_1/(n \cos(\theta)), \tag{5.26}$$

where $k_1 = 2\ (\sum^{n}_{i=2} (1+\eta^{(n-1)}))$

$$k_2 = (t_s/t_c) + t_c.$$

The energy absorption will be a function of the scales volume fraction (γ_S), which depends on the scales density, that is, the value of (p), scales length, degree of overlap, overlapping angle θ, *and* k_1 *and* k_2.

The low value of β and an overlapping angle greater than 10° redistribute the impact over a larger area.[75] The failure stress of the scales is a function of failure stress of the scales material, Young's modulus and the ratio α, β, η, k_2.

5.6 SOME DESIGN ASPECTS OF BULLETPROOF ARMOR

The modern armor features not only prevent the penetration of the bullet, but also prevent the penetration of fragments of projectiles. Body armor systems are multi-functional due to the change in the nature of threats. The protective armor performance (PAP) depends on fiber's, yarn's and fabric's mechanical properties, weave structure and design (2D or 3D), and the number of fabric layers.

However, the PAP of the flexible vest can be improved with the use of the additional elements, such as ceramic plates, hard plates, or composite structures, implanted in the tightly woven structures. In all armor designs, the comfort and ergonomic aspects must not inferior the wearer injury-related musculoskeletal conditions or not leave trauma or body syndrome.[76-80] The armor should be properly suitable to fit each wearer not creating pressure on the body. Its light weight is a final requirement, conducted pulmonary function test, the heart rate and oxygen uptake on wearing armor of 11.0 kg, showed a significant increase as compared to the values without wearing it in routine exercise, developing an early feeling of fatigue.[81] The armor should be evaluated for the freedom of movement, suitability of the design with other equipment worn, which presents a new challenge for the armor designer. The optimization of all the above contradicting factors affects the level of protection of the designed armor.

5.6.1 SUGGESTED DESIGNS OF THE BODY ARMOR

The design of body armor has been drawn attention all over the human history with unstopped efforts of the armor designer due to the progress and variability of the threats as well as the need for reducing the weight of the armor without reducing its PAP. The general design criteria are to increase the ratio of (energy absorption/impact energy). There are several tactics to attain this target:

- use of high-performance fibers,
- use fabrics with high weight per unit area,
- application of textile reinforced composite material, either flexible or hard,
- application of nanoparticles,
- application of a new weave structure,
- application of graphene sheets, and
- introduction of the innovative designs for the armor

5.6.2 SOFT PROTECTIVE ARMOR

In this section, several suggestions are discussed.

5.6.2.1 QUILTED FABRIC COMPONENT FOR BODY ARMOR

Quilting is the process of sewing two or more layers of fabric together with stitch line taking a diagonal or perpendicular direction to the fill yarns. The pitches between the stitch lines are equal. This process increases the stiffness of the fabric layers and the area participating in resistance of ballistic impact force. The amount of energy absorbed during ballistic impact over a range of velocities for a diamond quilted pack is between 14% and 22% higher compared to a non-quilted pack, depending on the number of fabric layers in the pack.[82] Fast impacts resulted in 45% more energy absorbed than slow impacts. The conical depths (back face signature) upon impact on the multiaccess stitched structures were small compared to those of the unstitched structures.[83]

5.6.2.2 WOVEN FABRIC DESIGN

The fabric structure of 2D or 3D structure is found to have an influence on the ballistic resistance. Several investigations indicate that the yarn pulling force depends on the inter-yarn friction and the way it is interlaced.

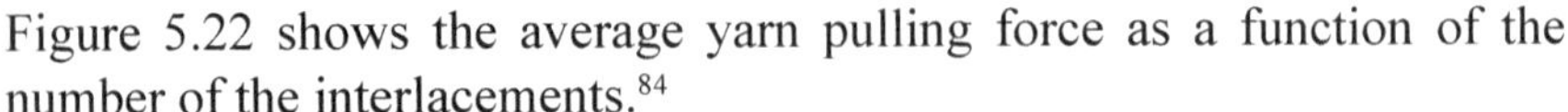

Figure 5.22 shows the average yarn pulling force as a function of the number of the interlacements.[84]

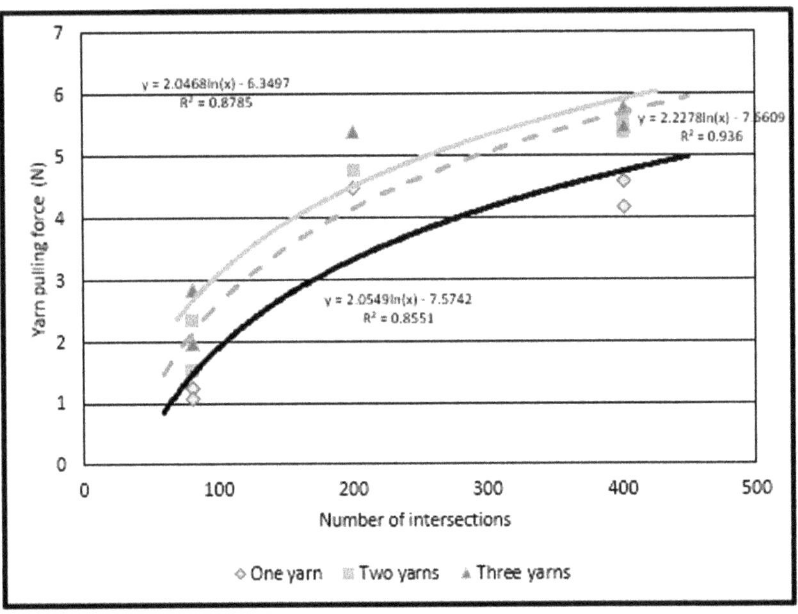

FIGURE 5.22 (See color insert.) Yarn pulling—off force versus the number of intersections.

The index to express the relation between the fabric properties and the yarn pulling force will be:

$$\text{FM} = S \cdot I \cdot T, \tag{5.27}$$

$$\text{assume yarn pulling force (PF)} = K\,(\text{FM})^{0.5}, \tag{5.28}$$

where K is a material constant. Then

$$\text{PF} = 0.03\,(\text{FM})^{0.5} \tag{5.29}$$

Fabric of high index FM results in higher energy absorption. This has been supported by several researches.[85,86] Fabric with high shear modulus, tightness, and between yarns friction contributes to the increase of the fabric absorption of the impact energy.[85-87]

In the cases of very low values of transverse and shear moduli or very high value of shear modulus, yarns are early damaged and failed.[88] The fabric is characterized by both, fabric areal density and the value of specific absorption energy. The choice of the areal density value of the protective fabric to stop penetration of the projectile is proportional to the impact of energy.[84,89] Bullet tip hits the fabric between the threads, hence sufficient space allows it to pass, as in the case of fabric with low density in fill and warp, so that the yarns will move in different directions permitting bullet tip to pass through without fiber or yarn damage, or bullet tip punches the fabric between the threads without cutting matrix, pushing the yarns aside without cutting it. The bullet will be subjected to the friction between the yarns and the bullet body, pressing the yarns with the increasing pressure. If the energy of the bullet is less than the energy to penetrate a fabric, the fabric absorbs bullet energy forming a cone. Otherwise, the yarns will break, and the bullet will pass through. The absorption energy depends on the fabric energy absorption capacity. For tight fabric near the jamming condition, the yarns start to be cut by bullet tip.[90,91]

5.6.2.3 HYBRID STRUCTURE

In the case of multi-layer woven fabric, the first layer is subjected to a higher impact velocity than the last one. The orientation of the fill or warp yarns in each layer relative to the next one and the angle-layering improves the extent of the isotropy of the entire structure and causes higher energy absorption. Again, a higher number of layers of low fabric areal density with small angle-layering is preferable.

Various armor structures from different materials are targeting to get high ballistic performance, such as Kevlar plain weave, Triaxial fabric layers from Kevlar 29, Polyester and Vectran.[89,90] The armor is structured by adding the micro rubber particles between layers or carbon yarn mat. The most effective is the Kevlar plain weave fabric because of its high strength, closed spacing, and heavyweight, 385 g/m^2. The structure shows the best performance when using the micro-particles of rubber, to disperse the energy of the projectile, between a combination of layers from both Kevlar and polyester triaxial fabrics.

Thus, a good understanding of the role of yarns mechanical properties in the impact behavior of woven fabrics is necessary. Longitudinal mechanical properties of yarns play a considerable role. The good ballistic materials, as implied by several investigators,[92,93] would be mainly characterized by a tensile failure. Therefore, the tensile properties of high-performance yarns have been considerably investigated. The analysis of various combinations of fabrics design, construction of fabric's layers layout and type of fibers indicate the importance of the proper choice of each element. Using particles of rubber between fabric layers had a significant effect on the absorption of the ballistic energy hence it increases the friction between the particles, and the bullet during penetration decreases its velocity.

5.6.3 HARD COMPOSITE PANELS

Conferring to the literature, hundreds of the hard-composite panels have been designed and tested using different materials, HPF composites, nanoparticles composites, ceramics, and so forth to maximize their absorbing energy to defeat the projectile. Since the development of new material and the innovation of new designs, a continuous challenge for the armor designer still exists.

5.6.3.1 CERAMIC PLATES PROTECTIVE ARMOR

The continued search for protective material that has a price efficiency ratio suggested ceramic materials for the design of the protective armor for the soldiers and military vehicles. For ballistic protection, the following materials were developed[94]: silicon nitride (SN), titanium boride (TiB_2), aluminum nitride (AlN), silicon aluminum oxynitride (SIALON), fiber-reinforced ceramic, ceramic-metal composite materials (CMC), silicon carbide (SiC), boron carbide ceramics, and aluminum oxide ceramics with different contents of Al_2O_3. The appropriate material can be selected based on situation and ballistic requirements. Alumina ceramics are widely used for plates formation of ballistic protection for its high performance and low cost.[95]

The general structure of ceramic plates, which can be the hard single plate or flexible, consists of small ceramic elements arrangement (Fig. 5.23), depending on the end use. To challenge the high energy of the projectile, the hard ceramic surface should absorb the most part of the energy through the fragmentation of the bullet. On penetration through the surface part of the ceramic plate, the energy of the fragments is completely absorbed.

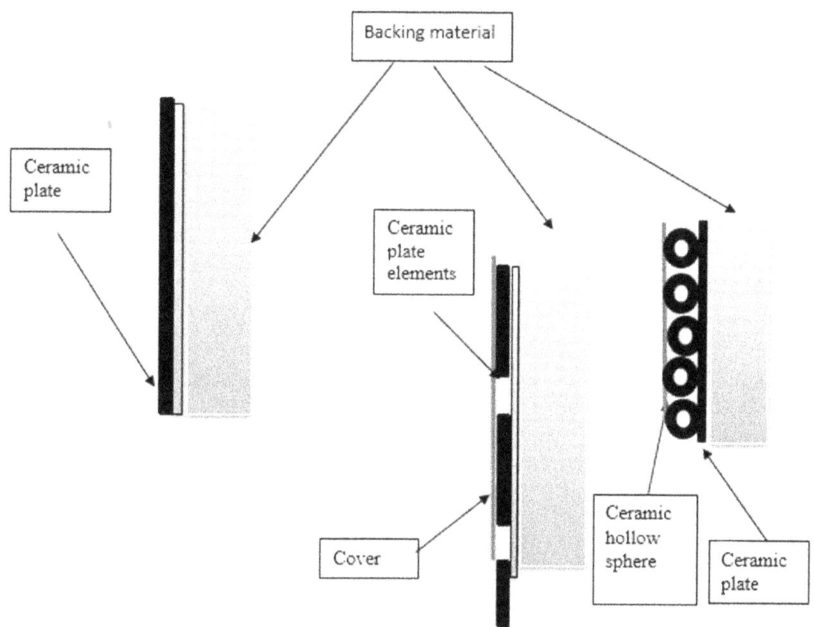

FIGURE 5.23 Ceramic armor systems.

The characteristic properties of the ceramic are created by the sintering process at temperatures of 1600°C for alumina, and more than 1900°C for carbide materials. This process is associated with a distinct reduction in the dimensions of the pre-shaped component.[94,96] The processing of the required shapes of the plates is done before the sintering process. The hardness of the plates reaches 14 GPa and bending stress 150 MPa, the density 3.19–3.95 g/cm^3, Modulus 304–380 GPa, strength 200–370 GPa with thickness depending on the type of threat and can reach 2–100 mm. The thickness is also controlled by the total weight of the armor required. Due to the high value of ceramic surface hardness, the bullet will strike the first ceramic surface, resulting in the absorption of its kinetic energy through ceramic plate cracking and friction between bullet and the ceramic materials, then it will shatter (Fig. 5.24), and the resultant fragments will penetrate the ceramic layer. The backing material layers will absorb the residual energy safely.

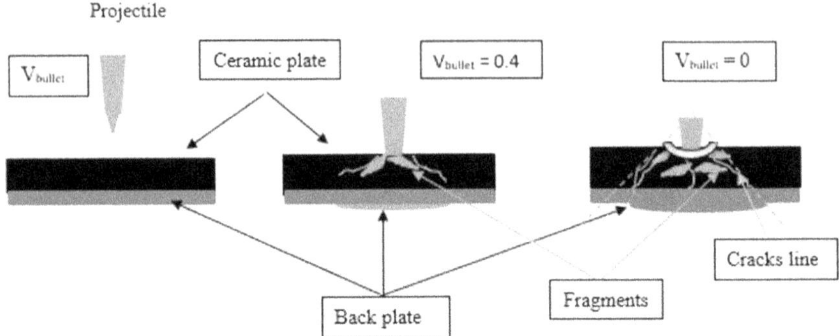

FIGURE 5.24 Profile of the ceramic structure under bullet impact.

The ballistic performance of the material is evaluated by the following parameters[97]:

1. Fracture toughness is calculated from the Vickers indentation. Brittleness (B) of homogeneous ceramic materials is calculated using formula:

$$B = (\mathrm{HV}\ E)/K^2. \tag{5.30}$$

The values HV is (1230–1560), E is (280–450 GPa), and K (3–3.4) for ceramic.

2. Ballistic energy dissipation ability (D-criterion) is calculated in accordance with the following formula:

$$B = 0.36\ (\mathrm{HV}\ E\ c)\ /\ K^2, \tag{5.31}$$

where c is the sonic velocity of the material.
It was revealed, that the brittleness of alumina ceramic varies between 340 and 545 × 10^{-6} and ballistic energy dissipation criterion "D" from 1.07 to 2.4 × 10^{-12}, and they are governed by the wetting of alumina particles with a liquid phase, interaction between the phases in sintering and by the firing conditions.[95] A proper selection of alumina starting materials, that allows optimizing ceramic microstructure with desirable grain sizes and physical properties, is considered as one of the imperative factors of the ceramic improvement.

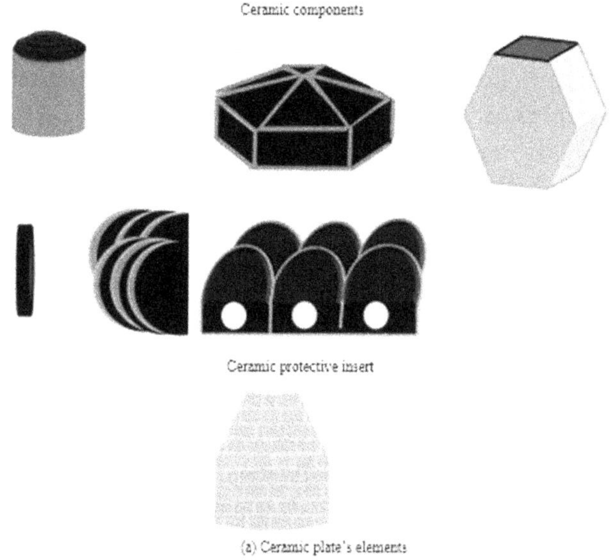

(b)

FIGURE 5.25 **(See color insert.)** Some elements of the ceramic plates.[101]

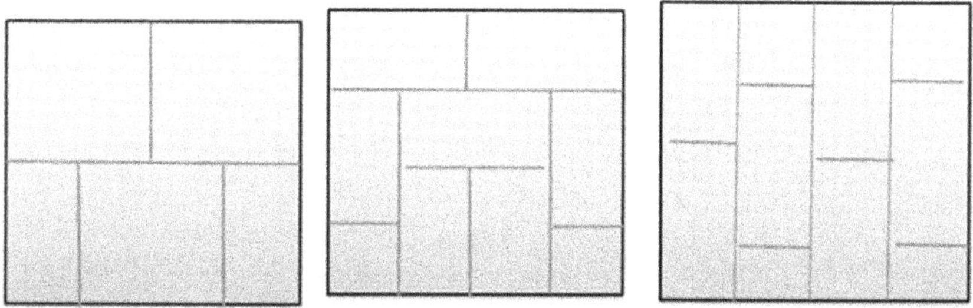

FIGURE 5.26 Layout of the ceramic tiles.

3. Depth of penetration method (DOP)
In this method, flat ceramic samples are fixed on a thick backing material.[98] The DOP corresponds to the depth at which the projectile embedded itself in the backing material. The results can be used to determine the following parameters:

$$\text{Mass efficiency } (E_m) = (\rho_{bac} P_{ac})/(\rho_{cer} t_{cer} + \rho_{bac.} \text{DOP}) \tag{5.32}$$

$$\text{Ballistic efficiency } (\eta_b) = \rho_{bac} (P_{bac} - \text{DOP})/\rho_{cer} t_{cer} \tag{5.33}$$

$$\text{Critical thickness } (t_{crit}) = t_{cer} P_{bac} (P_{bac} - \text{DOP}). \tag{5.34}$$

Several designs of the ceramic composite armor have been introduced to defeat the bullet through multilayer composite structure of different materials. In all the designs, the ceramic tile will be laid on the face to confront the bullet, followed by a combination of backing materials to reduce the trauma effect. A layer of rubber may be used to damp the residual energy of the projectile fragments after hitting the ceramic plates. The overall performance of each design is a function of the individual elements. The stiffness of the successive layers is gradually reduced in each layer, depending on the thickness of the layer, material properties and structure.[99] The last layer should be soft enough to prevent body discomfort. Nature and thickness of backing materials have a significant impact on crack propagation due to their own abilities to absorb the residual energy with limited deformation and capture the bullet fragments. Fabrics of Kevlar, Twaron, Spectra, KM, and Dyneema are commonly used as backing materials; the number of fabric layers is select to reduce the size of fragments.[96] The protective armor structure is covered by a layer of the textile composite to ensure the integrity of the final design. The optimization of the choice of the material thickness in each layer is a necessity to fulfill both: the degree of protection and the wearer comfort. The final weight of the armor to achieve the protection performance depends on the thickness of the used layers and their densities. The plate is usually inserted in a pocket in the flexible protective vest.

The total areal weight of the structure is given by:

$$W = \sum^{n} \rho_i t_i, \tag{5.35}$$

where n is the number of layers, ρ_i is the density thickness of ith layer, t_i is the thickness of ith layer, *i is equal to* 1, 2, 3, 4…n.

The thickness of the ceramic plate required to stop projectile was found to be proportional to its velocity. The depth of bullet penetration decreases as the ceramic tile thickness is increased. Consequently, the ceramic plate thickness should be augmented for higher ballistic limit.[100] The hard plate weight, in this case, will vary between 1.4 and 3.7 kg. Several designs of the ceramic plates were introduced, either to be one solid plate or assembly of small size ceramic plates to cover the protected area and fixed in a particular manner on a flexible sheet of material. Figure 5.25a, b shows different shapes of ceramic components which can be assembled in the final armor form.

The layout of the ceramic tiles influences the propagation and absorption of the impact energy. Figure 5.26 shows some layouts of the ceramic plates. The crack propagation depends on the width of the tile and neighborhood of impact zone which had a restricted damage.[102] Consequently, the width of the ceramic tile should be higher than the radius of the area of crack propagation, which is a function of the interaction between the ceramic material properties, the level of projectile energy, and shape of the bullet nose. The increase of the time between the moment a bullet hits the surface of the ceramic plate and starts to penetrate, delay time, is the key issue to improve the armor performance.

5.6.3.2 SOFT BODY ARMORS WITH CONCEALABLE VEST DESIGNS

The concealable structure consists of flexible lightweight armor with pockets for the insertion of ballistic packs or plates. These plates are designed to prevent the bullet penetration in specific areas; the design of these plates is made of nonwoven, woven or combination of them or made of hard ceramic, boron-carbide plates for a high level of threats. It should be mentioned, that the body vest is provided by several pockets to protect other parts of the body, such as side plates, shoulder plates or groin plates.[103] An advanced composite armor can be designed from several layers, as shown in Figure 5.27, of fabric, ceramic, anti-trauma layer and finally, textile fabric to improve the ballistic protection. The performance of each layer affects the overall performance of the armor. Some designs use a rubber layer to support the ceramic tile layer. The ceramic layer is the main defense line that will defeat the bullet and damp its fragments, while the other layers are responsible for the absorbing the residual of the impact energy and preventing the back signature from causing trauma. Several other designs of the back plate of the armor either use new high-performance fibers materials or different damping layered structures.[99] The backplate, functioning as a residual energy absorber, may have the different designs, for instance, Kevlar fabric reinforced polymer composite, should be sufficiently stiff and of lightweight. The textile fabric covers the ceramic plate just to prevent the ceramic fragments to move out of the armor surface. The use of a cover layer, which is bonded to the ceramic plate, improves the armor performance. The hard backing material absorbs the residual energy of the bullet after the impact with the ceramic plate. In some armor designs, the use of the rubber layer helps in the redistribution of the energy after impact on a larger area. The mechanism of fabric-adhesive-ceramic plate will keep the integrity of the area under the bullet nose, especially when multi-hit. More improvement is expected when the radial movement of the ceramic plate fragments is constrained.

5.6.3.3 LAMINATED AND HYBRID SOFT ARMOR FOR BALLISTIC APPLICATIONS

The formation of flexible laminate of woven or nonwoven fabrics is suggested to increase the energy absorption due to the delaminating of the layers under ballistic impact. In such a case, the percentage of the polymer matrix should be low, and the polymer gives a lower shear property. A hybrid system using nonwoven fabric layers on the impact side and relatively tighter woven fabrics between the layers were suggested for influencing the energy absorption mechanism indirectly by restricting the lateral movement action of the primary yarns during the ballistic impact event.[104] Another construction of the laminated armor is by the reorientation of the direction of the warp yarn relative to that at the subsequent fabric layers in order to spread the strain in all directions of the impact area, that results in increasing the absorption energy and reducing the size of the cone height.[105] The thickness of the composite multilayer laminates plays a significant role in the mechanism of penetration of the projectiles. It was found that to stop the bullet at velocity 850 m/s, the armor of thickness higher than 50 mm is needed (96 layers of Kevlar fabric).[106] The relation between the residual bullet velocity and the multi-layer thickness was found to reduce as the thickness increases, after a defined initial thickness t_0, the first thin layers are not effective in reducing the bullet velocity, Figure 5.28. A new trend is to use 3D fabric reinforced polymer composite with 3D orthogonal woven carbon/epoxy composite plates. Composites made of 2D weave architect are weak in the z-direction which is in the path of the bullet during impact. In 3D fabric, the architect of the yarns supposed to have yarns in the xyz-directions; consequently, it can create compactness of the weave structure increasing the bullet-resisting force and fabric deformation under the impact. The major failures in all plates were matrix and fiber breakages in the impacted area. However, there was not any internal yarn splitting observed between matrix and fibers around the impacted region. Damage propagation in the 3D woven structure is small.[107]

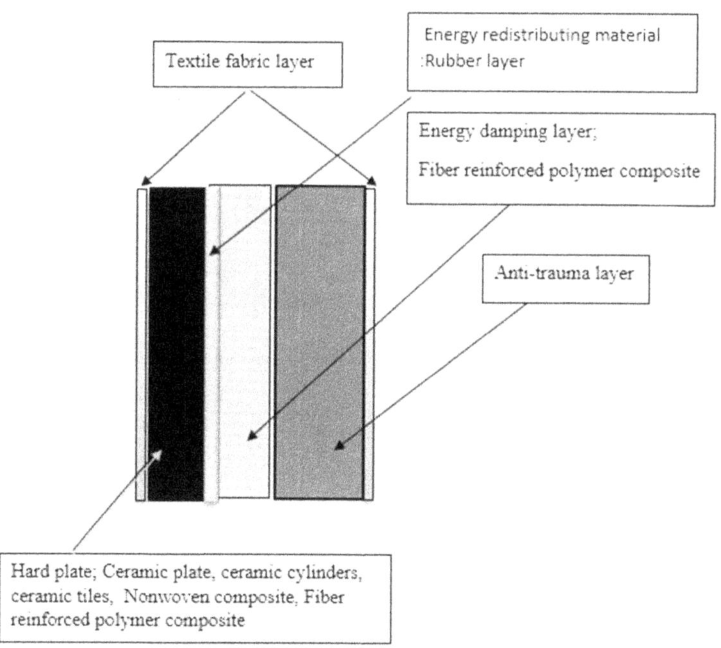

FIGURE 5.27 Concealable vest design.

5.6.3.4 NANOPARTICLES TO FABRICATE COMPOSITE PROTECTIVE PLATES

In present time, a great attention raised toward the use Nanoparticles to fabricate composite protective plates. Glass Nanoparticles composite is suggested for the formation of protective tiles with a multilayer of Triaxial fabric/Polyester/Glass nanoparticles. Nanoparticles composite with Vg 2.8% glass nanoparticles and thickness 2 mm increases significantly the resistance energy to impact.[108] Carbon nanotube–metal matrix composites (CNT-MMC) are an emerging class of new materials that are being developed to take advantage of the high tensile strength and electrical conductivity of carbon nanotube materials. Several CNT composites were designed, such as Aluminum–CNT composites, Copper–CNT composites, Nickel–CNT composites, Magnesium–CNT composites, Nano-crystal steel,[109,110] and suggested to form a lightweight upper armor.

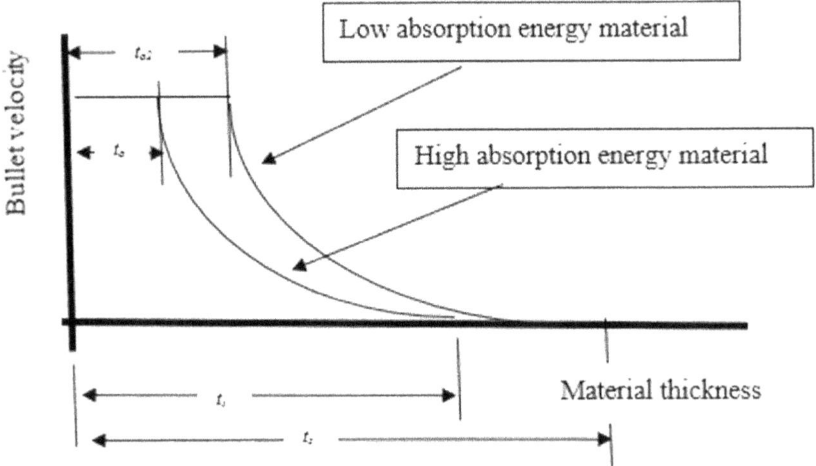

FIGURE 5.28 Effect of the material energy absorption capacity of the bullet speed.

The back-plate material properties, design and thickness influence the value of the thickness of the hard plate to defeat the bullet at a particular velocity. A multilayer fabric, fiber-reinforced polymer, etc. are designed to keep the integrity of the shield but also participate in the determination of the wearer comfort.[99]

The use of the certain type of protective armor is accomplished by risk level which depends on several factors, the risk relates to the probability of the occurrence of the threat, the intensity of the threat, and the intensity of the harm expected.[111] Several factors are involved in the body armor effectiveness such as:

I. Ballistic force

Mass of the bullet,
Shape of the bullet tip,
Bullet material,
Bullet velocity, and
Angle of incident.

II. Penetration dynamics

Armor material,
Armor thickness,
Armor design,
Weave design, and
Weave layers orientation.

III. Body–armor interaction

Comfort;
Body blunt trauma (blunt injury, nonpenetrating trauma ability);
Heat exchange between the human body and the environment; and
physiological and psychological reactions of the vest user.

The evaluation of the armor should satisfy the following other requirements[43, 112] for the acceptability of continuous-wear limited protection garments:

1. Garment performance ergonomics;
2. Garment effect on wearer psychological deterioration; and
3. Cost and ease of manufacturing.

Above this all, the armor should satisfy the ballistic limit. The most common parameter is 50% of nonperforated bullets (V_{50}) of the armor for a particular projectile at a certain level of protection; the test is conducted to determine a velocity at which a vest is failing by full penetration 50% of the time with a given round/velocity. If after 20 rounds fired for the V_{50} test, only 10% the armor, they pass to meet V_{50}

for that round/velocity for the selected level of protection. Figure 5.29 shows the results of testing different armors that satisfy the various levels of protection. Users might be more interested in the velocity at which the probability of perforation is 10% (estimating V_{10}), or even lower.[113]

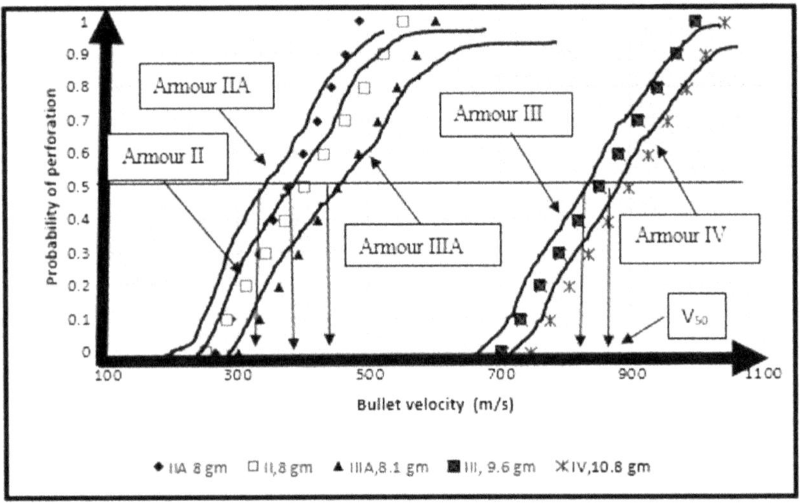

FIGURE 5.29 An illustration of ballistic limit test results.

There is an open research for the innovations of new structures of multi-layers laminated armor with improved performance to fulfill the flowing requirements:

1. High ballistic protection;
2. Lower weight;
3. Higher ratio of protection area to the total torso area;
4. Greater comfort index;
5. Easy body fit;
6. Longer serviceability; and
7. Lower cost.

KEYWORDS

- **classification of armor**
- **moisture management**
- **biomimicry**
- **seashell armor**
- **fish scales**
- **ceramic armor system**
- **laminate composite**

REFERENCES

1. Nath, K. Ballistic Protection Fabric and Bulletproof Vests. Textile Review https://www.scribd.com/document/284056221/ballistic-protection-fabrics-and-bulletproof-vests (accessed May 6, 2018).
2. https://upload.wikimedia.org/wikipedia/commons/c/cc/LAV3patrol.jpg (accessed May 10, 2018)
3. US Department of Justice, National Institute of Justice, Office of Science and Technology. Ballistic Resistance of Body Armor NIJ Standard–0101.06, 2008.
4. Body Armor. https://www.nij.gov/topics/technology/body-armor/Pages/welcome.aspx. (accessed May 6, 2018).
5. US Department of Justice Office of Justice Programs National Institute of Justice. Selection and Application Guide to Personal Body Armor. NIJ Guide 100–01, 2001. https://www.ncjrs.gov/pdffiles1/nij/189633.pdf. (accessed May 6, 2018).
6. NIJ guide Body Armor. Selection & Application Guide 0101.06 to Ballistic-Resistant Body Armor. 2014, NCJ 247281. https://www.hsdl.org/?abstract&did=761328. (accessed May 6, 2018).
7. US Army and Navy have Multiple Projects to Reduce Body Armor Weight by 20-50% While Maintaining or Improving Protection. https://www.nextbigfuture.com/2016/09/us-army-and-navy-have-multiple-projects.html (accessed May 6, 2018).
8. Milenković, L.; Škundrić, P.; Sokolović, R.; Nikolić, T. Comfort Properties of Defense Protective Clothing. *Univ NIŠ, Sci J. FACTA Univ., Ser.: Work. Liv. Environ. Protect.* **1999,** *1*, (4), 101–106. http://facta.junis.ni.ac.rs/walep/walep99/walep99-14.pdf (accessed May 4, 2018).
9. Matusiak M. Thermal Comfort Index as a Method of Assessing the Thermal Comfort of Textile Materials. *Fibres Text. East. Eur.* **2010,** *18*, 2 (79), 45–50.
10. Branson, D.; Sweeney, M. Conceptualization and Measurement of Clothing Comfort: Toward a Metatheory. In S. Kaiser & M. L. Damhorst (Eds.), Critical Linkages in Textiles and Clothing: Theory, Method and Practice, Monument, CO: International Textile and Apparel Association, 1991.

11. Kamalha, E.; Zeng, Y.; Mwasiagi, J.; Kyatuheire, S. The Comfort Dimension; a Review of Perception in Clothing. *J. Sens. Stud.* **2013,** *28*, 423–444. https://onlinelibrary.wiley.com/doi/pdf/10.1111/joss.12070. (accessed June 6, 2017).
12. Wong, A.; Li, Y. Clothing Sensory Comfort and Brand Preference. http://iffti.com.bh-in-10.webhostbox.net/downloads/papers-presented/iv-HKPU,%202002/F/1131-1135.pdf. (accessed May 6, 2018).
13. Sztandera, L. Tactile Fabric Comfort Prediction Using Regression Analysis. *WSEAS Trans. on Comput.* **2009,** *8* (2), 292–301.
14. Pan, N. Quantification and Evaluation of Human Tactile Sense Towards Fabrics. *Int. J. Design Nat.* **2007,** *1* (1), 48–60.
15. Bajzík, V. Some Approaches to Objective Evaluation of Fabric Hand. *WJET* **2015,** *1*, 48–58.
16. Cardello, A.; Winterhalter, C. Predicting the Handle and Comfort of Military Clothing Fabrics from Sensory and Instrumental Data: Development and Application of New Psychophysical Methods. *Text. Res. J.* **2003,** *73* (3), 221–237.
17. Bell, R.; Cardello, A.; Schutz, H. Relations Among Comfort of Fabric, Ratings of Comfort, and Visual Vigilance. *Percept. Motor Skil.* **2003,** *97*, 57–67.
18. Winakor, G.; Kim, C. Fabric Hand: Tactile Sensory Assessment, *Text. Res. Inst.* **1990,** 601–610.
19. Andreen, J.; Gibson, J.; Wetmore, O. Fabric Evaluations Based on Physiological Measurements of Comfort. *Text. Res. J.* **1953,** *23* (1), 11–22.
20. Li, M.; Wu, Y.; Chen, C.; Du, Z. Characterization of Plantar Press-Comfort Performance of Warp-Knitted Spacer Fabrics. *Text. Res. J.* **2018,** *88* (4), 357–366.
21. Bajzík, V. Some Approaches to Objective Evaluation of Fabric Hand. *World J. Text. Eng. Technol.* **2016,** *2*, 1–11.
22. Kawabata S. The standardization and analysis of fabric hand, 2nd ed. *Text. Mach. Soc. Jap.*: Osaka, 1980.
23. Park, S.; Hwang, Y. Comparison of Total Hand of Single Knitted Fabrics Made from Linc LITE and Conventional Wool Yarns. *Text. Res. J.* **2002,** *72* (10): 924–930.
24. Comparison of the Fabric Evaluation by Kawabata Evaluation System (KES) and Fabric Evaluation By Simple Testing (fast principle) http://shodhganga.inflibnet.ac.in/bitstream/10603/75122/18/18_chapter%2011.pdf (accessed May 4, 2018).
25. Kamalha, E.; Zeng, Y.; Mwasiagi, J.; Kyatuheire, S. The Comfort Dimension; a Review of Perception in Clothing. *J. Sens. Stud.* **2013,** *28*, 423–444.
26. Pan, N.; Zeronian, S. An Alternative Approach to the Objective Measurement of Fabrics. *Text. Res. J.* **1993,** *63* (1), 33–43.
27. Sztandera, L.; Cardello, A.; Winterhalter, C.; Schutz, H. Identification of the Most Significant Comfort Factors for Textiles from Processing Mechanical, Hand Feel, Fabric Construction, and Perceived Tactile Comfort Data. *Text. Res. J.* **2013,** *83* (1) 34–43.
28. Schutz, H.; Cardello, A.; Winterhalter, C. Perceptions of Fiber and Fabric Uses and the Factors Contributing to Military Clothing Comfort and Satisfaction. *Text. Res. J.* **2005,** *75* (3), 223–232.
29. El Messiry, M., El Ouffy, A.; Issa, M. Microcellulose Particles for Surface Modification to Enhance Moisture Management Properties of Polyester, and Polyester/Cotton Blend Fabrics. *AEJ* **2015,** *54*, 127–140.

30. Das, B.; Das, A.; Kothari, V.; Fanguiero, R.; de Arau′jo, M. Moisture Transmission Through Textiles, Part I; Processes Involved in Moisture Transmission and the Factors at Play. *AUTEX Res. J.* **2007,** *7* (2), 100–110.
31. Das, B.; Das, A.; Kothari, V.; Fanguiero, R.; Araujo, M. Moisture Flow Through Blended Fabrics—Effect of Hydrophobicity. *J. Eng. Fibers* **2009,** *4* (4), 20–28.
32. Yoon, N.; Buckley, A. Improved Comfort Polyester, Part I: Transport Properties and Thermal Comfort of Polyester/Cotton Blend Fabrics. *Text. Res. J.* **1984,** 289–298.
33. Hu, J.; Li, Y.; Yeung, K.; Wong, A.; Xu, W. Moisture Management Tester: A Method to Characterize Fabric Liquid Moisture Management Properties. *Text. Res. J.* **2005,** *75* (1), 57–62.
34. Barnes, J. C.; Holcombe, B. Moisture Sorption and Transport in Clothing During Wear. *Text. Res. J.* **1996,** *66* (12), 777–786.
35. Wong, A.; Li, Y.; Yeung, K. In *Comfort Perceptions and Preferences of Young Female Adults for Tight-fit Sportswear*, Proceedings of Textile Institute 82nd World Conference, Cairo, Egypt, 2002.
36. Li, Y., Zhu, Q., Simultaneous Heat and Moisture Transfer with Moisture Sorption, Condensation and Capillary Liquid Diffusion in Porous Textiles. *Text. Res. J.* **2003,** *73* (6), 515–524.
37. Scheurell, D.; Spivak, S.; Hollies, N. Dynamic Surface Wetness of Fabric in Relation to Clothing Comfort. *Text. Res. J.* **1985,** *6*, 394–399.
38. Sadikoglu, T. Effect on Comfort Properties of Using Superabsorbent Fibres in Nonwoven Interlinings. *Fibres Text. East. Eur.* **2005,** *13*, 3 (51), 54–57.
39. Raccuglia, M.; Hodder, S.; Havenith, G. Human Wetness Perception in Relation to Textile Water Absorption Parameters under Static Skin Contact. *Text. Res. J.* **2017,** *87* (20) 2449–2463.
40. Kim, J.; Spivak, M. Dynamic Moisture Vapor Transfer through Textiles. *Text. Res. J.* **1994,** *64*, (2), 112–121.
41. Schlader, Z.; Stannard, S.; Münde, T. Human Thermoregulatory Behavior during Rest and Exercise—A Prospective Review. *Physiol. Behav.* **2010,** *99*, 269–275.
42. Barnes, C.; Holcombe, V. Moisture Sorption and Transport in Clothing during Wear. *Text. Res. J.* **1996,** *66* (12), 777–786.
43. Bogdan, A.; Marszałek, A.; Majchrzycka, K.; Brochocka, A.; Luczak, A.; Zwolinska, M. Aspects of Applying Ergonomic Tests in the Evaluation of Ballistic Body Armors Using the Example of Ballistic Vests. *J Text Sci Eng* **2012,** *2*:7, 1–5.
44. Filingeri, D.; Havenith, G. Human Skin Wetness Perception: Psychophysical and Neurophysiological Bases. *Temp. Multidiscip. Biomed. J.* **2015,** *2* (1), 86–104.
45. Gagge, A.; Gonzalez, R. Mechanisms of Heat Exchange: Biophysics and Physiology. *Compr. Physiol.* **2011,** 45–84.; http://dx.doi.org/10.1002/cphy.cp040104.
46. Gagge, A.; Stolwijk, J.; Hardy, J. Comfort and Thermal Sensations and Associated Physiological Responses at Various Ambient Temperatures. *Environ. Res.* **1967,** *1*, 1–20.
47. USA National Institute of justice. Body armor. https://www.nij.gov/topics/technology/body-armor/Pages/welcome.aspx. (accessed May 20, 2018).
48. DeMaio, M; Onate, J.; Swain, D.; Ringle, S.; Morrison, S.; Naiak, D. Physical Performance Decrements in Military Personnel Wearing Personal Protective Equipment (PPE). http://www.dtic.mil/dtic/tr/fulltext/u2/a567881.pdf (accessed May 20, 2018).

49. Larsen, B.; Netto, K.; Aisbett, B. The Effect of Body Armor on Performance, Thermal Stress, and Exertion: A Critical Review. *Military Med.*, **2011,** *176*, (11), 1265–1273.
50. Taylor, N.; Burdon, C.; van den Heuvel, A.; Fogarty, A.; Notley, S.; Hunt, A.; Billing, D.; Drain, J.; Silk, A.; Patterson, M.; Peoples, G. Balancing Ballistic Protection Against Physiological Strain: Evidence from Laboratory and Field Trials. *Appl. Physiol. Nutr. Metab.* **2015,** *41* (2). https://tspace.library.utoronto.ca/bitstream/1807/70963/1/apnm-2015-0386.pdf
51. Choi, C. Seashell Armor Could Offer Transparent Protection for Troops. March 31, 2014 https://www.livescience.com/44484-seashell-armor-for-troops.html (accessed Apr 18, 2018).
52. MIT NEWS., http://news.mit.edu/2005/seashells. (accessed May 20, 2018).
53. Barthelat, F. Biomimetics for next-generation materials. *Philos. Trans. R. Soc. A* **2007,** *365*, 2907–2919.
54. Downing, E. Seashells Offer a Lesson in Armor Design. MIT, *50* (2), http://news.mit.edu//2005/techtalk50-2.pdf (accessed Apr 18, 2018).
55. Study: Transparent, Tough Sea Shell Armor Could Protect US Troops., http://washington.cbslocal.com/2014/03/31/study-transparent-tough-sea-shell-armor-could-protect-us-troops/ (accessed Apr 18, 2018).
56. Crystal twinning., https://en.wikipedia.org/wiki/Crystal_twinning (accessed Apr 18, 2018).
57. Deep-sea snail shell could inspire better body armor. https://www.wired.com/2010/01/snail-armor/ (accessed Apr 18, 2018).
58. Shimomura, M. The New Trends in Next Generation Biomimetics Material Technology: Learning from Biodiversity. *Sci. Technol. Trends*, Quart. *Rev*. **2010,** *37*, 53–75. http://citeseerx.ist.psu.edu/viewdoc/download?doi=10.1.1.463.9316&rep=rep1&type=pdf (accessed Apr 18, 2018).
59. Gu, G.; Takaffoli, M; Buehler, M. Hierarchically Enhanced Impact Resistance of Bioinspired Composites. *Adv. Mater.* **2017,** *29* (8), 150–155.
60. Sarikaya, M.; Aksay, I. (Eds). Biomimetics, Design and Processing of Materials. Polymers and Complex Materials. Woodbury, NY: American Institute of Physics, 1995.
61. Sanchez, C.; Arribart, H.; Giraud, G. Biomimetism and Bioinspiration as Tools for the Design of Innovative Materials and Systems. *Nat. Mater.* **2005,** *4*, 277–288.
62. Barthelat, F.; Li, C.; Comi, C.; Espinosa, H. Mechanical Properties of Nacre Constituents and Their Impact on Mechanical Performance. *J. Mater. Res.* **2006,** *21*, 1977–1986.
63. Hurst, N. This Conch-Shell Inspired Material Could Make Helmets and Body Armor Safer. https://www.smithsonianmag.com/innovation/conch-shell-inspired-material-could-make-helmets-and-body-armor-safer-180963612/ (accessed Apr 18, 2018).
64. Vernerey, F.; Barthelat, F. On the mechanics of fish scale structures. *Int. J. Solids Struct.* **2010,** *47*, 2268–2275.
65. Ikoma, T.; Kobayashi, H.; Tanaka, J.; Walsh, D.; Mann, S. Physical Properties of Type I Collagen Extracted from Fish Scales of Pagrus Major and Oreochromis Niloticas. *Int. J. Biol. Macromol.* **2003,** *32* (3–5), 199–204.
66. Vernerey, F.; Barthelat, F. Skin and Scales of Teleost Fish: Simple Structure but High Performance and Multiple Functions. *J. Mech. Phys. Solids* **2014,** *68*, 66–76.

67. Zhu, D.; Ortega, C; Motamedi, R.; Szewciw, L.; Franck Vernerey, F.; Barthelat, F. Structure and Mechanical Performance of a "Modern" Fish Scale. *Adv. Eng. Mater.* **2012,** *14*, (4), 185–185–194.
68. Li, R.; Zhong, Y.; Li, M. Analytic Bending Solutions of Free Rectangular Thin Plates Resting on Elastic Foundations by a New Simplistic Superposition Method. http://rspa.royalsocietypublishing.org/ (accessed Apr 24, 2018).
69. Shermana,V.; Quana, H.; Yangb, W.; Ritchiec, R.; Meyersa, M. A Comparative Study of Piscine Defense: The Scales of *Arapaima gigas*, *Latimeria chalumnae*, and *Atractosteus spatula*. *J. Mech. Behav. Biomed.* **2017,** *73*, 1–16.
70. Sherman, V.; Yaraghi, N.; Kisailus, D.; Meyers, M. Microstructural and Geometric Influences in the Protective Scales of *Atractosteus spatula*. *J. R. Soc. Interface* **2016,** *13*. http://dx.doi.org/10.1098/rsif.2016.0595. (accessed June 10, 2017).
71. Giraud, M.; Castanet, J.; Meunier, F.; Bouligand, Y. The Fibrous Structure of Coelacanth Scales: a Twisted "plywood." *Tissue cell* **1978,** *10* (4), 671–686.
72. Yang, W.; Sherman, V.; Gludovatz, B.; Mackeyc, M.; et al. Protective role of *Arapaima gigas* Fish Scales: Structure and Mechanical Behavior. *Acta Biomater.* **2014,** *10*, (8), 3599–3614.
73. Pangolin. https://commons.wikimedia.org/wiki/File:Pangolin_(Gir_Forest,_Gujarat,_India).jpg (accessed Mar 20, 2018).
74. Johnson, A. Establishing Design Characteristics for the Development of Stab-Resistant Laser Sintered Body Armor. PhD Thesis, Loughborough University, UK, November 2014.
75. Browning, A.; Ortiz, C.; Boyce, M. Mechanics of Composite Elasmoid Fish Scale Assemblies and Their Bioinspired Analogues. *J. Mech. Behav. Biomed.* **2013,** *19*, 75–86.
76. Nayak, R.; Crouch, I.; Kanesalingam, S.; Ding, J.; Tan, P.; Lee, B.; Miao, M.; Ganga, D.; Wang, L. Body Armor for Stab and Spike Protection, Part 1: Scientific Literature Review. *Text. Res. J.* **2018,** *88* (7) 812–832.
77. Park. H.; Branson, D; Petrova, A.; et al. Impact of Ballistic Body Armor and Load Carriage on Walking Patterns and Perceived Comfort. *Ergonomics* **2013,** *56*, 1167–1179.
78. Yoo, S.; Barker, R. Comfort Properties of Heat-Resistant Protective Workwear in Varying Conditions of Physical Activity and Environment. Part II: Perceived Comfort Response to Garments and Its Relationship to Fabric Properties. *Text. Res. J.* **2005,** *75*, 531–539.
79. Phillips, M.; Bazrgari, B.; Shapiro, R. The Effects of Military Body Armor on the Lower Back and Knee Mechanics during Toe-Touch and Two-Legged Squat Tasks. *Ergonomics* **2015,** *58*, 492–503.
80. Armstrong N.; Gay L. The Effect of Flexible Body Armor on Pulmonary Function. *Ergonomics* **2016,** *59* (5), 692–696.
81. Majumdar, D.; Srivastava, K.; Purkayastha, K.; Pichan, G.; Selvamurthy, W. Physiological Effects of Wearing Heavy Body Armor on Male Soldiers. *Int. J. Ind. Ergonom.* **1997,** *20*, (2), 155–161.
82. Carr, D.; Lankester, C.; Peare, A.; Fabri, N.; Gridley, N. Does Quilting Improve the Fragment Protective Performance of Body Armor? *Text. Res. J.* **2012,** *82* (9), 883–888.
83. Bilisik, A.; Turhan, Y. Multidirectional Stitched Layered Aramid Woven Fabric Structures and their Experimental Characterization of Ballistic Performance. *Text. Res J.* **2009,** *79* (14), 1331–1343.

84. El-Messiry, M; El-Tarfawy, S. Effect of Fabric Properties on Yarn Pulling Force for Stab Resistance Body Armor, Sixth World Conference on 3D Fabrics and their Applications, North Carolina State University (NCSU), Raleigh, NC, USA, May 26–28, 2015.
85. Rao, M.; Duan, Y.; Keefe, M.; Powers, B.; Bogettic, T. Modeling the Effects of Yarn Material Properties and Friction on the Ballistic Impact of a Plain-Weave Fabric. *Compo. Struct.* **2009,** *89*, (4), 556–566.
86. Duan, Y.; Keefe, M.; Travis. B.; Cheeseman, B. Modeling Friction Effects on the Ballistic Impact Behavior of a Single-Ply High-Strength Fabric. *Int. J. Impact Eng.* **2005,** *31* (8), 996–1012.
87. Chu, Y.; Min, S.; Chen, X.; Jianxin He, J.; Cui, S. Numerical and Experimental Investigations on Inter-yarn Friction in UHMWPE Fabrics Subjected to Ballistic Impact. Proceedings to the 8th World Conference on 3D Fabrics and Their Applications, Manchester, UK, 28–29 March, 2018.
88. Ha-Minh, C.; Imad, A.; Boussu, F.; Kanit, T.; Cre´pin, D. Numerical Study on the Effects of Yarn Mechanical Transverse Properties on the Ballistic Impact Behavior of Textile Fabric. *J. Strain Anal.* **2012,** *47* (7), 524–534.
89. Chen, X; Sun, D.; Wang, Y.; Zhou, Y. 2D/3D Woven Fabric for Ballistics Protection. In 4th World Conference on 3D Fabrics and Their Applications; Xiaogang, J. (Ed.); Aachen, Publisher TexEng/RWTH Aachen: Germany, 2012. https://www.escholar.manchester.ac.uk/api/datastream?publicationPid=uk-ac-man-scw:171241&datastreamId=FULL-TEXT.PDF (accessed Mar 20, 2018).
90. El Messiry, M.; El-Tarfawy, S. Performance of Weave Structure Multilayer Bulletproof Flexible Armor, In 3rd Conference of National Campaign of Textile Industry, NRC, Cairo Egypt, 9–10 March 2015.
91. Abdulghaffar, H. Study on the Arrangement of Fabric Materials for Multilayer Soft Body Armor Based on Their Mechanical Properties. *J. Fash. Tech. Text. Eng.* **2015,** 3, (3), 1–5.
92. Cunniff, P. Dimensionless Parameters for Optimization of Textile-Based Body Armor Systems. In Proceeding of the 18th International Symposium on Ballistics, San Antonio, TX, 1999.
93. El Messiry, M.; Eltahan, E. Stab Resistance of Triaxial Woven Fabrics for Soft Body Armor. *J. Ind. Text.* **2016,** *45* (5), 1062–1082.
94. Ceramic Materials for light-weight Ceramic Polymer Armor Systems https://pdfs.semanticscholar.org/230b/8e4a24557409ec4b08216fe0c5e878a94e8a.pdf (accessed Apr 24, 2018).
95. Medvedovski, E. Ballistic Performance of Armor Ceramics: Influence of Design and Structure. Part 1. *Ceram. Int.* **2010,** *36*, 2103–2115.
96. Medvedovski, E. Ballistic Performance of Armor Ceramics: Influence of Design and Structure. Part 2. *Ceram. Int.* **2010,** *36*, 2117–2127.
97. Quinn, J.; Quinn, G. On the Hardness and Brittleness of Ceramics, Key Engineering Materials. *Trans. Tech. Publ.* **1997,** *132–136*, 460–463.
98. Bolduc, M.; Anctil, B.; Lo, J.; Zhang, R.; Lin, S.; Simard, B.; Bosnick, K.; Bielawski, M.; Merati, A. Ballistic Evaluation of Nanocomposite Ceramic. http://cradpdf.drdc-rddc.gc.ca/PDFS/unc298/p804201_A1b.pdf (accessed May 3, 2018).

99. Ma, Z.; Hui Wang, H.; Cui, Y.; Rose, D.; Socks, A.; Ostberg, D. Designing an Innovative Composite Armor System for Affordable Ballistic Protection. Department of Mechanical Engineering, The University of Michigan, Ann Arbor, MI, USA, 1–8.
100. Liaghat, G.; Shanazari, H.; Tahmasebi, M.; Aboutorabi, A.; Hadavinia, A. Modified Analytical Model for Analysis of Perforation of Projectile into Ceramic Composite Targets. *Int. J. Compos. Mater.* **2013,** *3* (6B), 17–22.
101. Composite Armor. https://en.wikipedia.org/wiki/Composite_armor. (accessed May 3, 2018).
102. Evci, C.; Gu¨lgec, M. Effective Damage Mechanisms and Performance Evaluation of Ceramic Composite Armors Subjected to Impact Loading. *J. Compos. Mater.* **2014,** *48* (26), 3215–3236.
103. Cavallaro, P. Soft Body Armor: An Overview of Materials, Manufacturing, and Ballistic Impact Dynamics. NUWC-NPT Technical Report 12,057, 1 Aug 2011. http://www.dtic.mil/dtic/tr/fulltext/u2/a549097.pdf (accessed May 3, 2018).
104. Kocer, H. Laminated and Hybrid Soft Armor Systems for Ballistic Applications. MSc Thesis, Auburn University, Alabama, Dec 17, 2007.
105. Thomas, H. Ballistic resistant fabric. US Patent 2003/ 0022583, 2003.
106. Braga, F.; Lima Jr, E.; Lima, E.; Monteiro, S. The Effect of Thickness on Aramid Fabric Laminates Subjected to 7.62 MM Ammunition Ballistic. *Impact Mat. Res.* **2017,** *20* (Suppl. 2), 676–680.
107. Bilisik, K. Experimental Determination of Ballistic Performance of Newly Developed Multiaxis Non-Interlaced/Non-Z E-Glass/Polyester and 3D Woven Carbon/ Epoxy Composites with Soft Backing Aramid Fabric Structures. *Text. Res. J.* 2010, 81(5) 520–537.
108. Elmessiry, M.; Aloufy, A.; Abd-Ellatif, S.; Elmor, M. Enhancement of Quasi-Static Puncture Resistance Behaviors of Triaxial Fabric/Polyester Hybrid Composites. In 7th International Conference on Recent Trends in Science, Engineering and Techno logy (RTSET—2017), London (UK), June 29–30, 2017.
109. Bakshi, S.; Lahiri, D.; Agarwal, A. Carbon Nanotube Reinforced Metal Matrix Composites—A Review. *Int. Mater. Rev.* **2010,** *55*, 41–64.
110. Ceramic Materials for light-weight Ceramic Polymer Armor Systems. https://www.ceramtec.com/files/et_armor_systems.pdf (accessed May 20, 2018).
111. Struszczyk, M. Risk Analysis in Designing of Body Armor. TechniczneWyrobyWłókiennicze (Technical Textile Products), Instytut Technologii Bezpieczeństwa "Moratex", 2009, 66–73. http://yadda.icm.edu.pl/baztech/element/bwmeta1.element.baztech-journal-1230-7491-techniczne_wyroby_wlokiennicze. (accessed June 19, 2019).
112. National Institute of Law Enforcement and Criminal Justice Law Enforcement Assistance Administration U.S. Department of Justice. Body Armor Field Test and Evaluation, Final Report. https://www.ncjrs.gov/pdffiles1/Digitization/46837NCJRS.pdf (accessed May 20, 2018).
113. Johnson, T.; Freeman, L.; Hester, J.; Bell, J. Ballistic Resistance Testing Techniques. IDA Research Notes, 12–19. https://pdfs.semanticscholar.org/7a79/4c7a04191e6627 7266076c8cd0090ad14ff1.pdf (accessed May 22, 2018).

CHAPTER 6

Flexible Protective Armor: Modern Designs

ABSTRACT

The flexible soft body armors are preferable for their performance and the advantages against the hard ones in the thought of their low weight, comfort and the wearer mobility. The optimal design of an armor system should balance three conflicting characteristics: efficient protection, mobility, and comfort. The armor comfort is affected by all the parameters that influence both protection performance and mobility. The mechanism of failure of the armor under bullet impact was investigated taking into consideration the theories of strain propagation under bullet impact and the effect of weave structure in the case of single and multilayer armor structures. The effect of some parameters on the ballistic velocity are given. In some armor designs, the hard plates are used as an insert in the flexible armor to increase its protection capacity. The designs of such inserts are illustrated. A recommendation to enhance the ballistic penetration resistance of woven fabrics is given.

6.1 INTRODUCTION

Since the invention of the body armor, the flexible soft body armors targeted for their performance and the advantages against the hard ones in the concept of weight, comfort, and the wearer mobility. Basically, the flexible body armor consists of the multilayer fabric and or flexible fibrous composites. It was acknowledged that the number of variables that affect protective flexible body armor performance was found to be vast, besides the interactions that make the prediction of their impacts more complicated. The designer of the bulletproof armor starts with the choice of the basic design concepts and the modifications based on the body armor field test and evaluation.

The fabric, when hit by a bullet, is deformed at the area in contact with the bullet and around it (Fig. 6.1) and distorted causing a back-face signature of depth "δ."

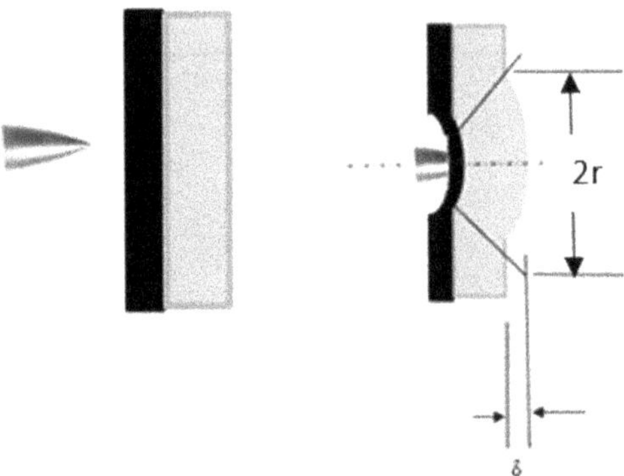

FIGURE 6.1 Flexible body armor deformation under hit by bullet.

The basic principle of bulletproof armor design recognizes the following:

1. The bullet kinetic energy should be fully absorbed by the armor material without penetration through the back layers.
2. A back-face signature should be of the minimum depth "δ" and wider radius "r."

In some cases of flexible armor, the increase of "δ" causes armor blunt trauma, even when it succeeds to completely stop the bullet through. The following may be considered:

1. Fabric material mechanical properties and structure are capable to redistribute the strain over large area at a high speed before failure.
2. The multilayer body of the armor matrix can enable the delamination to increase the energy absorption.
3. Using HPF.
4. Application of absorbing materials and structures between the HPF fabric layers acting as a shock absorber.

5. Using a hard or soft plate.
6. Increasing the coefficient of friction between fabric and bullet.
7. Increasing the coefficients of friction between yarns, between yarns and bullet surface, between the surface of the different fabric layers of the armor.
8. Applications of hybrid materials to produce an isotropic composite plate structure.
9. Using the application of a new material that changes its structure under the impact force.
10. Applications of graphite sheets as part of the armor.

The success to find out the optimum combination of the above factors results in better armor design. In all solutions, the problem of armor weight is permanently a decisive factor for any design.

6.2 CRITERIA FOR MODERN ARMOR DESIGN

The armor design criteria should satisfy the main condition—to protect the wearer against a certain threat level through the solving the problem of energy absorption and, at the same time, to ensure the free mobility and the wearer comfort requirements (Fig. 6.2).

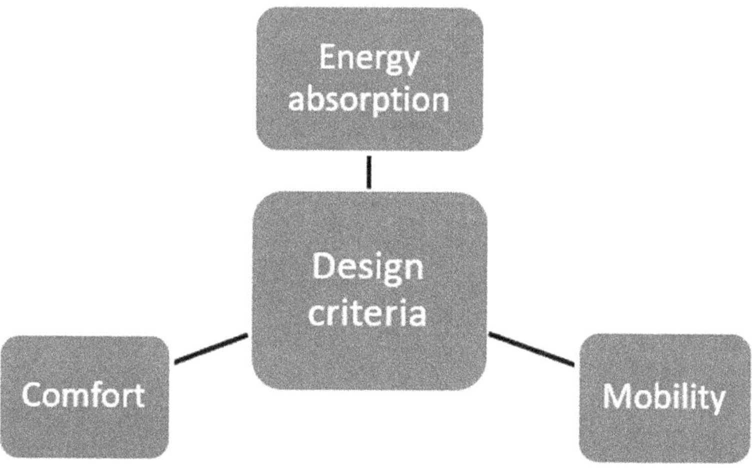

FIGURE 6.2 Elements design criteria.

For each element of the of design procedure, the steps, indicated in Figure 6.3, represent a closed loop design process to react on the unpredictability of the safety demands.

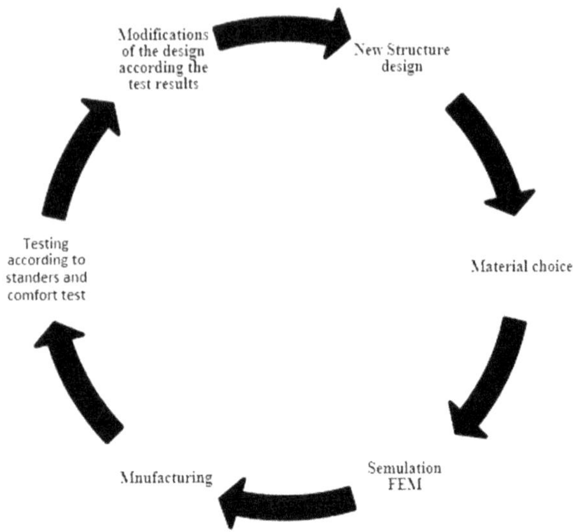

FIGURE 6.3 Steps of armor design.

The elements that influence the design of each component were discussed in Chapter 5. Accordingly, the optimal design of an armor system should balance three conflicting characteristics: efficient protection, mobility, and comfort. For instance, the efficiency of protection is directly responsible for the armor weight while the mobility is directly deteriorated by the armor weight and finally, the armor comfort is affected by all the parameters that influence both, protection performance and mobility. The close-fitting of the armor improves the mobility but has negative effect on the comfort. It was proved that wearing armor during different activities significantly reduced mobility during key task elements and resulted in greater physiological effort.[1] In view of the main armor design, the problems can be summarized into: weight, maximum ballistic absorption energy, comfort, flexibility, serviceability, and cost. The variables are and not limited to: fabric design, fabric thickness, number of layers, fabric areal density, fabric stiffness, fiber material properties, and system construction.

6.3 SOFT BODY ARMORS

Throughout the history, the individuals are always trying to seek better materials to protect themselves from the injuries by weapons of various types. For ballistic protection, soft body armors are constructed from multilayers of Kevlar fabric with at least 28 layers and up to 90 layers required to reach the armor thickness to defeat the bullet at the necessary protection level.[2] The PAP relies on fiber properties, fabric design, and fabric linear density. The armor ballistic energy upsurges with the high number of fabric layers and its thickness [number of layers (n = armor weight/m^2 W_t/fabric weight/m^2 W_{fabric})]. Assuming the total projectile energy to be defeated by the armor at a certain level of threat is E_s and the absorption energy per layer of the fabric is EF_{fabric}, then the approximate number of fabric layers required for the armor is:

$$n = K\,E_s/EF_{fabric}. \tag{6.1}$$

Then the armor areal density:

$$W_t = K\,(E_s/EF_{fabric})\,W_{fabric}. \tag{6.2}$$

The armor ballistic limit (ABL) is the utmost for no penetration of the bullet. The back-face signature BFS complies with the standard values (NIJ standards). Simply, the armor is a plate on flexible foundation, with maximum deformation:

$$\delta_{max} = f\,(V_{bullet}, EF_{fabric}, J, \gamma, b, W_t, W_{fabric}, V_{ballistic}, m, M, E_{fabric}, S_{fabric}), \tag{6.3}$$

δ_{max} depends basically on the dynamical properties of the armor material.

For convenience, the symbols used throughout are listed as follows in Table 6.1.

TABLE 6.1 List of Symbols and Abbreviations.

Symbols	Designation
PAP	protective armor performance
K	factor of safety
HPF	high performance fiber
KE	kinetic energy
C_f	constant of fabric

TABLE 6.1 *(Continued)*

Symbols	Designation
E_{muzzle}	muzzle energy
W_t	armor weight/m^2
W_{fabric}	fabric weight/m^2
E_s	level of the energy to be defeated by the armor
EF_{fabric}	energy absorbed by the one layer of the fabric
E_{bullet}	bullet kinetic energy
N	number of fabric layers of the armor
V_{bullet}	bullet ballistic velocity
E_{armor}	armor Young's modulus
J	area second moment of inertia of the armor
δ_{max}	maximum deflection of the armor under force
M	mass of the armor area participating in the resistance of the impact
$V_{balistic}$	fabric layer ballistic velocity
M	bullet mass
R_p	radius of the bullet
E_{fabric}	fabric Young's modulus
S_{fabric}	fabric strength
Γ	coefficient of stiffness of foundation per unit area
V_s	longitudinal wave speed
ρ	density of the material (ρ_f, ρ_y, ρ_{fabric} are density for fiber, yarns, fabric, respectively)
σ	breaking stress of the material (σ_f, σ_y, σ_{fabric} are breaking stress for fiber, yarns, fabric, respectively)
E_f	Young's modulus of fibers
V_c	cone wave velocity
V_{bullet}	bullet ballistic velocity
ε_{yarn}	maximum strain in the yarns
$R_{c\,max}$	maximum cone radius
R_c	cone radius
C_s	Euler transverse wave speed
CFF	cross-over firmness factor
FYF	floating yarn factor
I	interlacement index
i_{wp}	total number of interlacements in warp direction per repeat
i_{wf}	total number of interlacements in fill direction per repeat
f_{wp}	total number of floats in warp direction per repeat
f_{wf}	total number of floats in fill direction per repeat
R_{wp}	number of warp yarns per repeat

TABLE 6.1 *(Continued)*

Symbols	Designation
R_{wf}	number of fill yarns per repeat
W_{wf}	width of the fill yarns
W_{wp}	width of the warp yarns
D_{wf}	distance between two adjacent fill yarns
D_{wp}	distance between two adjacent warp yarns
σ_{cx-y}	compression stress of the material
$N_{compactness}$	normal force to the bullet body due to the compactness of the yarns
m_{wf}	number of the fill yarns involved in the strained zone
n_{wp}	number of the warp yarns involved in the strained zone
A_{wf}	cross section area of the fill yarn
A_{wp}	cross section area of the warp yarn
σ_{wf}	failure stress of fill yarns
σ_{wp}	failure stress of warp yarns
K	number of sheared yarns in the impact zone
T	yarn shear force
$V_{before\ impact}$	projectile initially travels at velocity
E	fiber material Young's modulus
ρ_m	fiber material density
F_i	compression forces of *i*th fabric layer
σ_{ci}	compression stress of *i*th layer
D_{bullet}	bullet diameter
$E_{absorbed}$	energy absorbed by the fabric
$V(T)$	velocity of the bullet after passing through the target
$V(0)$	velocity of the bullet before impact
ES	energy to shear the yarns
ED	energy to deform all yarns and yarn pull-out force
ET	energy of tensile failure of directly impacted yarns
EF	energy to overcome friction between fabric layers
EJ	energy to overcome friction between bullet body and yarns
EM	energy to deform the fabric during impact forming cone shape
$V_{bullet}(t)$	velocity of the bullet during penetration at a time t
$m(t)$	mass of the fabric attached with the bullet at a time t
$V_{bullet}(t)$	bullet speed at a time t

TABLE 6.1 *(Continued)*

Symbols	Designation
$V_{bullet}(T)$	bullet velocity when leaving the target
$V_{bullet}(0)$	bullet impact velocity
A_p	projectile presented area
V_{50}	ballistic limit
A_d	system areal density
ε_{max}	failure strain of the fabric
K_{max}	strain concentration constant
T_m	thickness of the material
A_{bullet}	bullet body area
SF	bullet shape factor = $\sin^2 \beta$
B	half angle of the conical tip of the bullet
ρ_{bullet}	bullet material density
R	radius of the bullet
$T_{ballistic}$	total ballistic impact time
$\rho_{laminate}$	density of laminate
σ_c	out-of-plane compressive strength of the laminate
S_{sp}	shear plugging strength
H	thickness of laminate
$E3$	elastic modulus through-thickness
σ_R	mean resistive pressure provided by the target material
α_0	(V_r/V_{bullet})
β_0	m/M
γ_0	$(V_{balistic}/V_{bullet})$
ζ_0	factor related to the effect of the bullet nose shape
η_0	factor related to the material and the armor structure
θ_i	the angle of inclination of the bullet to the surface of the layer
I	$1, \ldots, n$
n	the number of fabric layers
R_{ci}	radius of cone wave front normalized to bullet radius (R_p)
R_1	(R_{c1}/R_p)
R_c	cone wave front radius of the first layer
V_{si}	strain wave velocity in weave layer *i*th
Vsj	strain wave velocity in weave layer *j*th

6.4 BULLET VELOCITY

The bullet velocity of particular bullet design (muzzle velocity)[3] was found to depend on firing KE of penetrator ammunition and projectile mass and radius. An example of the bullets weight and their muzzle velocity (Fig. 6.4) indicates that the higher the mass of the bullet, the lower will be its muzzle velocity for the same arm type and ammunition, as given by equation:

$$V_{\text{bullet}} = (2\,E_{\text{muzzle}}/M)^{0.5}. \tag{6.4}$$

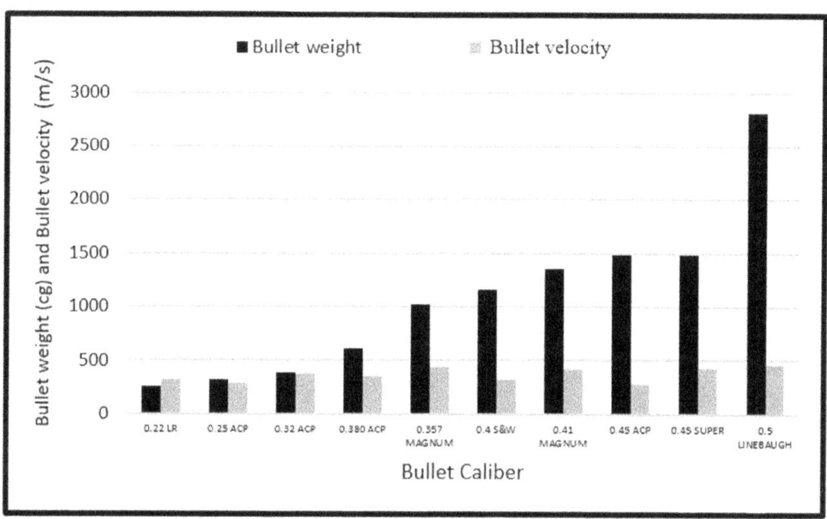

FIGURE 6.4 Bullet muzzle velocity versus bullet weight.

The bullet specifications depend on its caliber, muzzle velocity, and muzzle energy[4] (Fig. 6.5a, b). The bullet weight, its diameter, muzzle energy, and muzzle velocity are correlated. Bullet weight is a determining element of recoil, so as is bullet weight/velocity ratio. The ability of the armor to defeat the bullet is its ballistic velocity and it should be more than V_{bullet}. While the ballistic limit V_{50} expresses the velocity of the bullet when it totally penetrates ballistic armor material with about 50% probability.

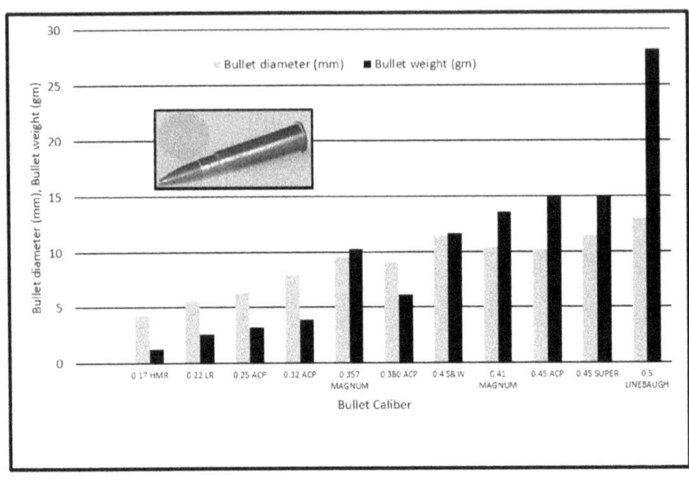

(a)

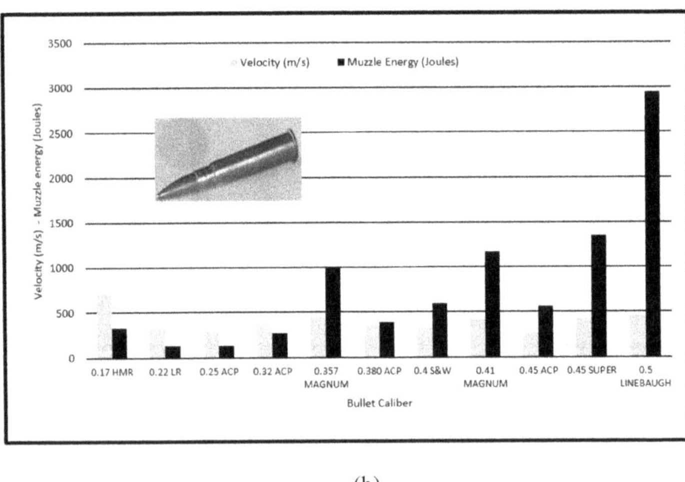

(b)

FIGURE 6.5 Examples of the bullet specifications: (a) weight and diameter; (b) velocity and muzzle energy of handgun calibers.

6.5 FABRIC ARCHITECTURE

The fabric architecture means the way the fibers, staple, or continuous filaments, are arranged to form a continuous homogenous sheet of fabric for the different applications. The way of the yarns arrangement influences

TABLE 6.2 Remarks on Fabrics Structural Properties.

Fabric type **2D fabrics**	**Remarks**	**Fabric type** **3D fabrics**	**Remarks**
Woven fabric	Thin, isotropic, high tenacity, Young's modulus and low flexural rigidity, and shear modulus	Multiaxial knitting fabrics	Thick, less flexibility, limited strength in *z*-direction
Knitted fabric	Anisotropic, high extensibility, low strength, high-featured rigidity, and high porosity	3D-contoured weaving	Limited shapes, profile with constant cross section, high flexural rigidity, and shear modulus
Triaxial fabric	Isotropic, low extensibility, low shear modulus, and high tear strength	Stitched assembly	Limited flexibility, Limited strength in *z*-direction, low heterogeneity in properties in *x*, *y*, *z* directions, and high compression stress
Braiding fabric	Isotropic, low extensibility, high shear modulus, and penetration resistance. High flexural rigidity	3D multiaxial weave	Anisotropic, high fractural rigidity, high shear modulus, and high compression stress
Nonwoven fabric	Anisotropic, high extensibility, low strength, and flexural rigidity, high porosity and penetration resistance, and heterogeneity in properties in *x*, *y*, *z* directions	Spacer fabric	Anisotropic in properties in *x*, *y*, *z* directions. Flexible, high porosity, low shear modulus, high compressibility in *z* directions, and high porosity
		3D Knitted fabric	• Anisotropic, flexible, high porosity, low shear modulus, and heterogeneity in properties in *x*, *y*, *z* directions
		3D Braiding fabric	• Flexible, high tenacity, low shear modulus, and heterogeneity in properties in *x*, *y*, *z* directions
		Multilayer nonwoven knitted interlacing fabric	• Anisotropic, high stiffness, less flexibility, high shear modulus, and heterogeneity in properties in *x*, *y*, *z* directions

the fabric mechanical properties. The yarn specifications, such as count, twist, evenness, yarn friction properties, mechanical properties, directly reflect on the fabric ballistic performance. When fabric is a concern, its structure, areal density, tightness, moisture management, and fabric dynamical mechanical properties should be specified. The fibers or yarns are forming fabrics with various architects as 2D or 3D (Table 6.2) which indicates that each architect has different final properties that define its suitability to be used in the bulletproof armors.

6.6 MECHANISM OF FAILURE OF THE ARMOR UNDER BULLET IMPACT

The flexible armor is designed to prevent the penetration of the bullet by a complete absorption of its E_{bullet} before its penetration through the full armor thickness. When the bullet hit armor, the fabric layers will be deformed and create a cone where the bullet touched the fabric layers. As the area of fabric strain propagation increases, the height of the cone *(h)* is increased too. The fabric's out-of-plane displacement increases as the bullet pushes the fabric, the rate of displacement is high at the first few milliseconds and declines by further penetration of the bullet into the armor thickness (Fig. 6.6).

During the impact of the bullet, the faster longitudinal strain wave will propagate through the yarns, and at the same time, a slower moving traverse wave will start in the other direction of the material from the point of its fixation. The higher the longitudinal strain wave speed, the higher will be the dissipation of the impact energy. The longitudinal wave velocity for isotropic solid material is given by:

$$V_s = (E/\rho)^{0.5}. \tag{6.5}$$

Equation 6.5 can be applied for solid fibers. For yarns, V_s is less and for fabric, it is much lesser. Several investigations have revealed that the randomly oriented fibers have low longitudinal wave, dynamic modulus, and some other properties. It was indicated that sonic modulus convinced interpretations are believed to be related to both the physical and chemical structures of synthetic fibers over a broad spectrum and is influenced by the elastic modulus of yarns, such as static and dynamic strength of the yarns, recovery parameters of yarns, bending stiffness of fabric, and abrasion resistance of the yarns and fabrics.[5–11] The investigation of the relationship between the acoustic dynamic modulus of yarns and their twist factor and fiber orientations found that sonic modulus is significantly affected by these parameters, as well as by the blending ratio of their different components.[9,12,13] Moreover, dynamic moduli are also dependent on the configuration of the individual fibers within a fabric.[14] The translation of fiber to yarn results into a reduction in the level of sonic modulus, a further reduction corresponds the transform of yarns to fabric, which is a function of a fabric structure.[15] The inter-relationship between sound velocity, orientation of fibers in a fabric, material anisotropy, and such performance properties as strength, stiffness, elasticity, and resilience of material have been described to be the main parameters affecting the sonic velocity.[16] In order to study the ballistic impact behavior of armor structures, a thorough understanding of stress wave propagation is critical. Figure 6.7 illustrates the factors that influence the strain propagation speed of the fabrics.[10]

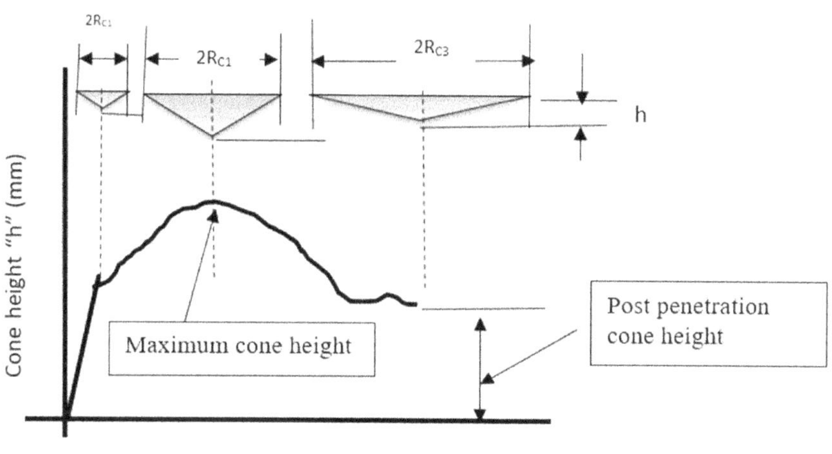

FIGURE 6.6 Fabric cone height propagation.

The fabric sonic velocity is as follows:

$$V_s = C_f\,(E/\rho)^{0.5}, \tag{6.6}$$

C_f depends on all mentioned above parameters.

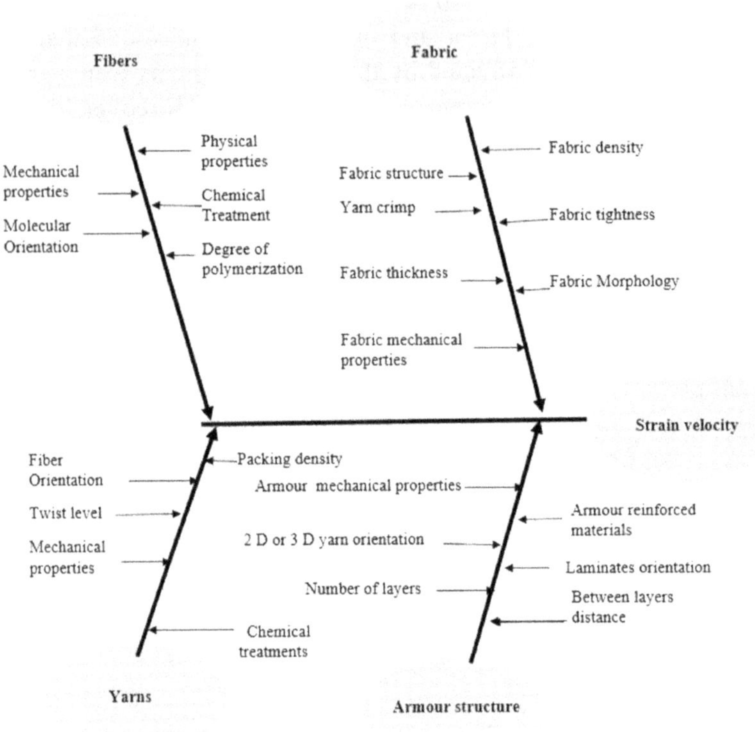

FIGURE 6.7 Strain propagation velocity.

6.6.1 MECHANISM OF STRAIN PROPAGATION UNDER BULLET IMPACT

An analytical model to predict penetration and perforation of armor impacted by rigid bullets was developed by several researchers.[17–19] The mechanism of deformation of a fabric under the impact by bullet can be summarized as[20–21]:

- Upon impact of the bullet on the single fabric layer, there is near instantaneous momentum transfer to a circular part of the fabric in the direct contact area, followed immediately by initiation of a radially growing tensile wave and a slower transverse wave to form a growing cone with the bullet at its tip.

- The yarns, in direct contact with the tip of the bullet in warp and fill directions, will be strained and result in a longitudinal strain wave away from the area of contact and traverse deflection.
- The other yarns, not in direct contact with bullet tip, will be pulled out at the point of intersections of the first group of yarns and out of the original fabric plane.
- The cone, formed due to the deformation of the fabric under the ballistic impact, will increase in its height with deacceleration rate, depending on the bullet decreasing velocity rate during impacting time. Consequently, there is on-in-plane and through-the-thickness wave propagation.
- When a fiber, yarn, or fabric is subjected to the external impact, a transverse wave is produced and propagates in the yarn or fabric. The speed of transverse wave has been found to be critical to the armor ballistic performance.[22]
- The longitudinal strain weave will propagate with the speed of the sound, both weaves are continuing until the strain in the yarns reaches the breaking strain.
- The yarns in the areas, not in contact with the bullet edge, will also experience the deformations due to transfer of the forces at the point of their interlacements, yarn to yarn friction, thus this phenomenon will be governed by the weave structure, the coefficient of the friction between yarns and the fabric tightness.
- A faster transverse wave speed is desired to dissipate the impact energy quickly. It was revealed that upon high-speed impact, not all yarns have been impacted instantaneously, which may result in progressive early failure of the yarn. This fact indicates that the energy absorbed by the target might be low at the first instance of impact.
- The fabric will be deformed, generating a cone at the point of penetration of the bullet. If force created by the bullet on the surrounding yarns is less than the compactness frictional energy, the bullet will penetrate till the total bullet kinetic energy will be less than the fabric compactness frictional energy and fabric deformation energy. Otherwise, the bullet will pass through.

Bullet tip may punch the fabric between the threads without cutting and pushing the yarns aside without cutting them. Compacting the yarns aside around the bullet outer surface, the bullet will be subjected to the friction between the yarns and the bullet body, pressing the yarns aside with the increase of the pressure on the bullet that increases the friction force as the bullet proceeds inside the fabric thickness.

The cone radius increases instantaneously after the impact and continues to grow at less rate during the time of cone formation, the higher the bullet velocity, the smaller will be the cone radius due to the lower impact time till the fabric failure.[23] The maximum cone height increases linearly as the impact velocity increases. The cone wave velocity V_c can be given by[24]:

$$V_c = constant\ ((V_s)^{0.333}\ (0.707\ V_{bullet})^{0.667}. \tag{6.7}$$

The constant can be taken as 1[25,26]:

$$V_c = ((E/\rho)^{0.1665}\ (0.707\ V_{bullet})^{0.667}. \tag{6.8}$$

For multilayer assembly of i layers:

$$Vc_i = ((E_i/\rho_i)^{0.1665}\ (0.707\ V_{bullet})^{0.66}. \tag{6.9}$$

The maximum strain in the yarns occurs when the cone radius R_c is equal to the maximum cone radius $R_{c\ max}$, which can be approximately calculated by[24]:

$$R_{c\ max} \approx R_p\ [(1-(m/M))/2\ (m/M)]^{0.5}. \tag{6.10}$$

Consequently, the maximum strain in the yarns is approximately calculated as:

$$\varepsilon_{yarn} = \mathcal{E}_p(0)\ K_1, \tag{6.11}$$

where K_1 is a function of the ratio of the armor material density to the density of the bullet and the strain propagation velocity V_s,[24,27] and $\mathcal{E}_p(0)$ is the strain at the bullet edge of radius R_p just after impact.

Euler transverse wave speed C_s was found to be[22]:

$$C_s = (0.5\ V_s)]^{0.333}\ V_{bullet}^{\ 0.667} \tag{6.12}$$

$$V_s = (E/\rho)^{0.5}. \tag{6.13}$$

In the case of twisted yarns, both strength and elastic modulus are changed as the twist increases. The Euler transverse wave propagates at higher speeds in the twisted fiber yarns than in normal ones. From the above analysis, it can be concluded that the fabric should have a high elastic energy per unit mass and the ability to transfer it away from the impact zone as fast as possible with the velocity V_s. The failure to do so will lead to stress concentration under the impact area and the bullet penetrates easily through the fabric. The specifications of the fabric and structure of the armor influence the energy absorbed which, in turn, depends on the bullet impact velocity. The higher the bullet velocity, the lower will be value of the fabric absorbed energy, as shown in Figure 6.8. It means that the ballistic velocity of the armor ($V_{ballistic}$), corresponding with the maximum absorption energy of the fabric, for fabric I is 240 m/s and for fabric II is 380 m/s. For higher bullet velocity, the absorption of the energy will be dropped abruptly.[23]

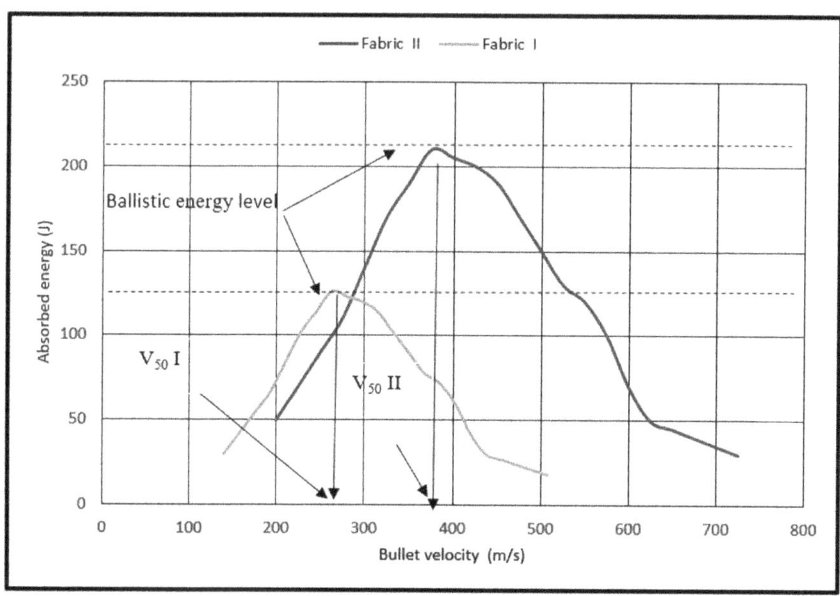

FIGURE 6.8 Fabric energy absorption–bullet velocity curve.

The failure of the yarn under impact by the bullet takes several phases: I. Decrimping; II. Pull-out under the increase of the strain during the cone formation; III. Yarns failure under tensile loading. The pull-out-force was found to be dependent on the inter-yarns friction and the number of yarns intersection. Phase I is characterized by weave structure, while phase II is dependent on the fabric firmness and the coefficient of friction between yarns. Finally, phase III is a function of yarn tenacity, toughness, and uniformity of the yarns. The mechanism of deformation of the fabric under impact is accomplished with the increase of the tension in the fabric, leading to the increase of the maximum pull-out-force.

It was revealed that the bullet will perforate the fabric when the recoil wave arrives, tensile stress vanishes, and the transverse wave stops.

It should be mentioned here that the dynamic properties of the yarns and fabrics should be considered since the impact duration is in micro-seconds. Consequently, the dynamic response of the fabric to the impact force should be high enough, otherwise, early failure of the fabric occurs, and bullet penetration is expected. The yarns elastic modulus increases as the strain rate increases while the tensile strength of the yarns reduces at the impact loading. Therefore, the yarn strain energy depends on the type of yarn material.[23–26] The irregularity of the yarn strength should be minimum, either for natural fiber yarn or synthetic ones, to prevent the early failure of some yarns under impact, that corresponds to the fabric failure at stress less than the average fabric stress.

6.6.2 EFFECT OF WEAVE STRUCTURE

Plain weave is the most 2D fabric structure used for formation of the armors due to its stability, higher density, number of intersections, friction between the interlaced set of warp and fill yarns, tightness, pull out yarn force, and shear modulus than the other fabric structures. The performance of the fabric under bullet impact depends on how much energy will be absorbed through its different components, which was investigated by several researchers.[23–27] Some of the above-mentioned parameters have contradicting effect on the fabric absorption energy. The loose structure with low number of intersections of warp and fill yarns have lower yarn pull-out force and reduces interference of the strain wave propagation upon ballistic impact. Higher fabric tightness gives higher areal density and at the same time increases the shear modules that increase the ballistic velocity. The increase of the inter-yarns friction improves the fabric rigidity and the yarn pull-out force, delaying the yarns breakage under the impact force through increase of their strain energy.[23]

Consequently, it was recommended to use a balanced fabric, that is, fabric with the same yarn count and number of yarns in both warp and fill directions. The following factors were suggested to explain the contact area of the yarns: cross-over firmness factor (CFF) and floating yarn factor (FYF).[28]

CFF = Number of cross-over lines in a complete repeat/ Number of interlacing points in a complete repeat (6.14)

FYF = Number of floating segments in a complete repeat/ Number of interlacing points in complete repeat. (6.15)

Another approach was introduced by dividing the fill or warp yarns into three parts: contact, interlacing, and space, to describe the effect of each part of the yarn on the fabric behavior under ballistic impact.[29] A modification to the eqs 6.14 and 6.15 is to replace the CFF and FYF parameters with an interlacement index (I) and float index (F).

The fabric firmness will be in warp and fill directions as:

$$I = (i_{wp} + i_{wf})/(R_{wp} - R_{wf}) \text{ and} \tag{6.16}$$

$$F = (f_{wp} + f_{wf})/(R_{wp} - R_{wf}). \tag{6.17}$$

For the entire fabric, the weave factor that explains the firmness is:

$$\text{Firmness} = (I/F)\,(i_{wp} \cdot W_{wf} \cdot D_{wp} + i_{wf} \cdot W_{wp} \cdot D_{wf}) + (f_{wp} \cdot W_{wf} \cdot D_{wp} + f_{wf} \cdot W_{wp} \cdot D_{wf})/(R_{wp} \cdot R_{wf} \cdot D_{wp} \cdot D_{wf}). \tag{6.18}$$

For balanced fabric[30]:

$$\text{Firmness} = (I/F)\,(2\,W_{wp}/D_{wp}). \tag{6.19}$$

Plain weave has the highest firmness comparing to other different structures.

As shown in Figure 6.9, the number of interlacements in both warp and fill directions varied consistently with the weave design. The strain propagation in the fabric on the bullet impact will vary in magnitude and direction as a function of the weave design. Thus, the absorbed energy was found to be higher for the plain weave than for the other fabric structures, and the residual bullet velocity due to a higher amount of yarn-bullet friction is less. However, in some cases of multilayer structures, ballistic resistance performance is less influenced by the weaving architecture and fabric firmness,[30] especially at high number of fabric layers. Despite that, using multilayer structure angle-layering fabric assembly improves the extent of isotropy that results in higher energy absorption. Consequently, it is recommended to increase the number of layers and reduce the angle between the warp yarns in the successive layers.[31] For example, the fabric assembly absorption energy increased linearly with the increase of the orientation angle in the case of eight fabric layers aligned as $[0/22.2/45/67.2]_2$ is higher than fabric aligned $[0/0/0/0]_2$. The impact energy absorption by these angled fabric panels showed an increase over the aligned counterpart, depending on the number of the plies.[32]

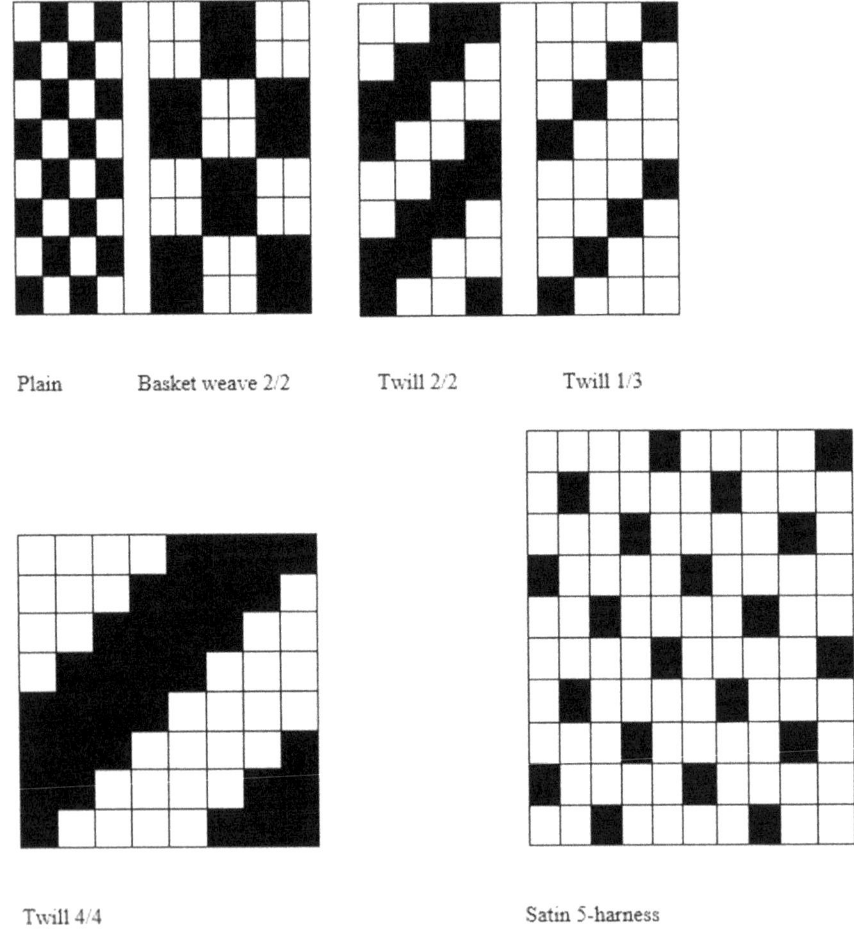

FIGURE 6.9 Different weave designs.

6.7 ANALYTICAL MODELS OF BULLET PENETRATION THROUGH THE FABRIC

6.7.1 SINGLE LAYER ARMOR

Analytical methods have been used to study ballistic impact. The fabric has a very complex geometry. Under the impact loading conditions, the deformation of fabric depends on the interaction between the fabric architect behavior under the dynamical forces and the change in the mechanical properties of the fibers. In most cases for bulletproof armors, the bullet will be faced by several resisting forces after touching the surface of the fabric, as indicated in Figure 6.10.

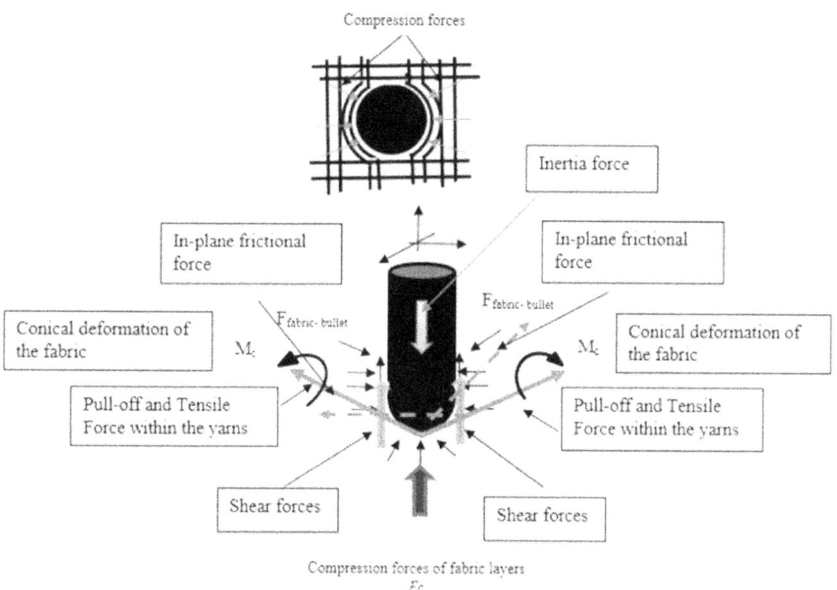

FIGURE 6.10 Forces acting on the bullet during its impact the primary strained zone.

The movement of the area under the bullet tip will cause a strain propagation along in-plane direction, compression wave, as well as shear wave, will be created at the same instance in the thickness direction. The compression stress will result in a formation of conical deformation under the bullet tip (Fig. 6.11) resulting in tension in the warp and fill yarns that

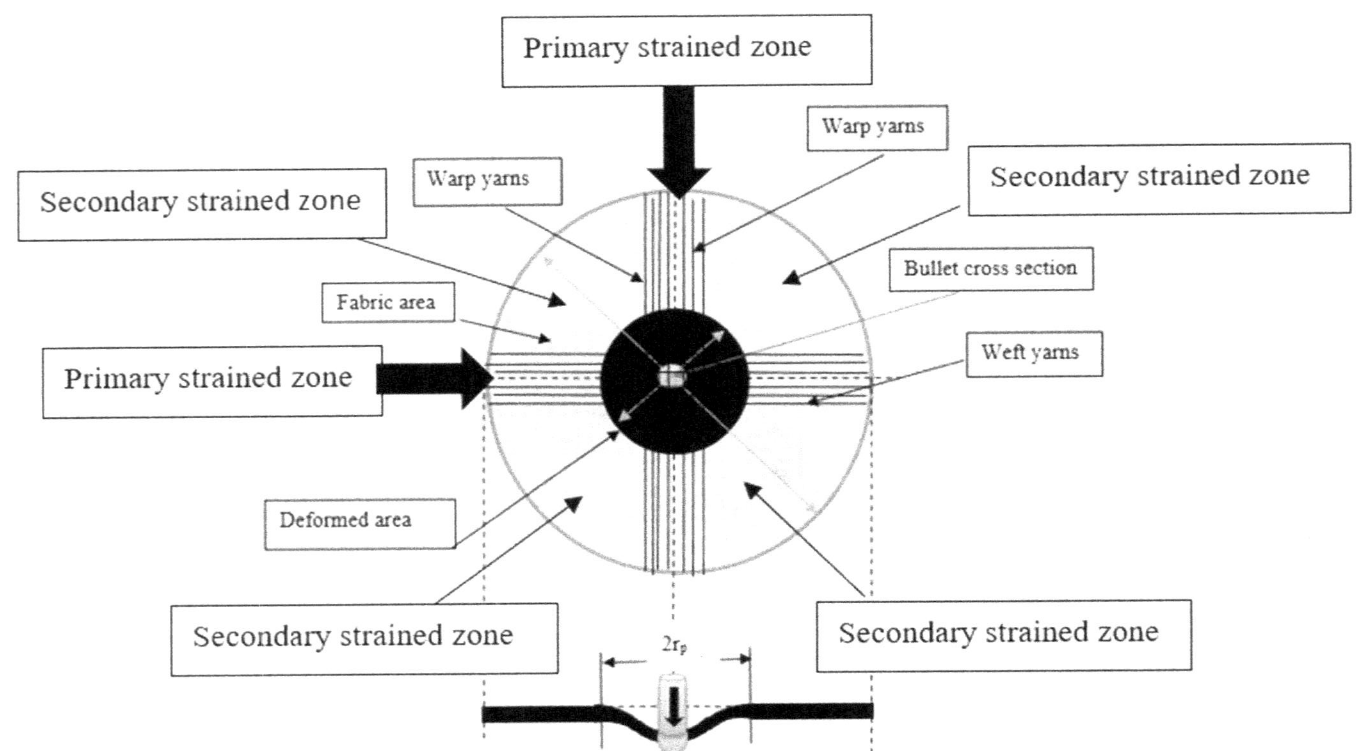

FIGURE 6.11 Schematic of a plain fabric deformation under bullet penetration.

forming the cone during their impacting in the primary strained zone. In this stage, in-plane friction forces between warp and fill yarns resist the formation of the cone. The failure of the yarns under the stresses, compression, tension, or shear, occurs if their strains exceed the corresponding failure strains. The yarns in the secondary strained zone will resist the movement of those in the primary zone due to the friction between the yarns at the point of intersections and between the adjacent yarns which increases the pull-out forces.

Resistance force, acting on the bullet during its penetration, is given by:

F = compressive force + $\sum$ shear force of the impacted zone + warp and fill yarns strain tension + force of the conical deformation of the fabric target + out-plane yarn frictional force (6.20)

$$F = N_{\text{compactness}} + \sum k\,\tau + \sum m_{\text{wf}} A_{\text{wf}}\, \sigma_{\text{wf}} + \sum n_{\text{wp}} A_{\text{wp}}\, \sigma_{\text{wp}} + F_{\text{out-plane yarn frictional force}} + F_{\text{conical deformation}}. \quad (6.21)$$

To stop the bullet, the value F (eq 6.21) should be higher or equal to F_{bullet} (0), and $F_{\text{conical deformation}}$ is the force required to bend the fabric under the inertia force of the bullet which is a function of the fabric flexural rigidity and the yarn tenacity. The cone size increases in diameter and height as the bullet punches the fabric area under the bullet tip (Fig. 6.7).

If bullet initially travels at a velocity $V_{\text{before impact}}$, then the strain of the yarns at the bullet edge (radius R_{p}) just after impact is:

$$\varepsilon_{\text{p0}} = \{V_{\text{before impact}}/(1.44\,(E/\rho_{\text{m}})^{0.5}\,(1 + m/M))\}^{1.333}. \quad (6.22)$$

The bullet velocity immediately after impact is:

$$V_0 = V_{\text{before impact}}/(1 + m/M). \quad (6.23)$$

The bullet velocity decreases due to the resisting force F, while the strain in the yarns increases as the cone height increases and reaching peak strain at its maximum value.[25] Figure 6.12 illustrates the change of the bullet velocity during penetration through the armor material. The slope of the V_{bullet}–t curve depends on several factors, such as target thickness, material properties, material structure, impact velocity, and bullet geometrical parameters.

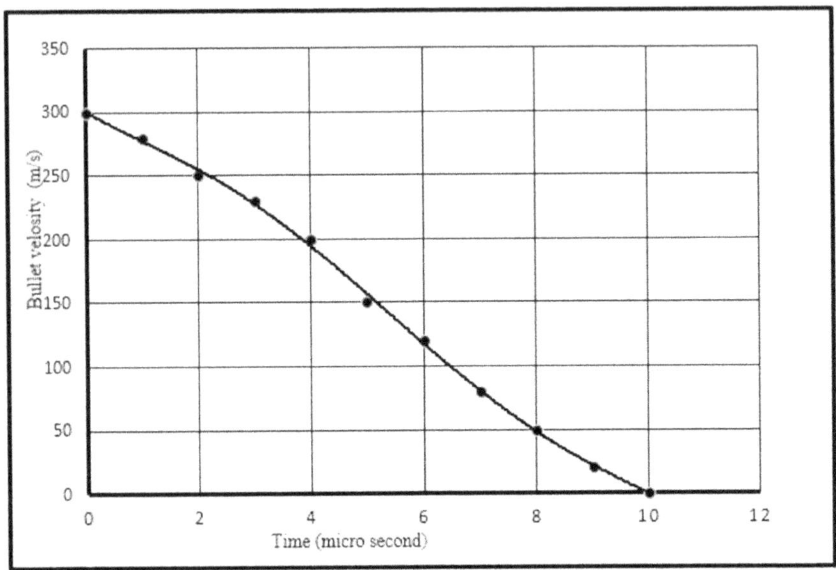

FIGURE 6.12 Bullet velocity versus the impact time through the armor.

6.7.2 *MULTILAYER STRUCTURE ARMOR*

From the above analysis, the energy lost by the bullet during its pass through the fabric is equal to the energy absorbed by the target material in the area targeted by the bullet. The absorbed energy of the multilayer structures depends on the orientation angle and the distance between the fabric layers. In the case of zero gap, the layers will be compressed together to act as a resisting media for penetration of the bullet. However, the failure of the layers will start successively as the bullet penetrates through the surface layer. In the case of multilayer system with space apart, the impact velocity of the second layer will be the exit velocity of the first one. Compression forces of fabric layers with no spaces can be given as:

The total compression force of layers:

$$F_c = \sum F_{ci} \tag{6.24}$$

The compression force for each layer can be expressed as:

$$F_{ci} = D_{bullet}{}^2 \sigma_{ci} / \sin \theta_i. \tag{6.25}$$

For layers with nonzero spacing, the absorbed energy by each layer varied affording to the bullet velocity (Fig. 6.8) which indicated that absorbed energy ($E_{absorbed}$) is low at high bullet velocity and increases in the successive layers but reduces again at lower bullet velocity. All these combined effects make the problem of ballistic impact through the multiply fibrous structures a complicated model. The deformation cone will be similar for all layers if the maximum height of the cone for a single layer is less than the space between layers. In the case of multilayer, the radius of the cone in any of layers is given by[31]:

$$R_c = 1 + (V_{si}/V_{sj})^{0.333}\,(R_1 - 1). \qquad (6.26)$$

If the space between the layers is more than that of maximum cone height of single layer, the cones for the first impacted layers will not interfere with the other layers. The value of V_{50} decreases as T' (the normalized total gap of the armor divided by the bullet radius) increases for high number of layers. In the multilayer armor with negligible spacing, the strain and cone waves in all the layers start almost simultaneously and almost all the layers fail together.[24] After the failure of the yarns in a certain layer, the bullet is continuously subjected to friction forces from the contacted yarns till its full penetration out of them. The width of the pull-out yarn zones, related to the energy transfer to the yarns in the case of multilayer, increased more than the width of the bullet diameter due to the inter-friction between layers. The width of strained yarns in pull-out zone is less for the first number of layers on almost constant, less than 35 layers, and will increase with the growing number of layers. This means that for higher number of layers, more resistance of the pull-out yarns is noticeable.[33] This may be due to the increase of friction between yarns in different layers that leads to broadening of the pull-out zone. The analysis of the strain in the yarns in the impact area has been theoretically studied by several researchers,[24,34–40] which revealed the complications in the reaching a concrete solution to satisfy the different materials and structures of the protective armor under impact by the bullet.

6.8 ENERGY-BALANCE MODEL OF BULLET PENETRATION THROUGH THE FABRIC

Several approaches considered the energy balance of the bullet during the contact with the armor body[41,42] involves all energy-dissipating mechanisms absorbing kinetic energy of the bullet. The kinetic energy of the bullet will equate the sum of the resisting to penetration energies due to contacts, such as strain, shear, fabric bending, surface friction, yarns pull-out, and compactness of the fabric. The energies, like the bullet deformation, damping of the material due to its plasticity, the vibration of the fabric surface, and the armor mass inertia effect, are not taking into the consideration.[42]

The maximum kinetic energy of the impactor before impact is:

$$KE_{\text{before impact}} = 0.5\, M\, V(0)^2 \text{ at } t = 0. \tag{6.27}$$

The kinetic energy of the bullet $t > 0$.

The energy balance at the end of ith time interval is obtained as:

$$E_{\text{residual}}(t) = KE_{\text{bullet}}(0) - E_{\text{absorbed}}(t). \tag{6.28}$$

The impact performance of a fabric is calculated by its energy absorption. Hence, the kinetic energy before impact of the bullet $E_{\text{before impact}}$ will be equal to the sum of the kinetic energy E_{absorbed} absorbed by the fabric and kinetic energy $E_{\text{after impact}}$ of the bullet after passing through the target.

Therefore,

$$E_{\text{absorbed}} = E_{\text{before impact}} + E_{\text{after impact}}. \tag{6.29}$$

The energy lost during impact E_{absorbed} is given by[24]:

$$E_{\text{absorbed}} = M\,(V(0)^2 - V(T)^2). \tag{6.30}$$

Kinetic energy absorbed by the fabric E_{absorbed} is defined by the following different components[43]:energy to shear the yarns, energy to deform all yarns, energy for yarn pull-out, energy of tensile failure of directly impacted yarns, energy to overcome friction between fabric layers, energy to overcome friction between bullet body and yarns, energy to deform the fabric to form a cone shape during impact.

Therefore, the total absorbed energy by the fabric can be expressed as:

$$E_{\text{absorbed}} = ES + ED + ET + EF + EJ + EM. \tag{6.31}$$

During the interval of impacting time, the value of impacting force will increase till full penetration, and then it will gradually decline and vanish after bullet passes through a fabric.

Figure 6.13 illustrates the variation of the different components of the absorption energy and the bullet energy during penetration through armor system which ended by a complete absorption of the bullet impact energy. The curve of the deacceleration of the bullet depends on the structure of the armor and properties of the fibrous material. In the case the armor absorption energy $E_{\text{before impact}} \leq E_{\text{absorbed}}$, then the $V(T) = 0$.

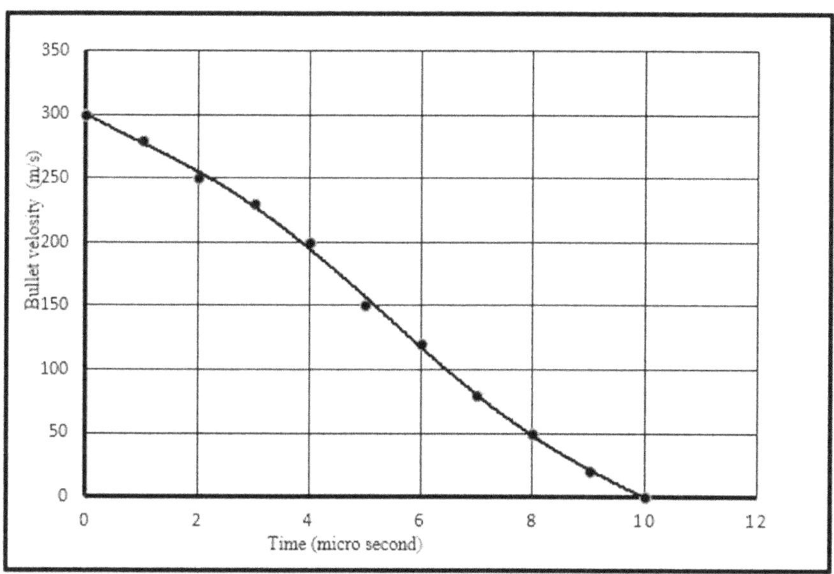

FIGURE 6.13 Energy–time curve throughout bullet penetration time.

The absorption energy will be:

$$E_{\text{absorbed}}(t) = 0.5\, M\, V_{\text{bullet}}(0)^2 - 0.5\,(M + m(t))\, V_{\text{bullet}}(t)^2 \tag{6.32}$$

$$= 0.5\, M\,[V_{\text{bullet}}(0)^2 - (1 + m(t)/M)\, V_{\text{bullet}}(t)^2].$$

Both $V_{\text{bullet}}(t)$ and $m(t)$ are also a function of t, then:

$$E_{\text{absorbed}}(t) = 0.5\, M\, V_{\text{bullet}}(0)^2 - \sum 0.5\,(M + m(t))\, V_{\text{bullet}}(t)^2. \tag{6.33}$$

If the change of the mass $m(t)$ is negligible relative to the bullet mass M, the change of the kinetic energy of the bullet due to the impact on the target will be:

$$E_{absorbed}(t) = \sum^{t} 0.5\,(M + m)\,[(V_{bullet}(0) - V_{bullet}(T))\,(1 - (t/T) + V_{bullet}(T)]^2. \quad (6.34)$$

Assuming the armor can stop the bullet, bullet velocity reaches zero m/s inside the armor, when the $V_{bullet}(0)$ is equal or less then $V_{ballistic}$ The value of the total $E_{absorbed}$ is found to be dependent on the function by which the bullet decelerates during its path through the armor thickness. $V_{bullet}(t)$ is a function of several factors such as the armor specifications and design which control the deacceleration of the bullet during its penetration in the target.[43] The components mentioned above of the absorbed energy depend on fabric structure and its specifications as well as on the yarn properties. The percentage of the different energy components varies and is consistent with the value of impact velocity, structure of the armor, fiber properties, fabric thickness, and its areal density (g/m^2). The areal density influences the absorbed energy components, with the increase of the areal density—the elastic energy, the fabric compactness energy, the frictional energy, and the shear energy rise, too. The value of V_{50}, generally, decreases as a result of heavy bullet weight, relying also on the armor structure and level of bullet impact velocity. At low impact velocities, the transverse deflection of the fabric has enough time to propagate, involving more yarns in fabric impacted area and therefore increasing energy dissipation. The initial elastic modulus, strength, and failure strain are the functions of increasing strain rate.[44] At high impact velocities to the fabric targets, yarn failure occurs and becomes more localized to the immediate area around the point of impact, and the transverse deflection of the fabric is minimal. Consequently, yarns fail under shear.[45] Temperature rises during the penetration of the bullet at high speed that may change the mechanical properties of the synthetic fibers, such as Spectra. This is another factor that differentiates low and high impact velocity mechanism, especially with fibers of low melting point.

6.9 EFFECT OF SOME PARAMETERS ON THE BALLISTIC VELOCITY

6.9.1 EFFECT OF THE FIBER PROPERTIES ON THE BALLISTIC VELOCITY

There are several formulae to evaluate the fibers toughness and strain weave velocity.

The velocity of strain propagation can be calculated by the following equation[27]:

$$V_s = \sqrt{E_f/\rho_f}, \tag{6.35}$$

where V_s is longitudinal wave speed in m/s, E_f is fiber modulus in Pa, and ρ is the fiber bulk density in g/m^3.

A dimensionless fiber property U is defined as the product of the specific fiber toughness and strain wave velocity, which can be used to qualitatively assess the performance of fibers, impacted by the bullet mass[46,47]:

$$U = (\sigma_f\, \mathcal{E}_f/2\rho_f)\,(E_f/\rho_f)^{0.5}. \tag{6.36}$$

A modified formula to express the fiber energy absorption capacity index (FCI) was driven as[43]:

$$FCI = (T_f\, V_s) \tag{6.37}$$

where T_f is the fiber tenacity in "GPa" and V_s is the longitudinal wave speed "m/s."

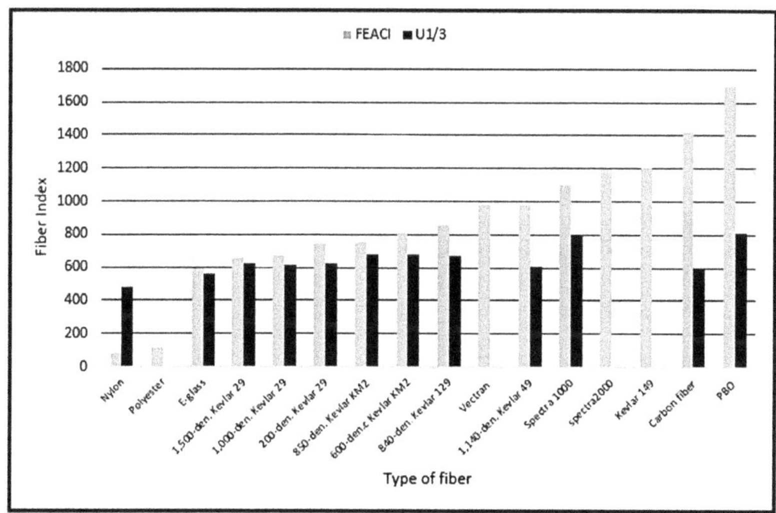

FIGURE 6.14 Fiber index (FCI and $U^{1/3}$) for different types of fibers.

Empirical work to relate ballistic impact performance to bullet and target characteristics has yielded dimensionally awkward relationships between perforation velocity and target and bullet quantities.[46,48] The dimensional ratios of bullet armor system are considered as:

$$\Phi\,[(V_{50}/U^{0.333}),\, A_d A_p/M)] = 0. \qquad (6.38)$$

For each type of material, the value of $(V_{50}(0)/U^{0.333})$ is proportional to $(A_d A_p/M)$ for angle of incident $\theta = 0°$. For angle θ, the ballistic limit $V_{50}(\theta)$ is:

$$V_{50}(\theta) = V_{50}(0)\, X^{\sec(\theta)-1}, \qquad (6.39)$$

where X is constant, determined from regression equation of experimental data, $V_{50}(0)$ is the ballistic limit at angle of incident $\theta = 0°$.

The ballistic limit V_{50} was calculated for different types of fibers and fabrics density to withstand the energy of different bullet calibers, as given in Table 6.3. It should be noticed that not all the fibers can defeat the bullet of the diverse calibers; consequently, the fiber type must be accurately selected.

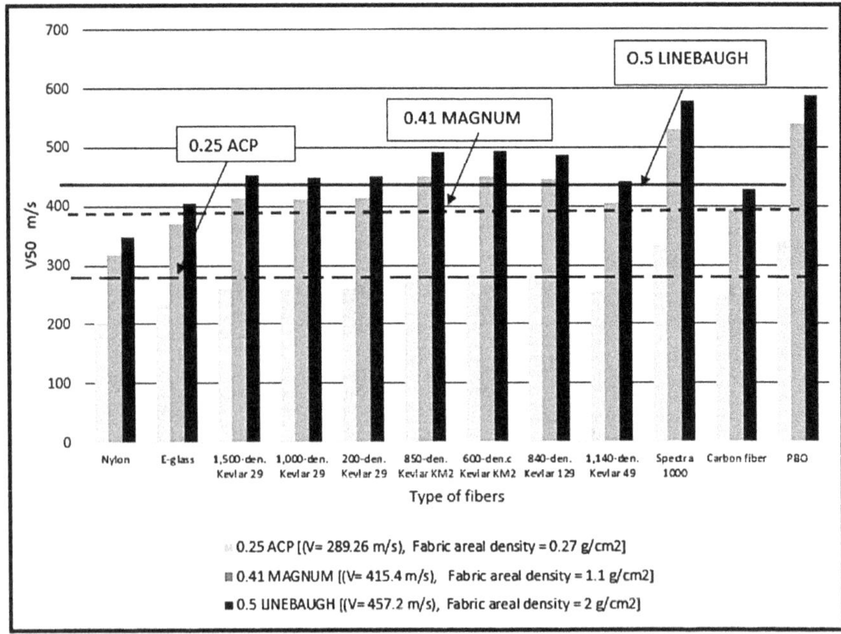

FIGURE 6.15 V_{50} versus the fabric areal density.

Several approaches to calculate value of V_{50} were tried, and it was found to be a function of: area density, ratio (m/M), strain to failure of the yarns, failure stress of the yarns, fiber-specific toughness, strain amplification factor, armor thickness, and cone wave velocity. In the case of multilayer, the V_{50} performance of the target is found to degrade progressively as the spacings between the layers are increased relatively to the sum of layer thicknesses without spacing.[24] For spherical and small caliber bullets, the V_{50} is inversely proportional to the bullet radius R_b. While for large calibers, it is deviated from the above relation.[49] The following relation may be applied considering the fabric as a membrane:

$$V_{50} = 2^{0.5} (1 + m/M)\, V_s (\varepsilon_{max}/K_{max})^{0.75}. \tag{6.40}$$

6.9.2 THE EFFECT OF FABRIC THICKNESS

The ballistic velocity or limit is the velocity required for a bullet to reliably (at least 50% of the times) penetrate a material. Bullet will, generally, not penetrate a given target when the bullet velocity is lower than the ballistic limit $V_{ballistic}$.

Assuming that the ballistic velocity of the armor is $V_{ballistic}$, then the general equation is:

$$V_{ballistic} = C\,(T_m D_{bullet}{}^2/M g)^{0.5}, \tag{6.41}$$

where C is constant of bullet–target interaction.

The relation between the ballistic energy and the thickness of the armor layers will take the shape of relation given in Figure 6.16.[50]

The ballistic limit energy of the armor $E_{ballistic}$[23] is:

$$V_{50} = (2\,E_{ballistic}/M)^{0.5} \tag{6.42}$$

$$E_{ballistic} = C'T_m. \tag{6.43}$$

Aramid fabric laminates can only stop 7.62 mm bullets (~850 m/s as impact velocity). The bullet easily perforates the thin armor; the armor starts to absorb higher impact energy at thickness over 37.5 mm till it can stop the bullet completely with the thickness higher than 50 mm.[50] To increase the thickness of the armor without increase the areal weight, fibrous body armor materials should have low density fibers, very high tensile strength and high stiffness.

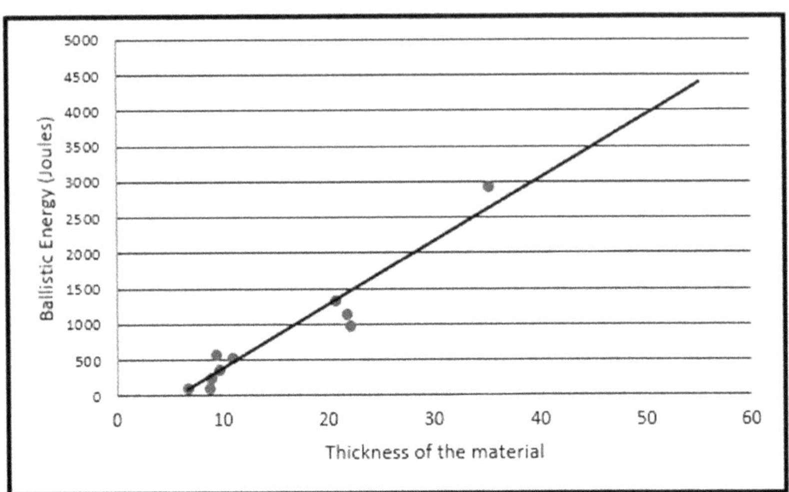

FIGURE 6.16 Ballistic energy versus target material thickness.

TABLE 6.3 Handgun Calibers.

Caliber	Bullet diameter (mm)	Bullet weight (g)	Velocity (m/s)	Muzzle energy (J)
0.25 ACP	6.38	3.24	289.26	135.55
0.22 LR	5.66	2.59	326	137.63
0.32 ACP	7.92	3.89	375.2	273.81
0.17 HMR	4.37	1.3	716.3	333.51
0.380 ACP	9.07	6.16	358.1	394.97
0.45 ACP	10.16	14.9	274.3	560.54
0.4 S&W	11.43	11.66	321	600.73
0.45 SUPER	11.46	14.9	426.7	1356.4
0.357 MAGNUM	9.53	10.24	442	1000.3
0.41 MAGNUM	10.41	13.61	415.4	1174.3
0.5 LINEBAUGH	12.95	28.19	457.2	2946.3

6.9.3 EFFECT OF BULLET INCIDENT IMPACT VELOCITY ON THE BALLISTIC VELOCITY

The model of failure of the armor under the impact indicated that if the bullet velocity at the end of the impact event V_r is > zero, then the bullet penetrates when:

$$V_r > [(0.5\, M\, V_{bullet}{}^2 - E_{absorption})/0.5\,(m + M)]^{0.5}.$$

If the bullet velocity at the end of the impact event is zero, then the bullet is stopped by the armor, $V_{bullet} \leq V_{ballistic}$. The behavior of the target at different bullet impact velocities is demonstrated in Figure 6.17. The bullet with impact velocity up to 200 m/s will not penetrate through the target material of thickness 10 or 20 mm. At higher velocity, the bullet perforates the target, exiting at different residual velocity, depending on the target thickness. At target thickness of 10 mm, the ballistic velocity is 200 m/s and with the increase of the target thickness to 20 mm, this velocity increases to 280 m/s. Most of the forces resisting the penetration of the bullet are a function of the bullet velocity, thus it will affect the value of the armor ballistic velocity.

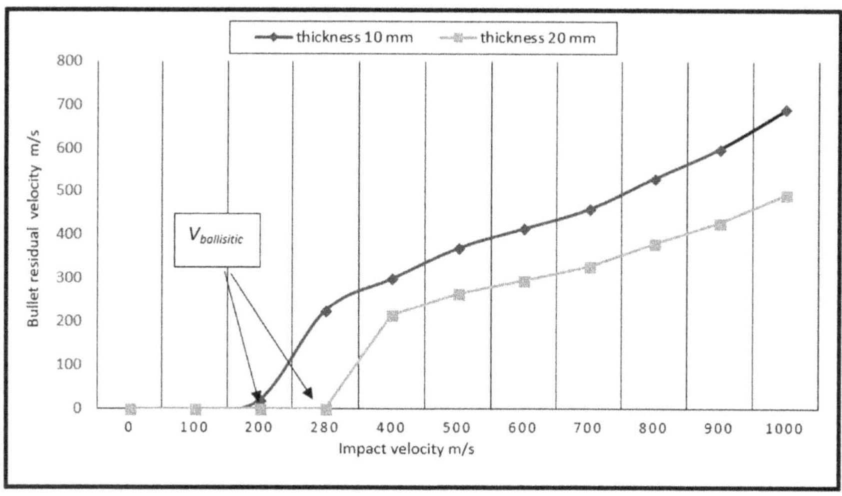

FIGURE 6.17 Bullet V_i–V_r curve.

6.9.4 *EFFECT OF BULLET GEOMETRY*

The key parameters, determining bullet resistance, are the shape of a bullet, friction between the yarns and the bullet, and friction between the yarns.[40] The governing factor in the armor damage is the tip's shape of the impacting bullet. The tip of the bullet has several profiles with different geometrical dimensions (Fig. 6.18).

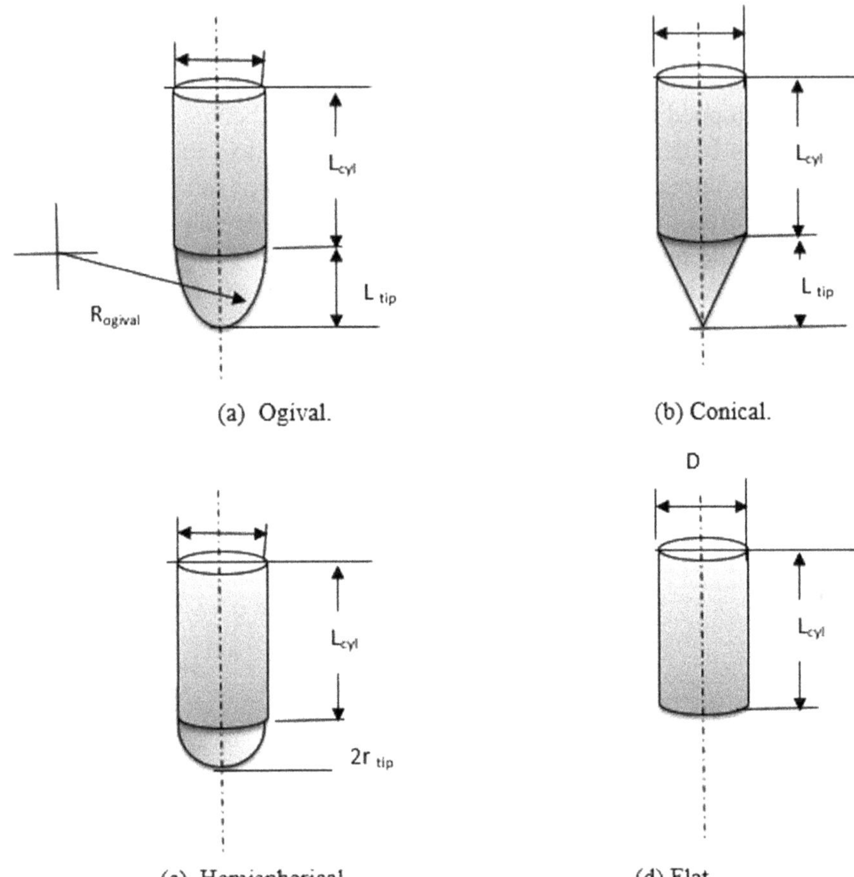

FIGURE 6.18 Bullet geometry.

The bullet geometry is defined by the bullet diameter D and the values of L_{cyl} and L_{tip}, all are varied based on the caliber and the bullet tip's shape.

The value of the inertia force of conical shape bullet can be expressed by the following equation:

$$F = 0.5\ SF\ \rho_{bullet}\ A_{bullet}\ V_{bullet}^{2}. \tag{6.44}$$

The use of the conical tip will increase the compression stress at the tip of the bullet, leading to faster penetration through the armor material.[31,51,52] The impact of the bullet on a hard plate showed that the tip of the bullet in contact with the target will be deformed. With further penetration, the shape of the tip will change to spherical shape or be fragmented, depending on the hardness of the target surface.[53] The bullet factor ($1/L^{0.5}$), where L is the ogive length of the bullet, can be used to characterize the bullet performance. A more pointed bullet is not decelerated as fast as a blunt bullet at higher velocities. At lower velocities, the pointed bullets decelerate faster,[54] but the back-plate deformation increases. More absorption energy is recorded of conical tip at a low impact velocity, however at high number of fabric's layers the difference between the tip's shapes is insignificant.[55] For plain weave single-ply Twaron CT716 fabric, conical and ogival bullets perforated the fabric with the least amount of yarns pull-out and the yarns stretched till failure. Flat-headed bullets sheared through the yarn thickness, and hemispherical bullets produced the most yarns pull-out[56,57] because of the large bullet-target contact area. The peak value of pull-out force is found either in-plane or out-of-plane and is linearly proportional to the value of pretension applied on the fabric. The ballistic limit for the same mass of the bullet increases as the diameter of the bullet increases, the relationship is nearly linear. Additionally, with the same diameter of the bullet, as the mass of the bullet increases, ballistic limit velocity decreases. But as the mass increases, the rate of decrease in ballistic limit velocity increases, too.[58]

All bullets that possess capabilities of wedging through the fabric, such as hemispherical, ogival, and conical shapes, cause a splitting of the fibers that show no significant apparent difference from single-ply systems of fibers.[59]

6.10 HARD PLATES FIBER REINFORCED POLYMER COMPOSITE

The total kinetic energy of the bullet lost during the ballistic impact event is equal to the total energy absorbed by the fiber reinforced polymer composite target until the time interval to pass through it. During the penetration of the bullet in the hard plate, it should have kinetic energy sufficient to overcome following resistance energies:

- Energy of compressive stress due to the compactness of the plate at the area of contact with the bullet tip.
- Frictional energy absorbed during interaction of the penetrator and laminate.

- Energy of delamination.
- Energy of matrix cracking.
- Energy of elastic deformation of the secondary yarns.
- Energy of tensile fiber failure of the primary yarns.
- Energy absorption because of shear plugging.
- Energy absorbed due to fiber-matrix debonding and pull-out.
- Energy absorption based on laminate parameters and
- Energy absorption due to cone formation.

The shear plugging force represents the highest percentage of absorption energy of the plate.

At the contact of the bullet with the plate, the bullet poses kinetic energy as:

$$E_{\text{bullet}}\ (\text{before impact}) = 0.5\ M\ V_{\text{before impact}}{}^{2}. \tag{6.45}$$

The velocity of the bullet will reduce by the penetration through the target, depending on the compressive stress along the thickness direction σ_c, density of the target material ρ, bullet constant SF dependent on the shape of the bullet, is equal to 1 for a flat-ended bullet; for cylindrical bullet tip, $SF = 0.5$; and $SF = (\sin \alpha)^2$ for a conical shape with apex angle α.

The ability of composite laminate to absorb impact energy depends on the numerous variables,[42,60] among them its thickness and the bullet speed are the dominated factors. The mechanism of failure of the composite plate is quite different than that of the multilayer fabric armor of the same thickness. Under high-speed impact, the composite structure is not able to withstand straining and delamination; when initiates, it will reduce composite's capability to absorb energy. Moreover, for composite material, the contribution of the yarns in the secondary zone and in cone formation is higher in energy absorption. Although the fiber matrix bonding strength increases laminate's strength, but also it decreases the energy dissipation. Consequently, low crimp woven fabric is recommended. The mechanism of failure of the composite is found to be different, depending on the velocity of bullet. At low velocity, the energy is absorbed through straining the elements of the composite, fibers, and matrix. While at high velocity, the fibers fraction and composite cracking are accomplished by the interfacial failure and finally, high-speed delamination of the laminates is predominated. Consequently, the failure mechanism of the fiber composite occurs either through the strain failure of the yarns or shear plugging or both. The cone formation energy increases at the early stage of the impact time, then it drops down as the bullet penetrates further. In the meantime, the energy of the deformation at the secondary zone increases until the complete penetration of the bullet through the target.

The percentage of the different components of the absorption energy differs, along with the composite structure and specifications, in particular, the thickness. Moreover, the increase in the different energy components has the direct effect on the bullet velocity $V_{bullet}(t)$.

The residual energy after impact E_r can be given by:

$$E_r = (E_{bullet} - E_{absorbed}).$$

The $E_{absorbed}$ increases linearly with the E_{bullet}.[61] During the bullet penetration through the composite material, the instant residual velocity $V_{bullet}(t)$ reduces till the bullet stops or passes through the whole thickness of the composite.

$$V_{bullet}(t) = [(M\,V_{bullet}{}^{2} - 2E_{absorbed}\,(t - \Delta t)/(M - m(t))]^{0.5}. \tag{6.46}$$

The distance traveled by the bullet during interval of time Δt till the target starts to resist the bullet:

$$dx = dV(i)\,dt,$$

$$T_{ballistic} = \sum dt.$$

The total ballistic time $T_{ballistic}$ is the time between the bullet touches the target to it completely stops or penetrates through.

Thus, the residual velocity V_r approximately can be calculated as:

$$(M + m(T_{ballistic}))V_r^2 = M(V_{bullet}{}^{2} - V_{ballistic}{}^{2}). \tag{6.47}$$

Assuming $\alpha_0 = (V_r/V_{bullet})$, $\beta_0 = m/M$, $\gamma_0 = (V_{ballistic}/V_{bullet})$, ζ_0 is a factor related to the effect of the bullet nose shape, η_0 is a factor related to the material and the armor structure, then:

$$\alpha_0 = [(1 - \zeta_0\,\gamma_0{}^{2})/(1 + \eta_0\,\beta_0)]^{0.5}. \tag{6.48}$$

Figure 6.19 ilustrates the approximate relation between the ratio α_0 and the ratio γ_0 for different values of β_0 (0.1 and 0.5) in the case of $V_{ballistic}$ 300 m/s. To reduce the value of $V_{residual}$, either the ratio β should be large or choose the armor with high value of (η_0).

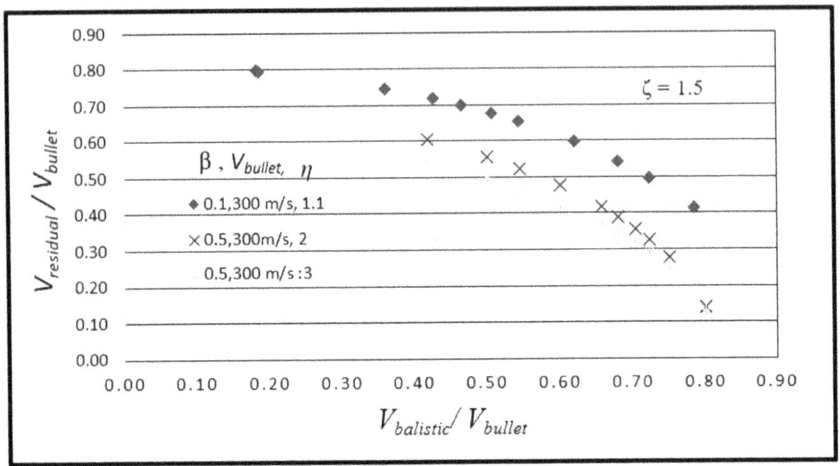

FIGURE 6.19 Bullet residual velocity ratio (α) versus the ratio of ($V_{balistic}$/ V_{bullet}).

In the case of hard plates, the question of what the required thickness is to stop the bullet has been intensively studied and several formulae are exiting concerned with the ballistic limit velocity of the plate. Assuming the entire volume of plate material in front of the bullet is resisting the penetration from the first instance, the relation between the diameter of the bullet, its weight and impact velocity can be given as[62]:

$$(T_m/D_{bullet}) = K_{material}\, g\, (M/D_{bullet}^{\ 3})\, [(V_{bullet}/Q) \cos(\theta)]^2, \qquad (6.49)$$

where $K_{material}$ is a constant dependent on material properties. The Q purposely had no form of scaling term, the coefficient "Q" was used as a measure of the required armor penetration energy and originally absorbed the effects of all factors, and θ is the angle of the bullet's direction of motion vector to the plate's normal line.

For thick composite hard plate, most of the energy is absorbed due to the shear plugging of the composite material[58,59] and is linearly proportional to the plate's thickness. In nonpenetration case, the shear plugging energy reaches 87–95% of the total absorption energy $V_{bullet} < V_{ballistic}$, and the ratio reduces as $V_{bullet} > V_{ballistic}$.

The ballistic velocity for composite hard plate has been tackled by several researchers,[41,61–64] resulting in numerous formulae proved for different types of composite laminates and fiber materials to calculate ballistic velocity for small and high caliber bullets at high and low velocities.[65–67] The ballistic velocity of composite plate can be predicted based on the different models: empirical models, approximate analytical models, and numerical models:

1. A proposed formula for the ballistic velocity of fiber reinforced polymer composite hard plate, based on general formula eq 6.49,[41] will be as:

$$V_{\text{ballistic}} = \alpha_0 \, [(T_m/D_{\text{bullet}})^{\beta_0} (D_{\text{bullet}}^3/M)]^{0.5}, \tag{6.50}$$

   where α_0 and β_0 are constants dependent on the fibers' material, properties and structure, structure of the composite plate laminates, and the fiber volume fraction.

2. Another modification of eq 6.49:

$$V_{\text{ballistic}} = 0.785A \, [((\rho_c \, \sigma_c)^{0.5} \, D_{\text{bullet}}^{\ 2} \, T_m/M) \\ (1 + (M/0.3925 \, A^2 . \, \rho_c . \, D_{\text{bullet}}^{\ 2} \, T_m)^{0.5}], \tag{6.51}$$

   where ρ_c is the density of the composite laminate, σ_c is the material elastic compression stress limit of the composite laminate, A is constant.

3. For fabric reinforced polymer composite material, assuming the fiber breaks at a failure strain ε_f and setting this be equal to the strain in the fabric, the computed rates of increase in the height and base diameter of the cone can be determined.[68] The value of the ballistic velocity is given as[69]:

$$V_{\text{ballistic}} = 4.5(1+\beta X) \, V_{sf} \, \varepsilon_f \, [(R_{bl}/R_{\text{bullet}})^{0.667} \\ - 2(R_{bl}/R_{\text{bullet}})^{0.333} + 3] \tag{6.52}$$

$$(R_{bl}/R_{\text{bullet}}) = [(9\pi/8)\,((1/x) + \beta_1)]^{0.5},$$

   where β_1 is a multiplier (experimental $\beta_1 \sim 2.56$), ε_f is the fabric failure strain, V_{sf} is the strain wave speed in the fiber, given by the square root of the elastic modulus E_f divided by the density ρ_f, $X = (\rho' A'/M)$ is dimensional parameter, M is the mass of the bullet, ρ' is the areal density of the fabric and resin, A' is the presented area of the bullet. The last formula was proved experimentally and the results agree to be sufficiently accurate.[69]

4. In this model, considering the bullet energy is absorbed through the following three energies: the energy of crushing, the energy of linear momentum, the energy of tensile fiber failure breakage, the ballistic velocity will be[70]:

$$V_{50} = \{[4(S_{sp}\, h/r\, \rho_{laminate})\, Xo - 4\,(S_{sp}/\pi\, r^3\, \rho_{laminate}{}^2)\, M)$$
$$+2(\sigma_c/\rho_{laminate}) - e^{-(\pi\, r2\, h\, \rho_{laminate}/M)\, Xo}\, (-\, 4\,(S_{sp}/\pi\, r^3\, \rho_{laminate}{}^2)$$
$$M + 2\,\sigma_c/\rho_{laminate})]/(e^{-(\pi\, r2\, h\, \rho_{laminate}/M)\, Xo})\} \quad (6.53)$$
$$Xo = 2\, h/(E3/\rho_{laminate})^{0.5}.$$

5. Fitting of the experimental data for the ballistic velocity of graphite/epoxy laminates, the ballistic velocity can be given by[41]:

$$V_{50} = 1550.4/(\rho_{bullet})^{0.5}. \quad (6.54)$$

6. The analysis of the fabric failure under impact indicates that the fabric system must store a lot of elastic energy per unit mass, as well as to transport that energy away from the impact region as quickly as possible. The value of V_{50} can be expressed by the following equation[37]:

$$V_{50} = (0.548/\mathcal{E}y_{max}{}^{0.0838})\, [(2\, E_y/\rho_y)^{0.5} - (\sigma_{ymax}/Ey)^{0.75}], \quad (6.55)$$

σ_{ymax}, ε_{ymax}, E_y, and ρ_y are the yarn's tensile stress, strain-to-failure, yarn elastic modulus, and yarn density, respectively.

7. In another approach for determination of ballistic velocity, taken into the consideration the properties of the composite material, its thickness and the bullet dimensions,[52] the ballistic velocity is given as:

$$V_{50} = [\pi\beta_2\, (\rho_t\, \sigma)^{0.5}\, D_{bullet}{}^2\, T_m/4\, M]\, [1 +$$
$$(8M/\pi\, \beta_2{}^2\, \rho_t\, D_{bullet}{}^2\, T_m)^{0.5}], \quad (6.56)$$

β_2 is determined experimentally, for a unidirectional laminate $\beta_2 = 1$, while for a (0/90) cross-piled laminate with equal proportions of the fibers in the two directions $\beta_2 = 0.5$, for a (±45, 90) composite with one quarter of the fibers in each direction $\beta_2 = 0.375$, for a three-dimensional random array $\beta_2 = 0.2$.

Fiber's elastic modulus at ballistic loading rates is higher compared to its values measured by the quasi-static tension tests, while its tenacity fells, the fiber's elastic modulus increase varied from none to an almost tripling of the quasi-static value.[26] The above formulae could be useful for preliminary assessment of the engineering design of the FRP laminated plate in order to reduce the cost of testing of the newly designed samples.

In most cases, the behavior of the composite under low-velocity impact is quite different from when impacted with high velocity. The structure of the fiber architect has a significant effect on the damaged area, energy absorbed, as well as the back-dentation depth. The response of the armor system is dominated by the inelastic impact. These results have the implications in body armor system design. Since the response of the first impacted area, the armor system when it has high flexural rigidity or stiffness, is dominated by inelastic impact.[27,71] The damage behavior of the plate indicated delamination, fiber-matrix debonding, and matrix failure as the possible modes of material damage. Local deformation and shear plugging are the major energy absorption mechanisms in impact perforation.[58,72,73] For 2D weave, the in-plane friction, primary and secondary yarns tension are the major absorption energy components.[20]

The cost of the multilayered armor composite plate depends on the absorbed energy of the plate and the cost of the fibers. It was revealed that the in some cases HPF composite plate has the higher cost than ceramic tiles or aluminum alloy sheet at the same $E_{absorption}$.[72] The Hornady index of terminal standards (HITS) is used to evaluate the effectiveness of the bullet/cartridge with the ballistic velocity $V_{ballistic}$ and its mass M. The HITS number uses the following formula to evaluate the effective bullet/cartage combinations to hit a specific target according to its weight[74,75]:

$$\text{HITS} = (Mg)^2/(V_{bullet}/D^2). \tag{6.57}$$

6.11 ENHANCEMENT OF THE BALLISTIC PENETRATION RESISTANCE OF WOVEN FABRICS

As the development of the weapons is continuously proceeding and represents a challenge for the armor designer, the following are the research directions to reduce the damage and enhance the armor performance[76–88]:

1. Development high-performance fibers of high toughness/weight ratio.
2. Development of material that resists the aging, wear, and exposure to various environmental factors.
3. Increase the surface friction of yarns and fabric through the different treatments (plasma treatment, coating of yarns or fabrics, etc.).
4. Use fabric structures that give better distribution of the strain energy via participation of all the yarns in the impact.
5. Impregnation of the fabric in the colloids shears thicker fluid (STF-fabric composites).
6. Blending of HPF and fine steel yarns.
7. Textile assemblies of CNTs (the fabrication of multiply yarns used in 3D braiding process as well as Z-yarns in 3D weaving process).
8. The self-locking 3D assemblies' structure.
9. Improving the armor comfort through the cooling systems.
10. Development of anti-multi-threat systems.
11. Development of hybrid armor system with elevated single-hit and multi-hit capacity.
12. Development of high cutting performance fibers.
13. Development of laminate composite of woven or nonwoven materials with the improved impact energy absorption.
14. Development of hybridization of the flexible armor using different materials or structures.
15. Better understanding of the transverse high-speed impact behavior of single yarns, single layer fabrics, and the multilayers structure.
16. Development of the mathematical modeling of the armor various designs under bullet impact and
17. Application of high-performance graphene sheets as bulletproof hard plate material.

To improve the ballistic properties of the bulletproof fabric, there are several materials for coating, such as silicon carbide, boron carbide, P2i nanocoating (water-repellent nanocoating)[76] and the coating by D3o, offers the thinnest and most advanced protection against impact that absorbs and dissipates energy during an impact, reduces the amount of force transmitted to wearer body,[77] which can be used for bulletproof vest. When such coating is used, the number of fabric layers required to form the armor decreases.

It is apparent that a considerable amount of research still requests to be done before the precise understanding of the mechanics of fabric-ballistic impact, especially fiber, yarn, and fabric interactions under the ballistic impact at different strain rate. The modeling of the fabric impact, particularly the 3D, needs a deeper approach in order to find out the solutions for the different situations that affect the ballistic penetration resistance of a fabric.

KEYWORDS

- **ballistic impact**
- **ballistic velocity**
- **energy absorbing mechanisms**
- **energy balance**
- **fabric target**
- **STF-fabric composites**
- **stress wave propagation**

REFERENCES

1. Dempsey, P.; Handcocka, P.; Rehrer, N. Impact of Police Body Armor and Equipment on Mobility. *Appl. Ergon* **2013,** *44* (6), 957–961.
2. Chu, Y. Surface Modification to Aramid and UHMWPE Fabrics to Increase Inter-yarn Friction for Improved Ballistic Performance. Ph.D. Thesis, School of Materials, University of Manchester, 2015.
3. Muzzle Velocity. https://en.wikipedia.org/wiki/Muzzle_velocity (accessed Sep 20, 2014).
4. Handgun-caliber-guide. https://gunnewsdaily.com/handgun-caliber-guide/ (accessed Sep 20, 2014).
5. Cybulska M. In *Dynamic Loading and Tensile Behavior of Staple Yarns*. The Fiber Society 2005 Spring Conference, Technical Textiles from Fiber to Composites, 25–27 May 2005.
6. Goetzendorf-Grabowska, B.; Karaszewska, A.; Vlasenko, V. et al. Bending Stiffness of Knitted Fabrics – Comparison of Test Methods. *Fibres Text. East. Eur.* **2014,** *22*, 43–50.
7. Hussain, G.; Iyer, K.; Patil, N. Sonic Modulus of Cotton Yarn and its Relationship with Recovery Parameters. *Text. Res. J.* **1984,** 54–61.

8. Rousselle, M.; Nelson, M. Sonic Pulse Velocity in Fabrics: Is it Related to Abrasion Resistance. *Text. Res. J.* **1980,** *50*, 211.
9. El-Shiekh, A. The Dynamic Modulus and Some Other Properties of Viscose-polyester Blends. *Text. Res. J.* **1974,** *44*, 343–343.
10. ElMessiry, M; Ibrahim, S. Investigation of Sonic Pulse Velocity in Evaluation of Knitted Fabrics. *J. Ind. Text.* **2016,** *46* (2) 455–472.
11. Subramaniam, V.; Muthukrishnan, P.; Nai, V. Sonic Modulus of Cotton Yarn and its Relationship to Abrasion Resistance, Friction, and Dynamic Loading Parameters. *Text. Res. J.* **1991,** *61*, 58.
12. Kremenakova, D.; Militky, J. In *Acoustic Dynamic Modulus of Staple Yarns*. The 4th RMUTP International Conference: Textiles & Fashion, 3–4 July 2012, Bangkok, Thailand, 2012.
13. Kremenakova, D.; Militky, J.; Pivonkova, D. In *Structure and Mechanical Behavior of Polypropylene Yarns*. 7th International Conference – TEXSCI 2010, 6–8 September, Liberec, Czech Republic, 2010.
14. Yan, A.; Postle, R. Application of Sonic Wave Theory to the Measurement of the Dynamic Elastic Moduli of Woven and Knitted Fabrics. *Text. Res. J.* **1981,** *51*, 732–740.
15. Blyth, G.; Postle, R. Measurement and Interpretation of Fabric Dynamic Modulus. *Text. Res. J.* **1979,** *49*, 601–608.
16. Charch, H.; Moseley, W. Structure-property Relationship Ships in Synthetic Fibers: Part I: Structure as Revealed by Sonic Observations. *Text. Res. J.* **1959,** *29*, 525.
17. Pandya, K. S.; Kumar, S.; Nair, N.; Patil, P.; Naik, N. Analytical and Experimental Studies on Ballistic Impact Behavior of 2D Woven Fabric Composites. *Int. J. Damage Mech.* **2015,** *24*4, 471–511.
18. Wen, M. Predicting the Penetration and Perforation of FRP Laminates Struck Normally by Bullets with Different Nose Shapes. *Compos. Struct.* **2000,** *49*, 321–329.
19. Wu, E.; Chang, L. Woven Glass/epoxy Laminates Subjected to Projectile Impact. *Int. J. Impact Eng.* **1995,** *16* (4), 607–619.
20. Shaktivesh, S.; Nair, S.; Naik, K. Ballistic Impact Behavior of 2D Plain Weave Fabric Targets with Multiple Layers: Analytical Formulation. *Int. J. Damage Mech.* **2015,** *24* (1), 116–150.
21. Shaktivesh, S.; Nair, S.; Kumar, S. et al. Ballistic Impact Performance of Composite Targets. *Mater. Des.* **2013,** *51*, 833–846.
22. Song, B.; Yang, W. Effect of Twist on Transverse Impact Response of Ballistic Fiber Yarns. Technical Brief, Luhttps://www.osti.gov/pages/servlets/purl/1236223. (accessed Sept 20, 2014).
23. Sun, D. Ballistic Performance Evaluation of Woven Fabrics Based on Experimental and Numerical Approaches. In *Advanced Fibrous Composite Materials for Ballistic Protection*; Chen, X., Ed.; Woodhead Publishing, 2016; pp 409–435.
24. Porwal, P.; Phoenix, S. Modeling System Effects in Ballistic Impact into Multi-layered Fibrous Materials for Soft Body Armor. *Int. J. Fract.* **2005,** 135, 217–249.
25. Phoenix, S.; Porwal, P. A New Membrane Model for the Ballistic Impact Response and V50 Performance of Multi-ply Fibrous System. *Int. J. Solids Struct.* **2003,** *40*, 6723–6765.

26. Phoenix, S.; Van der Werff, U.; Van der Jagt-Deutekom, M. Modeling and Experiments on Ballistic Impact into UHMWPE Yarns Using Flat and Saddle-Nosed Projectiles. *Fibers* **2017,** *5* (1), 8, 1–41.
27. Cunniff, P. Decoupled Response of Textile Body Armor. https://www.researchgate.net/profile/Philip_Cunniff/publication/ (accessed Sep 20, 2014).
28. Morino, H.; M. Matsudaira, M. Furutani, Predicting Mechanical Properties and Hand Values from the Parameters of Weave Structures. *Text. Res. J.* **2005,** *75* (3), 252–257.
29. Skliannikov, V. About Coefficient of Surface Cover of Woven Fabrics (in Russian Tekstiljnaja Promyshlennostj). *Text. Ind.* **1964,** 6, 32–36.
30. Yang, C.; Ngo, T.; Tran, P. Influences of Weaving Architectures on the Impact Resistance of Multi-layer Fabrics. *Mater. Des.* **2015,** *85*, 282–295.
31. Chen, X.; Wang, Y.; Zhou, Y.; Sun, D. 2D/3D Woven Fabrics for Ballistic Protection. JEC *Compos. Mag.* **2013,** *79*, 89–96.
32. Wang, Y.; Chen, X.; Young, R.; Kinloch, I.; Garry, W. An Experimental Study of the Effect of Ply Orientation on Ballistic Impact Performance of Multi-ply Fabric Panels. *Text. Res. J.* **2016,** *86* (1), 34–43.
33. Bazhenov. S. Dissipation of Energy by Bulletproof Aramid Fabric. *J. Mater. Sci.* **1997,** *32*, 4167–4173.
34. Wei, Li. Models for Projectile Impact into Hybrid Multi-Layer Armor Systems With Axisymmetric Or Biaxial Layers With Gaps. Ph.D. Thesis, Faculty of The Graduate School, 2011.
35. Cunniff, P. An Analysis of the System Effects in Woven Fabrics Under Ballistic Impact. *Text. Res. J.* **1992,** *62* (9), 495–509.
36. Carr, D. J. Failure Mechanisms of Yarns Subjected to Ballistic Impact. *J. Mater. Sci. Lett.* **1999,** *18*, 585–588.
37. Cunniff, P. In *Vs–Vr Relationships in Textile System Impact*. Proceedings of the 18th International Symposium of Ballistics, San Antonio, TX, 1999.
38. Cunniff, P. In *Dimensional Parameters for Optimization of Textile-Based Body Armor Systems*. Proceedings of the 18th International Symposium of Ballistics, San Antonio, TX, 1999; 1303–1310. https://www.researchgate.net/publication/256816090 (accessed July 22, 2018).
39. Hearle, J.; Leech, C.; Adeyefa, A.; Cork, C. Ballistic Impact Resistance of Multi-layer Textile Fabrics. Final Technical Report AD-A128064, University of Manchester Institute of Science and Technology, Manchester, England, 1981.
40. Heru-Utomo, B.; Ernst, L. In *Detailed Modeling of Projectile Impact on Dyneema Composite Using Dynamic Properties*. Proceedings of the ATEM'07, JSME-MMD, Fukuoka, Japan, 12–14 September 2007.
41. Abrate, S. In *Ballistic Impact on Composites. 16th International Conference on Composite Materials*. July 2007, Kyoto, Japan.
42. Shivakumar K.; Elber, W.; Illg, W. Prediction of Impact Force and Duration During Low Velocity Impact a on Circular Composite Laminate. NASA Technical Memorandum 85703, 1–31. https://ntrs.nasa.gov/archive/nasa/casi.ntrs.nasa.gov/19840003151.pdf (accessed Jun 28, 2018).
43. Elmessiry, M.; Eltahan, E. Stab Resistance of Triaxial Woven Fabrics for Soft Body Armor. *J. Ind. Text.* **2016,** *45* (5), 1062–1082.

44. Wang, Y.; Xia, Y. The Effects of Strain Rate on the Mechanical Behavior of Kevlar Fiber Bundles: An Experimental and Theoretical Study. *Compos. Part A* **1998,** *29A*, 1411–1415.
45. Prevorsek, D.; Kwon, Y.; Chin, H. Analysis of the Temperature Rise in the Projectile and Extended Chain Polyethylene Fiber Composite Armor During Ballistic Impact and Penetration. *Polym. Eng. Sci.* **1994,** (2), 141–152.
46. Cunniff, P. High Performance "m5" Fiber for Ballistics/Structural Composites. https://www.researchgate.net/publication/256839266_HIGH_PERFORMANCE_M5_FIBER_FOR_BALLISTICS_STRUCTURAL_COMPOSITES. (accessed July 22, 2018).
47. Anctil, B.; Keown, M.; Pageau, G. et al. Performance Evaluation of Multi-Threat Body Armor Systems. In 22nd International Symposium on Ballistics, Flis, W.; Scott, B., Eds.; Vancouver: Canada, 14–18 November 2005. http://www.dtic.mil/ndia/22ndISB2005/ Friday/anctil.pdf (accessed Sep 4, 2014).
48. Cunniff, P. A Semiempirical Model for the Ballistic Impact Performance of Textile-Based Personnel Armor. *Text. Res. J.* **1996,** *66* (1), 45–58.
49. Meng, Z.; Singh, A.; Qin, X.; Keten, S. Reduced Ballistic Limit Velocity of Graphene Membranes due to Cone Wave Reflection. *Extr. Mech. Lett.* **2017,** *15*, 70–77.
50. De Oliveira, F.; Lima Jr, E.; Lima, E.; Monteiro, S. The Effect of Thickness on Aramid Fabric Laminates Subjected to 7.62 mm Ammunition Ballistic Impact. *Mater. Res.* **2017,** *20* (2), 676–680.
51. Pierson, M. Molding the Impact Behavior of Fiber Reinforced Composite Materials. M.Sc. Thesis, University of Waterloo, 1991.
52. Onyechi, P.; Edelugo, S.; Chukwumuanya, E.; Obuka, S. Ballistic Penetration Response of Glass Fibre Reinforced Polyester (Gfrp) Composites: Body Armor. *IJSTR* **2014,** *3* (8), 226–237.
53. Khoda-rahmi, H.; Fallahi, A.; Liaghat, G. Incremental Deformation and Penetration Analysis of Deformable Projectile into Semi-infinite Target. *Int. J. Solids Struct.* **2006,** *43*, 569–582.
54. Montgomery, T.; Grady, L.; Tomasino, C. The Effects of Projectile Geometry on the Performance of Ballistic Fabrics. *Text. Res. J.* **1982,** *52* (7), 442–450.
55. Tabiei, A.; Gaurav, N. Ballistic Impact of Dry Woven Fabric Composites: A Review. *Trans. ASME* **2008,** *61*, 1–10.
56. Tan, V.; Lim, C.; Cheong, C. Perforation of High Strength Fabric by Projectiles of Different Geometry. *Int. J. Impact Eng.* **2003,** *28*, 207–222.
57. Dimeski, D.; Srebrenkoska, V.; Mirceska, N. Ballistic Impact Resistance Mechanism of Woven Fabrics and their Composites. *Int. J. Eng. Res. Tech.* **2015,** *4* (12), 107–111.
58. Naik, N; Doshi, A. Ballistic Impact Behavior of Thick Composites: Parametric Studies. *Compos. Struct.* **2008,** *82* (3), 104–116.
59. Rajagopal, A.; Naik, N.; Naik, K. Oblique Ballistic Impact Behavior of Composites. *Int. J. Damage Mech.* **2014,** *23* (4), 453–482.
60. Bartus, S. Simultaneous and Sequential Multi-site Impact Response of Composite Laminates. Ph.D. Thesis, University of Alabama at Birmingham, UK, 2006.
61. Gupta, B.; Davids, N. Penetration Experiments with Fiber-reinforced Plastics. *Exp. Mech.* **1966,** 445–450.

62. Okun, N. Major Historical Naval Armor Penetration Formulae. http://www.combinedfleet.com/formula.htm (accessed Jun 25, 2018).
63. Caprino, G.; Lopresto, V. On the Penetration Energy for Fibre-reinforced Plastics Under Low Velocity Impact Conditions. *Compos. Sci. Technol.* **2001,** *61* (1), 65–73, 38.
64. Caprino, G.; Langella, A.; Lopresto V. Indentation and Penetration of Carbon Fibre Reinforced Plastic Laminates. *Compos. Part B* Eng. **2003,** *34* (4), 319–325.
65. Reid, S.; Wen, H. Perforation of FRP Laminates and Sandwich Panels Subjected to Missile Impact. In *Impact Behavior of Fiber-reinforced Composite Materials and Structures*; Reid, S.; Zhou, G., Eds.;Woodhead Publishers Ltd.: Cambridge, 2000.
66. Villanueva, G.; Cantwell, W. J. The High Velocity Impact Response of Composite and FML-reinforced Sandwich Structures *Compos. Sci. Technol.* ***2004,*** *64*, 35–54.
67. Hazell, P.; Appleby-Thomas, G. J. The Impact of Structural Composite Materials. Part 1: Ballistic Impact. *J. Strain Anal. Eng.* **2012,** *47* (7), 396–405.
68. Walker, J. In *Ballistic Limit of Fabrics with Resin.* 19th International Symposium of Ballistics, Interlaken, Switzerland, 2001.
69. Sorrentinoa, L.; Bellinia, C.; Corradoa, A.; Polinia, W.; Aricò, R. Ballistic Performance Evaluation of Composite Laminates in Kevlar 29. *Proc. Eng.* **2015,** *88*, 255–262.
70. Fangyu, C.; Li, Z.; Yihao, T.; Xiaoming, Z. An Analytical Modeling for High-velocity Impacts on Woven Kevlar Composite Laminates. https://www.jvejournals.com/article/19253/pdf (accessed Jun 25, 2018).
71. Chen, F.; Hodgkinson, J. Impact Behavior of Composites with Different Fiber Architecture. *J. Aerospace Eng.* Proc. *I Mech. E* **2009,** *223* (G), 1009–1017.
72. da Luza, F.; Lima, E.; Louroa, L.; Monteiro, S. Ballistic Test of Multilayered Armor with Intermediate Epoxy Composite Reinforced with Jute Fabric. http://www.scielo.br/pdf/mr/2015nahead/1516-1439-mr-1516-1439358914.pdf (accessed Jun 25, 2018).
73. Kumar, S.; Gupta, D.; Singh, I.; Sharma, A. Behavior of Kevlar/Epoxy Composite Plates Under Ballistic Impact. *J. Reinf. Plast. Compos.* **2010,** *29* (13), 2010.
74. Sciencing. https://sciencing.com/calculate-bullet-impact-6951380.html (accessed Jun 1, 2018).
75. Blanton, B. A Physics Professor's View of Ballistics Part V (Terminal Ballistics including handguns. Rifle Ballistics-Part V-a-bu.doc. 1-5. http://poncarpc.org/rifle%20ballistics-part%20v.pdf (accessed Jun 1, 2018).
76. Repellency Treatments. http://www.aculon.com/repellency-treatments/?gclid=Cj0KCQjwpcLZBRCnARIsAMPBgF0wqwayR55HG1_t8S0Q7E7u2vs4_XL1fwADWR9rvVd5KuQHASkb7I4aAr_GEALw_wcB (accessed Jun 25, 2018).
77. What-is-d3o. https://www.d3o.com/what-is-d3o/ (accessed Jun 25, 2018).
78. Mitchell, C.; Post, C. Failure Examination of a New Body Armor Textile by Use of an Environmental Scanning Electron Microscope. *Electr. Inst. Phys. Conf. Ser.* **1999,** *161* (103–106), 103–106.
79. Bogdanovich, A. In *Advancements in Manufacturing and Applications of 3-d Woven Preforms and Composites*. Proceedings of the 37th International SAMPE Fall Technological Conference, Seattle, WA, 2005.
80. Bogdanovich, A.; Mungalov, D.; Baughman, R.; Fang, S.; Zhang, M. In *3-D Braided Material Made of Carbon Nanotube*. Proceedings of the 27th International SAMPE Europe Conference, Paris, France, 2006, 455–460.

81. Chou, T.; Gao, L.; Thostenson, E.; Zhang, Z.; Byun, J. An Assessment of the Science and Technology of Carbon Nanotube-based Fibers and Composites. *Compos. Sci. Technol.* **2010,** *70* (1), 1–19.
82. Thostenson, E.; Rez, Z.; Chou, T. Advances in the Science and Technology of Carbon Nanotubes and Their Composites: A Review. *Compos. Sci. Technol.* **2001,** *61*, 1899–1912.
83. Cheeseman, B.; Bogetti, T. Ballistic Impact into Fabric and Compliant Composite Laminates *CompoS. Struct.* **2003,** *61*, 161–173.
84. Lee, Y.; Wetzel, E.; Wagner, N. The ballistic impact characteristics of Kevlar. *J Mater Sci* 2003, 38, 2825 – 2833.
85. Dyskin, A.; Estrin, Y.; Kanel-Belov, A.; Pasternak, E. A New Concept in Design of Materials and Structures: Assemblies of Interlocked Tetrahedron-shaped Elements. *Scripta Mater.* **2001,** *44*, 2689–2694.
86. National Institute of Justice, NIJ Current and Future Research on Body Armor. https://www.nij.gov/topics/technology/body-armor/pages/research.aspx (accessed Jun 22, 2018).
87. Rahbek, D.; Johnsen, B. Dynamic Behavior of Ceramic Armor Systems. FFI-rapport 2015/01485. https://www.ffi.no/no/Rapporter/15-01485.pdf (accessed Jun 22, 2018).
88. Naik, N.; Shrirao, P.; Reddy, B. Ballistic Impact Behavior of Woven Fabric Composites: Parametric Studies. http://dspace.library.iitb.ac.in/xmlui/bitstream/handle/10054/798/5588-2.pdf?sequence=1 (accessed Jun 22, 2018).

PART IV

Protective Armor Testing Methods: Fibers, Yarns, Fabrics, and Protective Vests

CHAPTER 7

Testing Methods for Materials and Protective Vests: Different Components

ABSTRACT

Body armor system may consist of several components: fibers, yarns, woven fabric, nonwoven fabric, hard plates composite plates, and ceramic plates. Each component's properties have a specific impact on the armor performance. Consequently, the properties (physical, mechanical, thermal, etc.) of each component should be tested according to the international standard methods (ASTM, ISO, etc.). In this chapter, a summary of the principle testing methods of each competent is given. The testing for the ballistic and stab resistance armors as a final product are also mentioned according to the standard methods such asBallistic Resistance of Body Armor NIJ Standard-0101.06; UK Body Armor Standard (2017); Technische Richtlinie March 2008, Germany; RUSSIA - GOST R 50744-95.

7.1 INTRODUCTION

Body armor system must pass specific tests before it can be used, so all elements of the armor should provide the minimum level of performance. The armor, generally, consists of different layers of textile fabrics, each has a special function that represents a part of the armor system requirements. Synthetic fibers can be designed for numerous end uses for protection against various threats, starting from ballistic threat to UV protection. New developments in MMF have permitted a wide spectrum of properties to be imparted to protective fabric. For instance, the firefighter protective vest (Fig. 7.1) consists of three thermal layers, the moisture barrier in the middle, and the outer shell. Each layer is made from different fiber type and fabric structure.

Consequently, each armor component shall be tested separately as well as the entire armor system. Table 7.1 shows the different testing methods.

FIGURE 7.1 A firefighter vest.[1]

TABLE 7.1 Essential Tests for Armor.

I. Armor components	
a. Fibers	
1. Physical	Diameter, length, crimp, friction, density, moisture absorption, melting temperature, glass transit temperature "Tag"
2. Mechanical	Tensile properties of single fiber, bundle fibers, creep, torsion
3. Chemical resistance	Resistance against attack by acids, bases, salt solutions, organic solvents, and other miscellaneous chemicals
4. Electrical properties	Capacitance, dielectric constant, dielectric strength, electrical resistivity and conductivity, permittivity, piezoelectric constants
5. Thermal properties	Specific heat capacity, thermal conductivity, thermal diffusivity, coefficient of thermal expansion, emittance/emissivity
6. Radiation resistance	Stability of properties after exposure to different types of radiation
7. Flammability	TPP ratings, oxygen index, ignition temperature
b. Yarns	
1. Physical	Count, number of filaments, diameter, twist, number of ply, friction, density, moisture absorption
2. Mechanical	Tensile properties, creep, ballistic resistance, fatigue

TABLE 7.1 *(Continued)*

3. Chemical resistance	Resistance against attack by acids, bases, sat solution, organic solvents, and other miscellaneous chemicals
4. Electrical resistance	Capacitance, dielectric constant, dielectric strength, electrical resistivity and conductivity, permittivity, piezoelectric constants
5. Thermal properties	Specific heat capacity, thermal conductivity, thermal diffusivity, coefficient thermal expansion, emittance/emissivity
c. Fabric	
1. Physical	Fabric structure, areal density, thickness, fiber alignment, porosity and pores distributions, air permeability, moisture absorption, vapor permeability
2. Mechanical	Tensile properties, creep, tear strength, stiffness, fatigue strength, compression, strain propagation velocity, shear strength, cutting force
3. Thermal properties	Specific heat capacity, thermal conductivity, thermal diffusivity, thermal expansion, emittance/emissivity
4. Flammability	Ignition temperature, decomposition temperature, maximum flame temperature, melting drip, flame spread rate
5. Electrical properties	Volume resistivity, surface resistivity, dielectric constant, dielectric strength, dissipation factor, arc resistance
6. Dynamic parameters	Punching force, stabbing force, impact absorption energy
II. Protective Armor	
1. Ballistic	Ballistic protection level, V_{50}, V_0
2. Stabbing resistance	Punching force, stabbing force
3. Chemical resistance	Resistance against attack by acids, bases, salt solutions, organic solvents, and other miscellaneous chemicals
4. Flam resistant	Ignition temperature, decomposition temperature, maximum flame temperature, melting drip, flame spread rate
5. Heat shielding	Thermal resistance, thermal protective performance, arc thermal performance value, thermal conductivity, thermal absorptivity, thermal diffusivity

The scheme to design the protective armor system is a closed loop and follows the steps given in Figure 7.2:

- set the required parameters;
- choose design concepts and ideas;
- select armor system construction;
- define the specifications of new parts;

- test the armor performance; and
- define the new armor deficiency.

The assessment of each component of the protective armor should follow one of the recognized testing standards. There are several international and local testing standards for testing methods for each component of the protective armor for physical, mechanical, chemical, electrical, thermal, and so on Properties such as:

- ASTM;
- BS (The British Standards Institution);
- ISO;
- AATCC (American Association of Textile Chemists and Colorists);
- DIN (German Institute for Standardization);
- GOST R (The Russian Federal Agency on Technical Regulating and Metrology);
- EN standards (European Standard);
- NFPA (The National Fire Protection Association); and
- GB (Chinese for national standard).

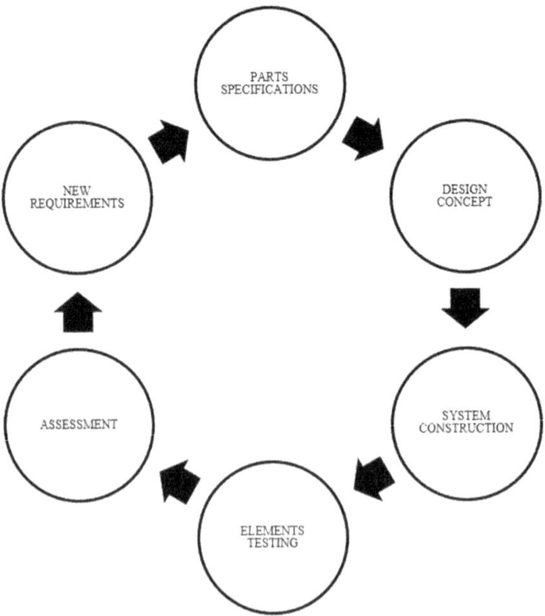

FIGURE 7.2 Design cycle.

7.2 TESTING OF THE DIFFERENT ARMOR'S COMPONENTS

7.2.1 MECHANICAL TESTING

Modern protective fabrics are manufactured from continuous filaments. The main mechanical parameters are given in Table 7.2.

TABLE 7.2 Some Mechanical Parameters of Fibers, Yarns, and Fabrics.[2–32]

Material	Dimensions	Remarks
Fibers		
Count	tex_f (g/km) or den_f (g/9 km)	Defined by: tex_f = weight of fiber "g"/ length of fiber in "km" or den_f = weight of fiber "g"/ length of fiber in "9 km"
Strength (tenacity)	*g/tex* or *g/den*	**FIGURE 7.3** Fibers stress–strain curve of selected fibers.
Yarns		
Count	tex_y (g/km) or den_y (g/9 km)	Defined by: tex_y = weight of yarn in "g"/ length of yarn in "km" or den_y = weight of yarn in "g"/ length of yarn in "9 km"

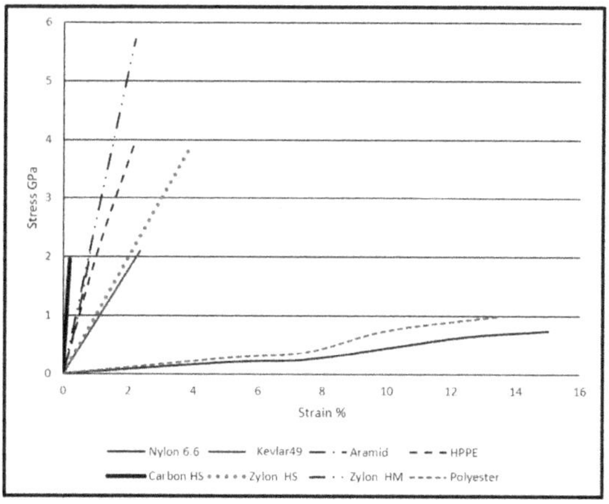

TABLE 7.2 *(Continued)*

Material	Dimensions	Remarks
Number of filaments	$n = tex_y/tex_f$	**FIGURE 7.4** The relation between number of filaments and yarn count.
Twist	T (turns/m)	For twist measurement of continuous filament yarns, untwist–twist method.
Twist factor (α)	$(tex)^{0.5}$(t/cm)	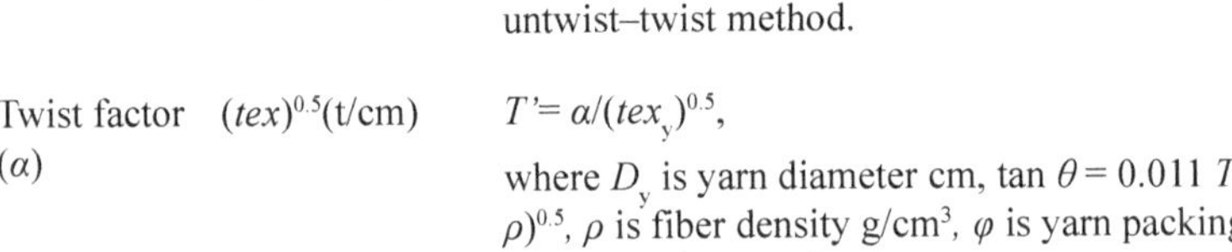 $T' = \alpha/(tex_y)^{0.5}$, where D_y is yarn diameter cm, $\tan\theta = 0.011\, T(tex_y/\varphi\rho)^{0.5}$, ρ is fiber density g/cm^3, φ is yarn packing density, and T is twist turns/cm.
Yarn evenness	Yarn mass coefficient of variation for various cut lengths (CVm) 1, 3, 10, and 50 m	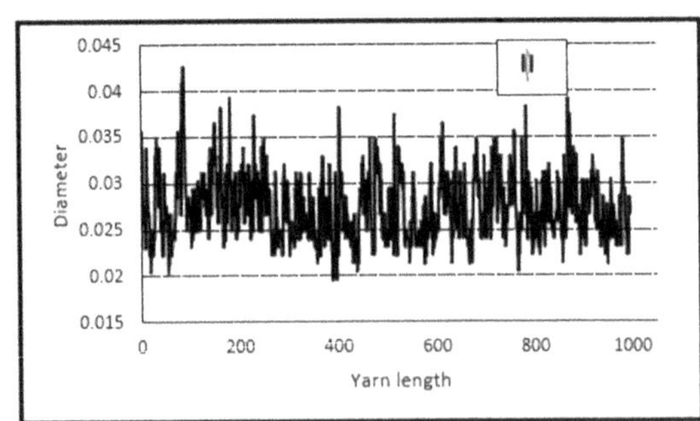 **FIGURE 7.5** Yarn diameter variation.

TABLE 7.2 *(Continued)*

Material	Dimensions	Remarks
Yarn strength (tenacity)	MPa	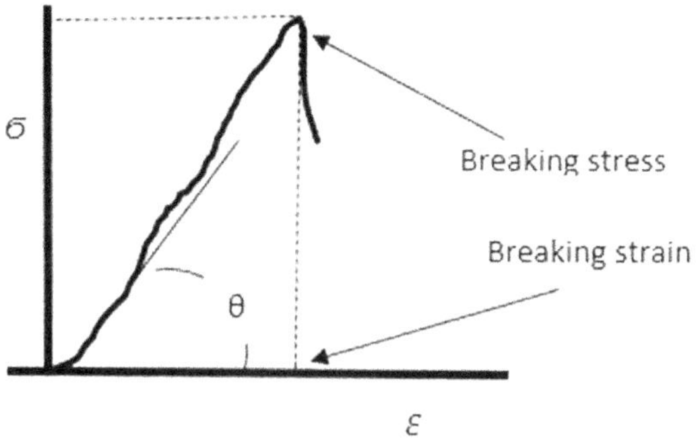 **FIGURE 7.6** Yarn stress–strain curve. $\sigma = 10^3\ \varphi\ \rho_f\ (N_{yarn}/tex)$ *tex* is yarn count, φ is yarn packing density, and ρ_f is fiber density (g/cm^3) (N_{yarn}) is yarn breaking load in Newton
Initial modulus, E_0	MPa	$E_0 = \sigma/(\varepsilon_o)$ (MPa), where E_0 is the initial modulus, σ is the stress, and ε_o is the corresponding strain
Work of rupture	Joule	Breaking load (N_{yarn}) × Breaking elongation (m)
Yarn creep	Elongation of the yarn under constant stress	
Fabric		
Fabric design		plain, twill, satin, basket, and so on.
Warp and weft crimp		$C\% = ((l_{yarn} - l_{fabric})/l_{fabric})\ 100$
Fabric GMS	g/m^2	GMS = (ends per cm) tex_{weft} (1 + $crimp_{weft}$ %)/10 + (warps per cm) tex_{warp} (1 + $crimp_{warp}$ %)/10
Fabric cover factor	%	= 0.444 {(tex(warp)/fiber density)$^{0.5}$ (number of warps per cm) + (tex(weft)/fiber density)$^{0.5}$ (number of weft per cm) − (0.444 (tex(warp)/fiber density)$^{0.5}$ (number of warps per cm)) × ((tex(weft)/fiber density)$^{0.5}$ (number of weft per cm))}
Fabric packing density	%	$\phi = (V_{fiber}/V_{fabric})$

TABLE 7.2 *(Continued)*

Material	Dimensions	Remarks
Fabric porosity	%	$= 1 - \phi$
Fabric air permeability	cm³/min/cm²	= rate of air flow (cm³/min) / sample area (cm²) × pressure drop (cm water)
Fabric strength	MPa	**FIGURE 7.7** Fabric stress–strain curve. $\sigma = 10^3 (\Phi \rho_f / n B)(N_{fabric}/tex_{yarn})$ tex is yarn count "tex_y," Φ is yarn packing coefficient, ρ_f is fiber density (g/cm³), n is number of ends/cm in the fabric, B is width of the tested fabric specimen (cm).
Fabric tear strength 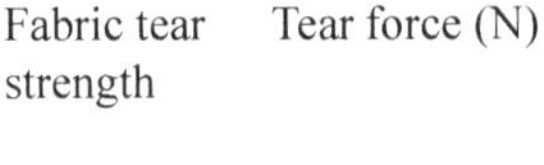	Tear force (N)	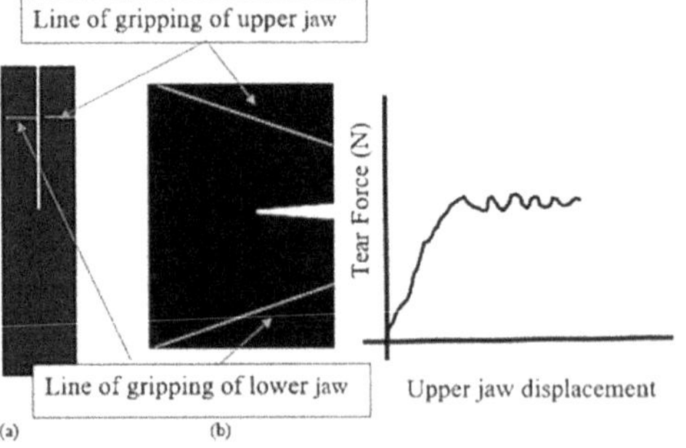 **FIGURE 7.8** (a) Tear test samples and (b) tear force-displacement curve.

TABLE 7.2 *(Continued)*

<table>
<tr><th>Material</th><th>Dimensions</th><th>Remarks</th></tr>
<tr><td>Fabric flexural rigidity</td><td>Micro N.m</td><td>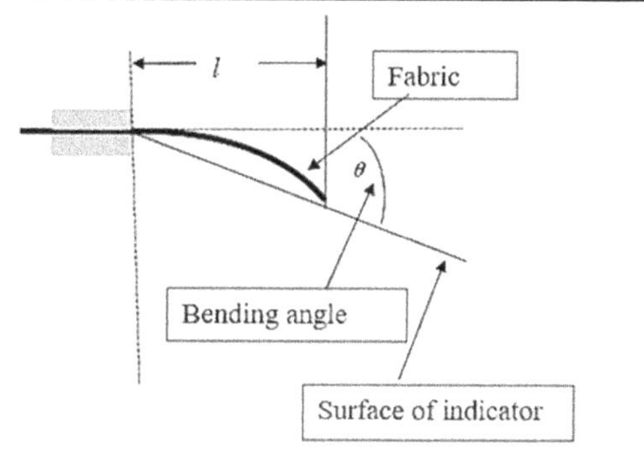

FIGURE 7.9 Principle of fabric stiffness measurement.
$C = M l^3 [\cos (0.5\theta)/8 \tan \theta]$,
where l is fabric length mm, θ is bend angle of fabric, M is fabric mass per unit area (g/m^2).</td></tr>
<tr><td>Fabric shear</td><td></td><td>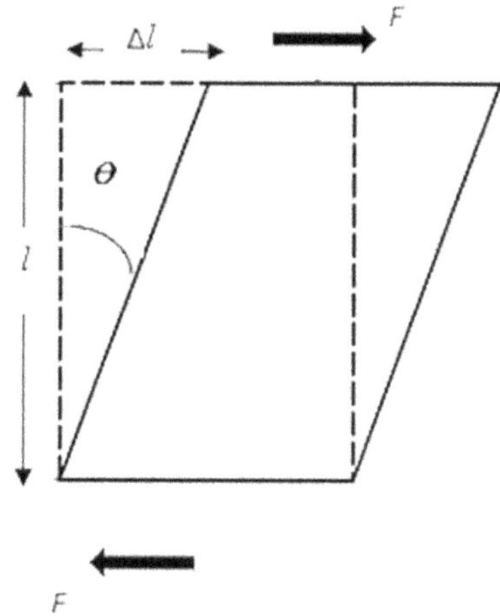

FIGURE 7.10 Fabric under shear force.
Shear stress $\tau = F/A$,
Shear strain $= \Delta l/L$,
Shear modulus $= (F/A)/\tan \Theta$,
F is shear force, A is area of the fabric cross section, and τ is shear stress.</td></tr>
</table>

TABLE 7.2 *(Continued)*

Material	Dimensions	Remarks
Dropped weight-impact stabbing energy	Stabbing energy (J)	**FIGURE 7.11** Dropped weight-impact stab tester. $E_{stabbing} = 0.5\, M\, V^2$, M is mass of the impactor.
Bursting strength	MPa	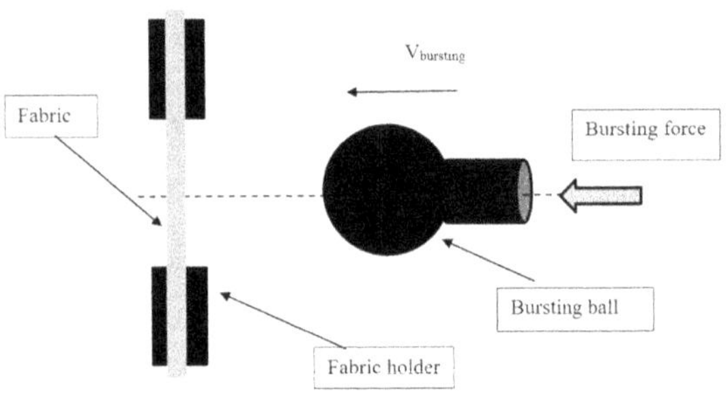 **FIGURE 7.12** Sketch of ball-bursting strength tester. Bursting strength (MPa) = 0.001974 $F_{bursting}$, $F_{bursting}$ is applied force (N).

TABLE 7.3 Fabric Structure.

Type of fabric design	Presentation	Type of fabric design	Presentation
Plain weave		2/2 rib weft weave	
2 × 2 mat weave	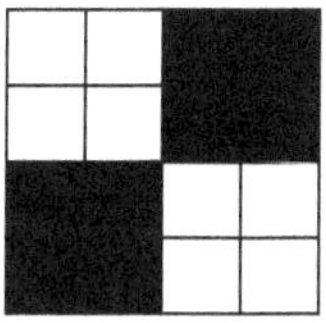	2/2 rib warp weave	
1/3 twill weave	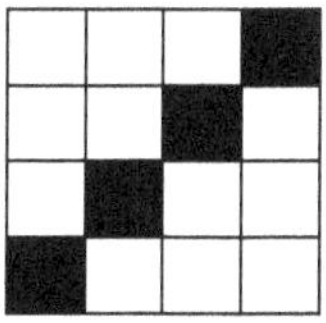	Compound design	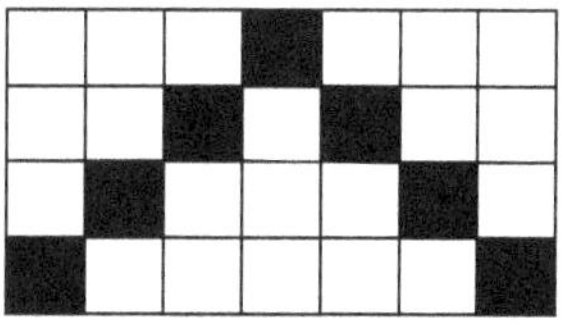
1/4 satin weave	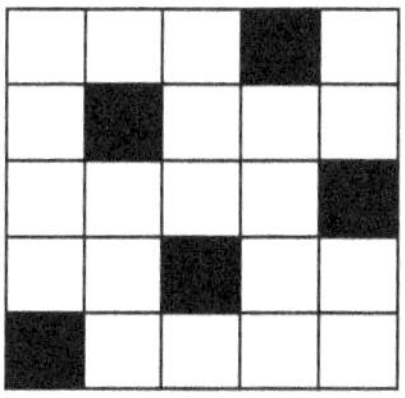	Crow fort satin weave	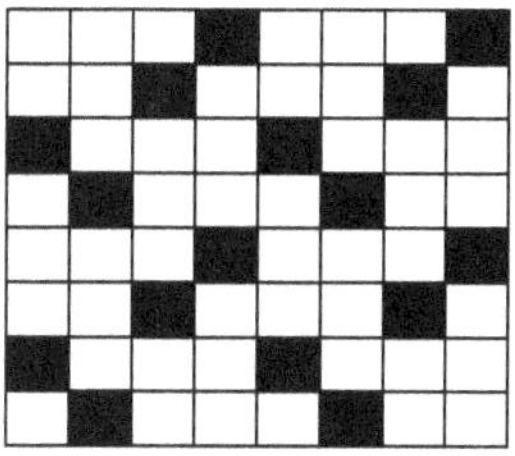
2/1 twill		2/2 twill	
2/2 matt weave	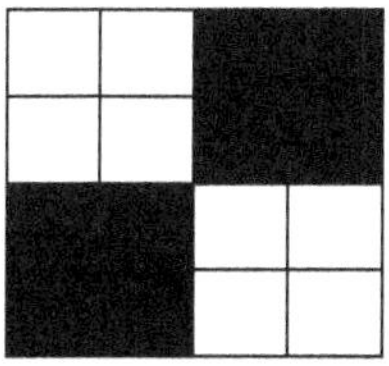	5/5 twill	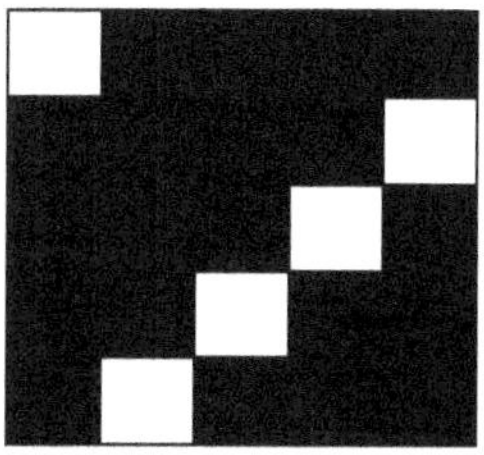

TABLE 7.3 *(Continued)*

Material	Dimensions	Remarks
Puncture strength	*N*	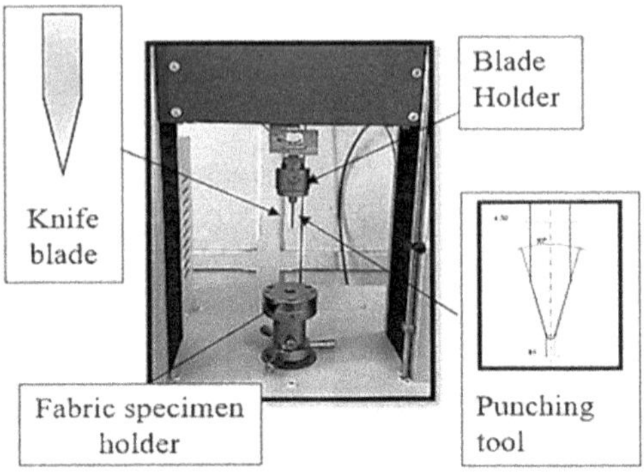 Knife blade; Blade Holder; Fabric specimen holder; Punching tool — **FIGURE 7.13** Punching attachment on universal testing machine.

7.2.2 ANALYSIS OF THE EFFECT OF SOME FABRIC SPECIFICATIONS ON FABRIC PROPERTIES

7.2.2.1 FABRIC STRUCTURE

Table 7.3 presents some of the woven fabric structures which are different mainly in the number of the interlacements per repeat. The fabric design plays a significant role in determining the behavior of the fabric under the different types of loads and environmental conditions.

Fabric mechanical properties are controlled by the yarns ability to move and resist the applied load and the number of warp and fill yarns intersections in the fabric design and the yarn to yarn coefficient of friction. For instance, the fabric shear modulus is contingent on the shear angle θ. Figure 7.14 illustrates the sketch of the 2D fabric under shear force. As value of the shear force increases, the shear angle increases till the yarns are jammed and any further deformation results in fabric buckling. For the same force, the higher the shear angle, the lower will be fabric shear modulus.

The higher number of warp and fill intersections are in the plain weave than in the other designs, satin designs result in higher shear modulus. The mechanism of fabric shear indicates that at small values of shear force, which cannot overcome the friction at the fill and warp yarns intersections, the shear strain depends on the deformation at the intersections. As the shear force increases, the slippages start at the yarns intersection, followed by the compression deformation till the complete fabric jamming.

The jamming angle α is a function of

1. number of the warp and fill yarns per cm;
2. yarn diameter;
3. fiber packing density;
4. jamming angle α;
5. fabric specifications; and
6. fabric design.

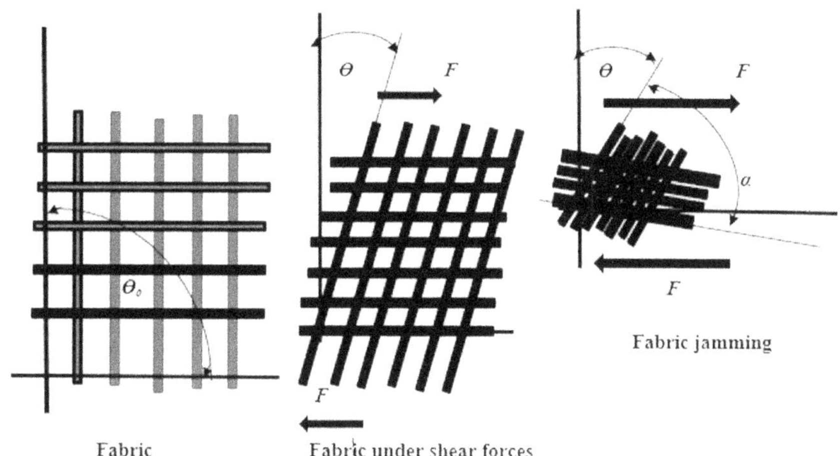

FIGURE 7.14 Fabric shear mechanism.

The illustration of the relation between shear force and shear angle for different fabric designs (Fig. 7.15) indicates that plain weave has the highest shearing modulus than the other weave designs due to the higher number of yarns interlacement and high fabric tightness.

The shear characteristic of fabric determines several properties as follows:

1. a fabric to conform to a double-curvature surface[26];
2. biaxial state of stress in the fabric;

3. resistance to the piercing of a bullet, knife, or spike into a fabric during impact; and
4. the cone height formation upon the impact.

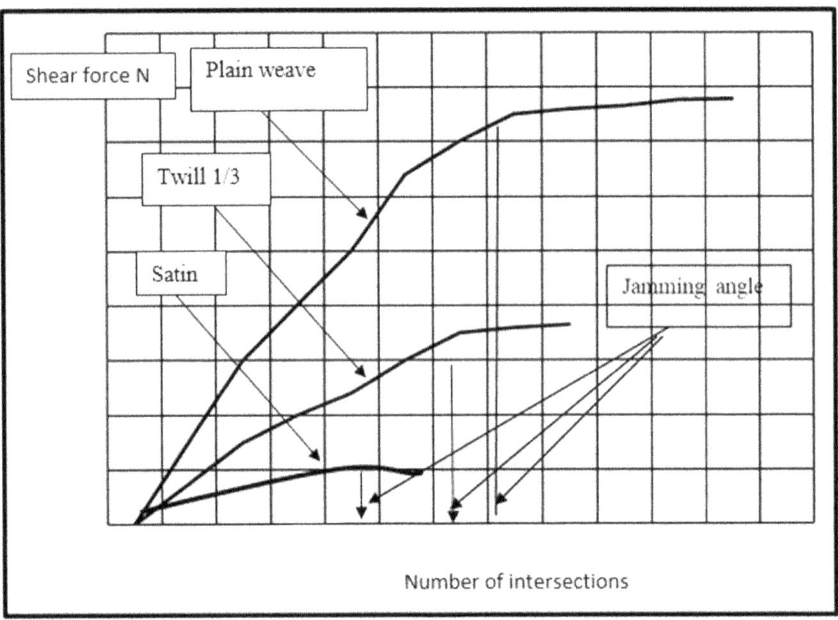

FIGURE 7.15 An illustration of the relationship between shear modulus and number of intersections.

7.2.2.2 FABRIC GMS

The fabric weight per square meter GMS plays a decisive role in determining all the mechanical, physical, electrical, thermal, heat reattendance, and chemical resistance of the fabrics, and also determines all the parameters in conjugation with comfort. Moreover, the V_{50} of the armor is proportional to the material thickness. The fabric thermal conductivity and absorptivity is in direct proportion to its thickness. It was revealed that the thermal energy input to a fabric over a certain time, that is required to result in a heat transfer through the fabric thickness to cause a second-degree burn in human tissue, is directly related to fabric thickness. In the lightweight fabrics, flame spreads faster than in the heavyweight thick fabrics.[28] The stab resistance and bulletproof properties of the protective fabric are directly proportional to the fabric thickness.[29–31]

7.2.2.3 FABRIC TEAR STRENGTH

The tear strength of the fabric is one of the properties that is involved in the resistance to the bullet or its fragments after impact to penetrate through it, increasing the fabric stopping power.

7.2.2.4 FABRIC STRENGTH AND MODULUS

The fabric strength and modulus are considered as the main parameters that determine the bullet penetration into ballistic fabric. It was revealed that the fabric cone formation, when impacted by projectile, depends on the velocity of strain propagation.

7.2.3 FABRIC CHEMICAL RESISTANCE

The chemical resistance of the protective fabrics (PFCR) is the effect of the different chemicals on the degradation of their physical and mechanical properties, in particular the tenacity.

The PFCR are tested using acids, bases, salt solution, solvents, and other miscellaneous chemicals. The chemical resistance depends on chemical concentration, temperature, and time of contact. The material assessment of the chemical resistance (chemical protective cloth) can be rated into the following categories consistent with the loss of its strength:

1. none (0–10% strength loss);
2. slight (11–20% strength loss);
3. moderate (21–40% strength loss);
4. appreciable (41–80% strength loss); and
5. degraded (81–100% strength loss).

The exposure to chemical hazards, necessitates the protective cloth of special material (CPC) cannot protect from all chemical hazards, but it protects against a limited number of chemicals.[54] The FCR is correlated with fiber chemical resistance, fabric structure, fabric specifications, fabric properties, including thickness, and weight/m^2. The fabric air permeability, porosity, and areal density are the key factors for CPC. The chemical effect on the fibers, fabric, and armor system can be influenced by chemical concentration and temperature, method of the application of the chemical—splashes or chemical vapor.

Ratings of chemical effect can be classified as[55,56]:

1. no effect—excellent,
2. minor effect—good,
3. moderate effect—fair, and
4. severe effect—not recommended.

The rating as stated by advanced materials group (AMG)[2]:

1. essentially no effect, less than 5% change in mechanical properties and less than 2% change in weight and dimensions;
2. some change; and
3. not recommended.

MPE (*Modern Plastics Encyclopedia 1986–87*) Rating System[2] suggests another classification:

1. No significant effect—1.0%, >0.5%, >20% change in weight, dimension, and strength, respectively;
2. Significant, but usually not conclusive—0.5–1.0%, 0.2–0.5%, 10–20% change in weight, dimension, and strength, respectively; and
3. Usually significant—>1.0%, >0.5%, >20% change in weight, dimension, and strength, respectively; distorted, warped, softened, or crazed.

Due to the damaging effect of some chemicals on fiber properties, the data sheet of each type of fibers contains a great attention to FCR of the MMF.

The FCR is tested using some chemicals only such as:

1. *Acids*: hydrochloric acid, sulfuric acid, and so on,
2. *Bases*: sodium hydroxide, calcium hydroxide, and so on, and
3. *Organic Solvents*: acetone, benzene, and so on with different concentrations varied 37–100% at temperature of 20–200°C.

The time of chemicals treatment varies, in relation to the end use, temperature and concentration, from 1 to 10,000 h. Several standards concerning the PFCR are given in Appendix I.

Personal protective equipment (PPE), protective armor system allows moisture vapor from the body to escape through the protective clothing layers while protecting the wearer from the passage of chemical agents in liquid, vapor, and aerosol forms.[57] NFPA Standards established performance for chemical protective clothing, such as NFPA 1991, NFPA 1992, which provide minimum levels of protection for emergency response of PPE against vapor, liquid splash, and particulate materials.

7.2.4 FABRIC UV LIGHT RESISTANCE

UV protection is essential for protective fabrics. Ultraviolet (UV) resistance of the protective fabric is a degree of degradation of the fabric when exposed to UV light. Fabrics with a higher weight, thickness, and cover factor absorb a higher amount of UV rays thus, these fabrics offer better protection against UV radiation.[33] AATCC Test Method 183-2004 standard determines the ultraviolet protection factor (UPF) of fabrics either in dry or wet states.[34] The UV spectrum lays between 200 and 400 nm and is commonly divided into three regions: UV-A—320–400 nm, UV-B—280–320 nm, and UV-C—200–280 nm. Most of the polymeric fibers are subjected to structural degradation due to break of the chemical bonds when exposed for a long period to UV, depending also on the wavelength. Each type of fiber has its critical wavelength region, for instance for Kevlar UV wavelength 300–490 nm is more critical. The samples of fabric shall be tested for UV evaluation, such as tensile, impact, abrasion, and color. The time–degradation curve should be plotted to calculate the properties degradation over time using one of the following methods:

1. weather meter, (exposure method - AATCC 16E); and
2. accelerator UV tester (accelerated weathering), (ASTM 4329, ASTM G154-16)

7.2.5 FABRIC FLAME RESISTANCE

The flammability of fire protective fabric (FPF) is dictated by the following properties: physical properties, chemical properties, thermal properties, and fabric specifications. Tight heavy weight fabric has higher flame resistance; FPF is tested applying ASTM D4108-87. Thermal protective performance (TPP) test measures the amount of thermal protection a fabric would give to a wearer in the event of a flash fire (flame resistance or flammability). To measure the relative flammability of a fabric, it is rigidly held in a three-sided frame and methane burner provides a small igniting flame which is allowed to impinge on the lower side of the fabric for 12 s, and the following data is measured:

1. the char length, after flame (the number of time flames continue on the fabric surface after the burner is turned off or removed);
2. afterglow (the amount of time the fabric continues to glow after any flame stops);
3. presence of drips; and
4. afterburn time.

This test results are qualitative pass/fail indicator of fabric flammability. For example, Kevlar is inherently flame resistant but can be ignited and burning usually stops when the burner is removed.[2] with none drips and no afterburn time. Flame resistance fibers have flame retardancy in their chemical structures (non-flammable).

NFPA 701 testing method (standard methods of fire tests for flame propagation of textiles and films) uses loss of weight as the main criterion of FPF. The flame from a gas burner is applied to each specimen for 45 sec and then removed. The sample will fall if: not stop burning within two seconds; weight loss is more than three standard deviations from the mean for the 10 specimens, the specimens on average lose no more than 40% of their weight during the test.[35]

For protection against flash fire simulation of flame resistance of clothing, an instrumented manikin is used, and test is performed by the standards (Appendix I).

The NFPA 2112, cloth is subjected to a 3 s engulfment fire at 2 cal/cm^2 and calculated for the total predicted body burn. Also, it should not have melt or drip or have more than 2 s burning after flame or 10 cm char length when tested[36,37].

7.2.6 FABRIC THERMAL PROTECTIVE PERFORMANCE TEST

The FPF shall be tested according to Standards (Appendix I) for the heat transmitted through the fabric specimen. The energy obtained through the time determines the possibilities of reaching second-degree burn (2nd DB). The TPP tester has two burners, as a convective/radiant heat source, and a number of quartz tubes to provide additional radiant heat. The 15 cm square protective fabric sample is heated at a rate of 84 ± 2 kW/m² (2.0 ± 0.05 cal/cm²)[38] and is measured by a copper calorimeter. The principle of the apparatus is given in Figure 7.16.[39]

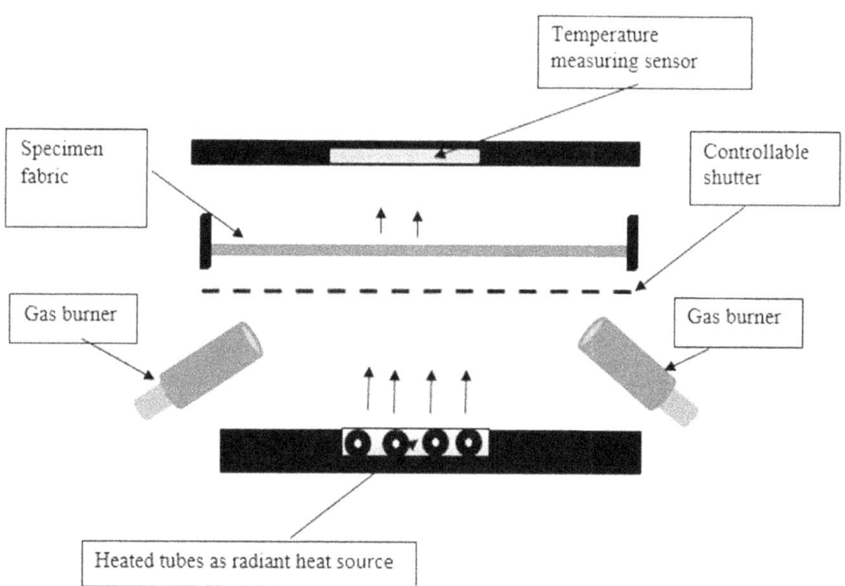

FIGURE 7.16 Sketch of TPP tester.

The heat, penetrating through the sample, is measured and compared with that enough to cause a second-degree burn using Stoll's curve (Fig. 7.17). Stoll's curve gives the measured energy J/cm²- exposure time curve to have a second-degree burn (when skin starts to blister).[40,41] The longer the time to have a certain value of heat energy, the higher is value of TPP. For TPP value of 6, the approximate time to second-degree burn is 3 s (Total heat energy 7.5 J/cm²) and for TPP value of 60, the approximate time is 30 s (total heat energy 13.5 J/cm²).[42,43]

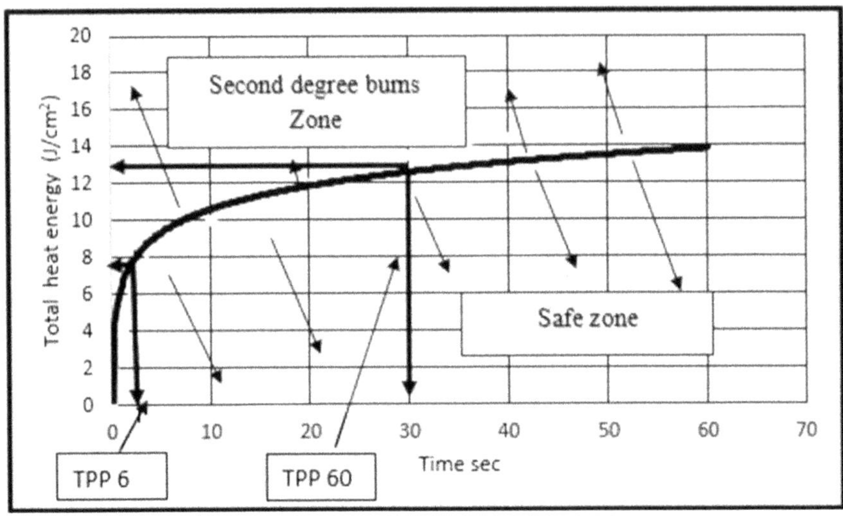

FIGURE 7.17 Stoll's curve.

To calculate (TPP) rating by NFPA (National Fire Protection Association) is presented by the following equation[44]:

$$\text{TPP} = F.\ T, \tag{7.1}$$

where F is total heat flux in cal/cm², T is time in seconds until the Stoll's burn criterion is met.

For NFPA, the minimum TPP score is 35, based on the formula used, which means it would take 17.5 s. Fabric, which has TPP 80, provides 40 s protection against fire.

7.2.7 ELECTRIC ARC FLASH (EAF) TEST

An EAF happens as electrical current passes through the air and generating heat of very high levels and the protective armor (PA) is required for safe work practices when exposed to the electric arc hazard. The PA performance in an electric arc exposure of various fabrics and/or systems is tested by either of two procedures:

- For non-flame-resistant materials, on the mannequins. Mannequin electric arc simulation (Thermo-Man; Appendix I; Fig. 7.18) measures how much heat a certain fabric will block from an electric arc before the reach of (second DB) for the wearer.[45] Heat attenuation factor (HAF; the percentage of total heat blocked by the fabric from reaching the sensor) is also determined. The mannequin is provided with sensors to measure the temperature at the different locations of its body when protective fabric is used as a barrier for protection from the effects of electrical arcs. Sensors are constantly sending measurements to data acquisition system for analysis. The mannequin must be equipped with more than 100 sensors to measure incident heat fluxes of up to 167 kW/m^2 at a high time response of ≤ 0.1 s. Heat flux transducer on the mannequin measures the total incident heat flux with and without protective clothing. Heat that is transferred through a tested garment is measured and used to model the skin's response and predict possibility for burn damage.[46]

- NEPA 70 E Standards; the electric arc protective clothing is classified, depending on the maximum capability of a cloth for arc-flash protection, into:

  I—5 cal/cm^2, II—8 cal/cm^2, III—25 cal/cm^2, and IV—40 cal/cm^2.[47–50]

- The ASTM Standards; the arc thermal performance value (ATPV) of a material is the incident energy on a material or a multilayer system of materials that results in a 50% probability that sufficient heat transfer through the tested specimen is predicted to cause the onset of a second-degree skin burn injury based on the Stoll's curve.

7.2.8 FABRIC THERMAL CONDUCTIVITY

The principle of measuring the fabric thermal conductivity (*k*) relies on the convection of heat emitted by the hot upper plate in one direction through the sample being examined to the cold bottom plate adjoined to the sample [ASTM F1060–16-2008, ASTM D1518]. The instrument directly measures the stationary heat flow, the temperature difference between the upper and bottom fabric surface to calculate the thermal conductivity of the fabric, rating the fabrics for their thermal insulating properties and the thermal protection time, ASTM F1060–16. Fabric areal density, thickness, porosity, and the weave design influence its thermal insulation properties; plain fabrics have a higher (k).[51,52] The thickness of the fabric/thermal conductivity ratio are related to the fiber's conductivity, fabric porosity, and weave design.[53]

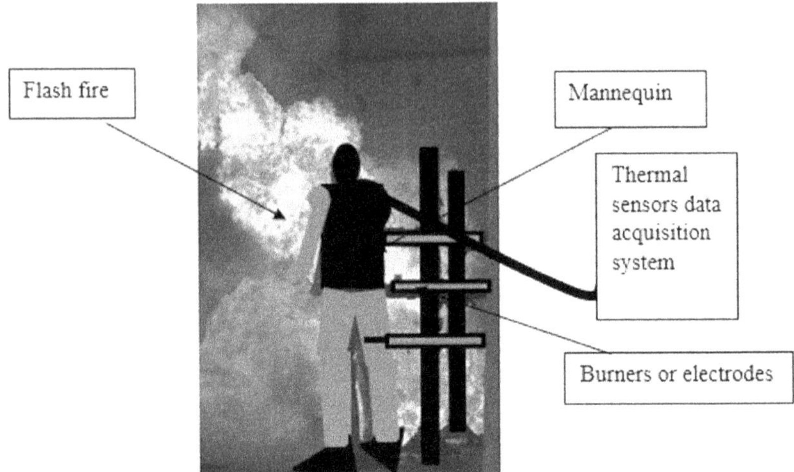

FIGURE 7.18 Flash fire or electric arc mannequin simulation.

7.2.9 ABRASION RESISTANCE

The abrasion resistance (AR) is the capability of fabric to resist surface wear due to abrasion by different materials in contact with under pressure. The value of AR measured in the laboratory represents only one of several factors contributing to wear performance in the actual use of fabric. The principle of the abrasion tester of the fabric is to subject the specimen to cyclic rubbing with a standard abrasive material in laboratory condition. Testing conditions, such as change in humidity, temperature, testing abrading, type of testing apparatus, can cause the significant variations in the test results. The fabric design, fabric thickness, fabric specification, yarn structure, twist factor, fiber fineness, fiber strength, and amount of finishing treatment are the main parameters affecting the fabric AR. The testing apparatus has several designs as follows:

- Martindale: where the circular sample is fixed and rubbed on the top with abrading cloth under a specific load. The abrading material holder is rotated in an enlarging elliptical shape.

- Flexing and abrasion method: the fabric is rubbed in flexing form against the abrading bar till its failure.
- Rotary platform double head abrader: the sample is fixed to the rotating platform and abraded by double-head abrading cylinder pressed to the rotating sample.

The AR of the fabric can be evaluated after abraded for specific number of cycles[58]:

- the number of cycles to develop a hole in a knitted fabric or cause rupture two or more yarns in a woven fabric;
- the number of cycles to rupture;
- percentage loss in fabric strength; and
- evaluation for visual changes after abrasion.

The standards for the testing of fibers, yarns, and fabric are listed in Appendix I.

7.3 ARMOR ESSENTIAL TESTING

The requirement for the performance of the final armor design defines the testing methods that are selected to ensure its reliability and certifies that the body armor is able to defeat the expected threat. By applying go/no go test, the armor will be rated to comply with particular standards limits.

7.3.1 BALLISTIC AND STAB RESISTANCE ARMORS

The armors are either soft or hard armor. Soft armor is sleeveless under-garment or lining of a jacket or outer garment.[59] Hard armor is used when expecting an unusual high risk of rifle fire or stabbing. It may include panels of different materials and constructions. Most of the standards concern with a flexible vest and hard armor plate or both together as a protective system.

The testing of the new design protective armor is needed to reveal performance spiral, since once protection against the existing offensive weapons is found, then the performance of the latter is immediately increased, which in turn entails new developments in the protection.[60]

7.3.1.1 STANDARDS AND GUIDELINES FOR KNIFE AND SPIKE STAB RESISTANCE

There are several standards specifying the body armor performance (BAP) requirements for knife and spike.[61,62]

7.3.1.2 STAB-PROOF TEST PRINCIPLE

The main principle of the stab resistance test is when the armor test sample or plate is subjected to strike with a standard shape knife blade or spike at a certain level of striking energy with standard impactor of weight 1.9 kg. The impactor kinetic energy is:

$$E = 0.5\, M\, V^2_{\text{impactor}},$$

where M is the total mass of impactor, V_{impactor} is the knife velocity at 0° angle of incidence.

The protection level determines the impactor drop heights, the penetration depth of the knife or spike will be measured on the backing material, the relation between the depth of penetration, and the witness paper cut length is shown in Figure 7.19,[61] in-cut-length is converted to depth of penetration as[63]:

$$\textit{Depth of penetration (mm)} = \textit{cut length (mm)} \times 2.396 \tag{7.2}$$

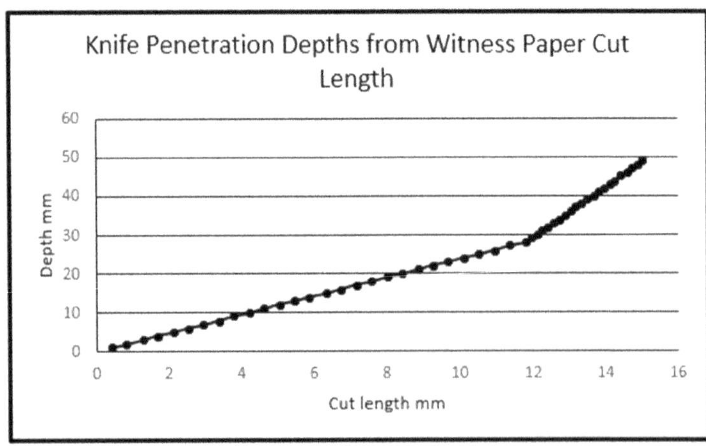

FIGURE 7.19 Knife blade penetration depths from witness paper cut length for 0° angle of incidence.

The sample must be tested for either knife resistance or spike resistance for one or more protection levels.[61,62]

7.3.1.3 BULLETPROOF ARMOR TESTING PRINCIPLE

The bulletproof armor protection level relates to the bullet mass M_{bullet} and velocity V_{bullet} as well as the design of the bullet shape. The muzzle energy of the bullet is the kinetic energy of a bullet as it is expelled from the muzzle of a firearm and equal to:

$$E = 0.5\, M_{\text{bullet}}\, V_{\text{bullet}}^{2}. \tag{7.3}$$

The velocity of bullet is measured by a chronograph; Figure 7.20 shows the layout of the stand.

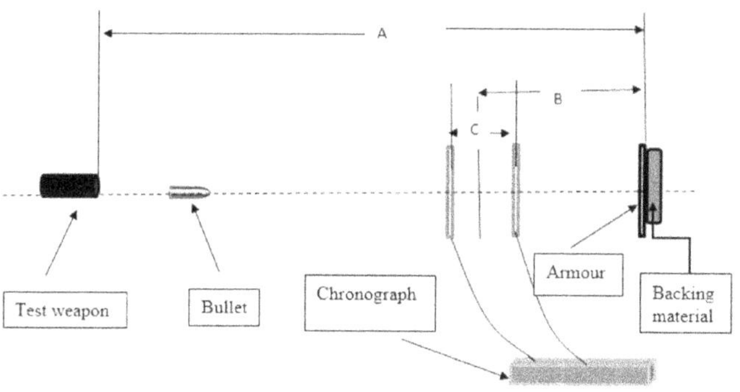

A- 5 ± 1m for type IIA, II and IIIA ballistic levels (Handgun Rounds) and 15 ± 1m for III and IV ballistic levels (Rife Rounds), B- 2.5 ± 0.025 m . C- adjusted according to the design of chronograph used.

FIGURE 7.20 Principles of testing stand for bulletproof armor ballistic test.

7.4 OVERVIEW OF COMPLIANCE TESTING FOR STAB RESISTANCE ARMOR

7.4.1 NIJ STANDARD–0115.00

The stab resistance test, NIJ standard–0115.00,[61] has two classes for knife or spike protection. The protection class includes an overall picture of the threats for which a body armor is designed to defeat, while the protection levels of the armor have three classes stating the graveness of a stab for which the armor is designed. The strike energy is different (Fig. 7.21) for knives and spikes.[61]

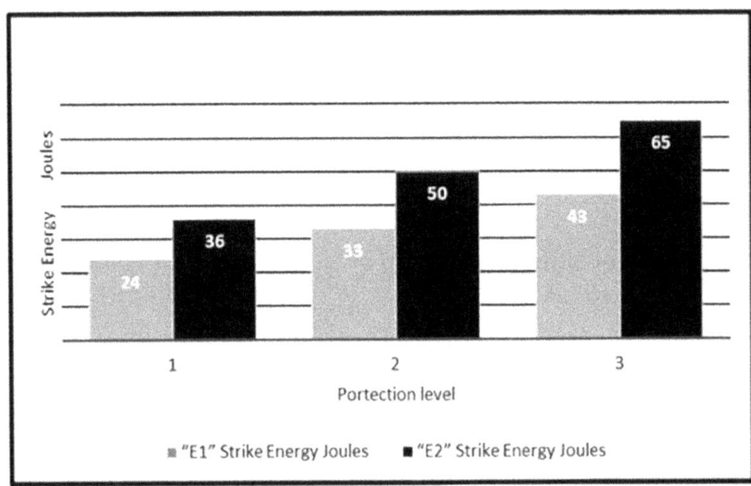

FIGURE 7.21 Levels of strike energy.

When the armor is tested under strike energy E1, the maximum allowable penetration is 7 mm and under overstrike energy E2 it is 20 mm.[61]

For a compliance tests, three armors are chosen: two of them are randomly collected for testing for stab penetration, using the two standard designs of knives[61] with two drop tests for front and back panels carried out for each knife at angle of incidence $\theta = 0°$ at energy levels E1 and E2 and one test, using standard knives at angle of incidence 45° at energy level E1.

Stab resistance drop tests for "Spike" Protection Class are carried out on both front and back of two armors: one drop test at angle $\theta = 0°$ at energy level E1 and E2 and one test at $\theta = 45°$ at energy level E1.

The stabbing points are distributed on the area of tested armor, with strike point of stab locations between distance 51 mm and far from the edge of the armor by distance at least 51 mm. The numbers of strikes are 24 (Fig. 7.22) which shows test sequence for values of E1 and E2 and for knives and spikes tests.

The armor drop test is recommended by NIJ to investigate the effect of fallen weight impact on the armor laid on hard surface.[64]

TABLE 7.4 HOSDB (2017) Knife and Spike Standards.[65]

Protection level	Energy level E1 (J)	Maximum penetration at energy level E1 (mm)	Single penetration limit SPL at energy level E1 (mm)	Energy level E2 (J)	Maximum penetration at energy level E2 (mm)	Single penetration limit SPL at energy level E2 (mm)
KR1	24	8	9	36	20	30
KR1 + SP1	24	KR1 = 8, SP1 = 0	KR1 = 9, SP1 = 0	36	KR1= 20, SP1 = N/A	20
KR2	33	8	9	50	20	30
KR2 + SP2	33	KR2 = 8, SP2 = 0	KR2 = 9, SP2 = 0	50	KR2 = 20, SP2 = N/A	KR2 = 30, SP2 = N/A 7.3

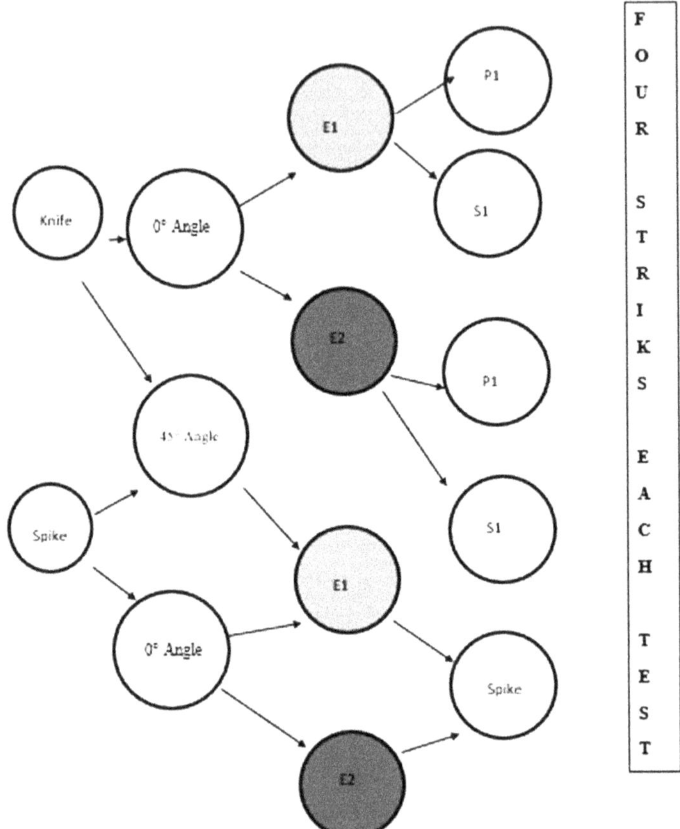

FIGURE 7.22 The tests for knife and spike at impact angle of incidence of 0° and 45°.

7.4.2 HOSDB BODY ARMOR STANDARDS

The performance standard test (PST) for stab resistance of the armors incompliance with HOSDB (2017)[62] is classified into three levels of protection: KR1 & KR1 + SP1, KR2 & KR2 + SP2, KR3 & KR3 + SP3.[62] The sample is strike by the knife or spike attached to mass, vertically freely falling to strike the fixed armor in a drop tester. The total impactor weight with the knife or spike is 1.9 kg.

Table 7.4 indicates that the maximum allowed penetration for different energy levels E1 and E2.

The knife and spike have a standard shape with specially tapered edge of length 100 mm. The knife width and thickness are 15 and 2 mm, respectively, while the spike has a round cross section of a diameter 4.5 mm.

To determine the depth of penetration of the falling tools, a composite backing material, consisting of seven different layers of predetermined specification, is placed at the back of the specimen. Calibrating the composite backing material is required to attain the consistent results,[62] at the same time the knife or spike must pass the tip sharpness test. The number of strikes and the angle of inclination of the knife is shown in Figure 7.23, while Figure 7.24 demonstrates the spike testing.

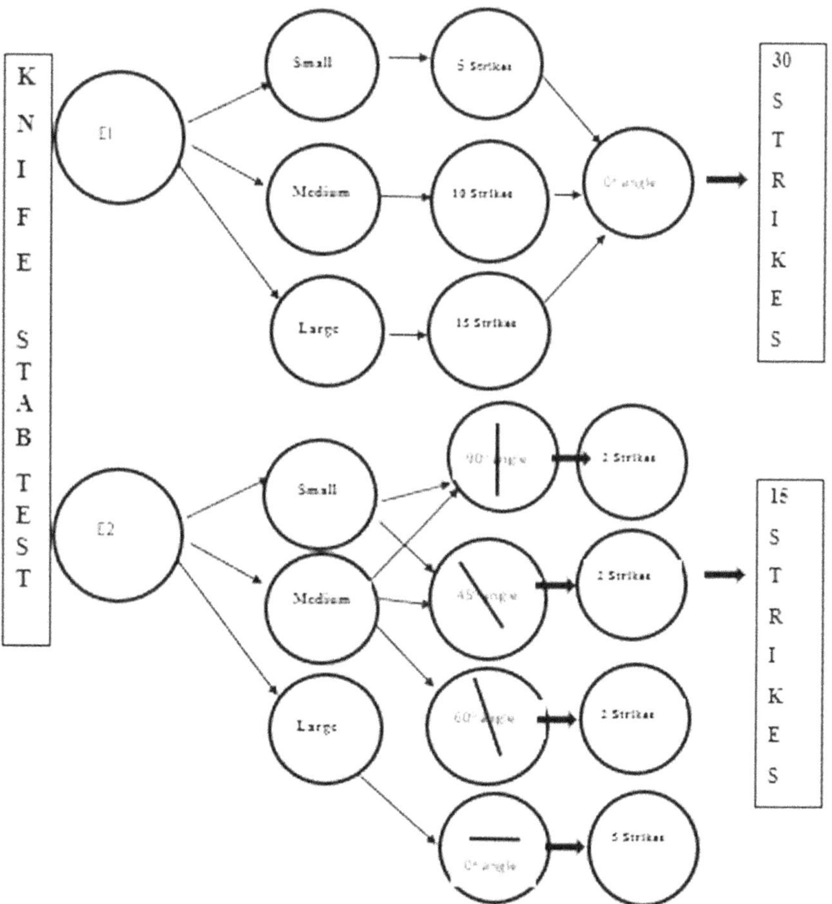

FIGURE 7.23 Unformed armor knife stab test quantities.

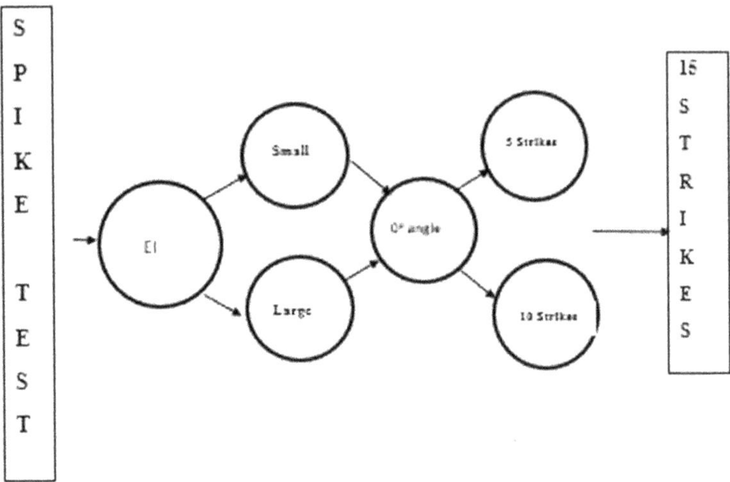

FIGURE 7.24 Unformed armor spike stab test quantities.

For formed armor, small and large panels are tested at energy level E1 for knife or spike impacts, four strikes for each size should be performed. For seams or joints, an additional test at energy level E1 is considered at each seam as well as for female armor, strike tests at $\theta = 0°$ at each line or area of weakness level within 5 mm of the cup, no penetration for spike or knife is permitted.

One penetration in a test series is permitted up to 9.0 mm in the case of panels, knife strikes, for energy level E1, are performed at angle of incident 0°, 30°, 45°, and 90°, four strikes at left, right, top, and bottom of bust tip, and for spike assessed test only at energy level E1 and no penetration is permitted. One penetration not exceeding 30 mm is permissible at E2 knife testing.

Several other standards for stab resistance of knives, spikes, or needles are existing such as: Technische richtlinie April 2003, Germany, Technische Richtlinie March 2008, Germany, RUSSIA—GOST R 50744-95.

7.5 OVERVIEW OF COMPLIANCE STANDARDS FOR BALLISTIC BODY PROTECTION

Every type of protective vest must pass several ballistics tests before it is certified. The armor is subjected to a series of shootings, from different calibers weapons, and inspected for the "penetration and backface signature." Clay

TABLE 7.5 Body Armor Classification Incompliance with NIJ Standard 0101.06.[66]

Ballistic level	Caliber	Bullet mass (g)	Velocity conditioned testing (m/s)	Velocity new and unworn body armor(m/s)	Distance to ballistic insert	Maximum BFS[a] (mm)	Bullet description	Armor
IIA	9 mm Luger [9 mm Dia.]; 40	8.0	355 ± 9.1	373 ± 9.1	5 m ± 1	44	Full metal jacketed round nose (FMJ RN) bullets	Vest
	S&W [9 mm Dia.]	11.7	325 ± 9.1	352 ± 9.1	5 m ± 1	44	Full metal jacketed (FMJ) bullets	Vest
II	9 mm Luger [9 mm Dia.];	8.0	379 ± 9.1	398 ± 9.1	5 m ± 1	44	FMJ RN bullets	Vest
	.357 Magnum [9.1 mm Dia.]	10.2	408 ± 9.1	436 ± 9.1	5 m ± 1	44	.357 Magnum jacketed soft point (JSP) bullets	Vest
IIIA	.357 SIG, [9 mm Dia.]	8.1	430 ± 9.1	448 ± 9.1	5 m ± 1	44	.357 SIG FMJ flat nose (FN) bullets	Vest
	.44 Magnum [10.9 mm Dia.]	15.6	408 ± 9.1	436 ± 9.1	5 m ± 1	44	.44 Magnum semi jacketed hollow point (SJHP) bullets	Vest
III	Rifles [7.62 mm Dia.]	9.6	847 ± 9.1	–	15 m ± 1	44	7.62 mm FMJ, steel jacketed bullets	Hard armor or plate inserts
			847 ± 9.1	847 ± 9.1			7.62 mm FMJ, steel jacketed bullets	flexible armor
IV	Armor Piercing Rifle [7.62 mm Dia.]	10.8 g	878 ± 9.1		15 m ± 1	44	.30 caliber armor piercing (AP) bullets	Hard armor or plate inserts
							.30 caliber AP bullets	Flexible armor
Special	Special threats need to be specified by the manufacturer for determining the correct reference velocity for other threats.							

[a]BFS (Perforation-Backface Signature)

TABLE 7.6 Summary Flexible Vest Standard for Tests.[66]

Information	NIJ standard 0101.06	Information	NIJ standard 0101.06
Submersion, new flexible vests	Armor are submersed tested wet, 100 ± 25 mm under water for 30 min, hanged and allowed to be dry for 10 min.	Min. shot-to-shot or shot-to-edge distance	51 mm
Sample applying the conditioned protocol	Preconditioning: temperature of 25°C ± 10°C relative humidity of 20–50 % for at least 24 h 30 min complete submersion and trembling for 10 days 72,000 cycles at 65°C at 80% rh. Perform drop testing on the armors. Conditioned armor tested dry.	Number of inserts to be tested	28 inserts (front and back)
Types IIA through IIIA armor s	Two test threats on 28 armors (11 Large, three of them conditioned and three small one of them conditioned).	Shot placements	Three shot near edge (19 mm) and three closed placements inside a circle of 100 mm.
Types III and IV	One test threats on 14 armors	Hits per panel at 0° angle	For all levels of threats
Back face signature (P-BFS)[a]	Samples: large conditioned and one small conditioned, new samples two small sized armors and two large sized armors.	Hits per panel at 30° or 45° angle	For all levels IIA, II, IIIA only. For P-BFS testing, each test panel must be shot with one hit at 30° and another hit at 45° angles
Ballistic Limit (BL)[b] test	Samples: large conditioned, five new samples large sized.	V_{50}[c] test	Both calibers (Type IIA through IIIA)
Total numbers of shots for soft armor	144/72 each cal. (Type IIA through IIIA) and only 24 for (Type III and IV)	Black face deformation requirements	Three measured above 44 mm and all other below 44 mm.
Total numbers of BFS tests	48 shot (24 shots each caliber)	Bullets each insert	Six shots

[a]P-BFS (Perforation-Backface Signature) [66]

[b]Ballistic Limit (BL): This is a test for determination of "limit" for perforation [66].

[c]V_{50} test is conducted to determine a velocity at which a vest is failing by full penetration (complete penetration) 50% of the time with a given round/velocity [67].

TABLE 7.7 Summary of NIJ Standard for Hard Armors and Plate Inserts.[66]

Information	NIJ standard 0101.06	Information	NIJ standard 0101.06
Submersion, all hard armors and plate inserts	Armor are submersed tested wet 100 ± 25 mm under water for 30 min, hanged and allowed to be dry for 10 min.	All hard armors and plate	Appling the conditioned protocol
Hard armor NIJ III	Nine test plates with six shots each panel	Type III. P-BFS test BL test	Nine armor panels Four armor panels, 24 shots Four armor panels, 24 shots
Hard armor NIJ IV	7–37 test plates with 1–6 shots each panel	Type IV P-BFS test BL test	Six armor panels Four armor panels, 24 shots Two armor panels, 12 shots
Total number of shots for Level II and IV hard armor	24 shots	Test duration	30 min, first shot shall be fired within 10 min
P-BFS test	24 shots	BL test	12–24 shots

tablet is used to measure the backface signature as a measure how deeply the bullet penetrates the armor. If any penetration of the vest or dent in the clay is not more than 44 mm and the depth is not too large and no part of the projectile penetrated the armor, the test is passed. If the armor is perforated or the divot is too deep, the test is failed. Each bulletproof vest is shot six times per test. Another testing to measure the armor ballistic limit (BL) is "V_{50}" (the velocity at which a bullet will penetrate the armor with probability 50%). There are several standard procedures to carry out the compliance test to the bulletproof vest or plates, for instance NIJ Standard-0101.06, to give the requirements and test methods for the ballistic resistance of body armor, rigid armor systems, such as plates, inserts, accessories, or semirigid armor systems. The conjunction armor is a combination of either two: flexible armor and panel or a flexible armor with a hard plate insert that is designed to provide increased stab or ballistic protection. In this case, the flexible armor is tested first to be compliant as a stand-alone armor and the combination of the flexible armor and hard armor/plate, then be tested as a system.

7.6 COMPLIANCE TESTS

7.6.1 NIJ STANDARD-0101.06

7.6.1.1 BODY ARMOR

Table 7.5 gives the tests conditions and parameters for protective armor which are classified into five levels (IIA, II, IIIA, III, and IV).

The summary of the body armor testing measures for incompliance with NIJ standard 0101.06 is given in Table 7.6. BL, V50 testing procedures are given in Tables II.2 and II.3.

7.6.1.2 BALLISTIC PLATES

Ballistic plate is inserted into a protective vest with purpose to absorb and redistribute the projectile energy and prevent the blunt trauma. A ballistic plate is requested to be tested separately and in combination with the vest which passes the incompliant test firstly. Armors and plates are passed condition procedure before testing.[66] Summary of the standard conditions for hard armors and plate inserts are given in Table 7.7.

TABLE 7.8 Bulletproof Vests Levels HOSDB Standard 2017.[63,65]

Protection level	Test weapon	Test caliber	Bullet mass (g)	Distance tile panel range (m)	Max. mean BFS (mm)	Single shot BFS limit (mm)	Velocity (m/s)
HO1	9 mm FMJ	MEN 9 mm FMJ DM11A1B2	8.0	5	–	44.0	365 ± 10
	9 mm JHP	Federal Premium 9 mm JFP P9HST1	8.0	5	–	44.0	365 ± 10
HO2	9 mm FMJ	MEN 9 mm FMJ DM11A1B2	8.0	5	–	44.0	430 ± 10
	9 mm JHP	Federal Premium 9 mm JFP P9HST1	8.0	5	–	44.0	430 ± 10
HO3	Rifle 7.62 caliber	Radway Green 7.62 mm NATO Ball L44A1 or L2A2	9.3	10	25.0	30.0	830 ± 15
	Rifle 7.62 caliber	7.62 × 39 mm surrogate	7.9	10	25.0	30.0	705 ± 15
HO4	7.62 rifle caliber	SAKO .308 Win 480A Powerhead or Barnes .308 TSX BT	10.7	10	25.0	30.0	820 ± 15
SG1	Shotgun 12 gauge true cylinder	Winchester 1 oz. Rifled 12RSE	28.4 g	10	25.0	30.0	435 ± 25
Special[a]	Depends on the threat						

[a]Depends on the threat level.

TABLE 7.9 Quantity of Armor Required for Ballistic Body Armor HOSDB Standard 2017.[63,65]

Protection level	Level panel type	Ballistic					
		Small size		Medium size		Large size	
		Front panel	Back panel	Front panel	Back panel	Front panel	Back panel
HO1 or **HO2**	Unformed	2	2	3	3	1	1
	Formed	4				4	
HO3	Plate			2	2	1	1
HO4[a]	Plate			3	3	1	1
SG1	Plate	1				1	

[a]when the armor's ballistic limit is sufficiently high.

TABLE 7.10 Bulletproof Vest Levels Classification German Schutzklasse Standard 2008.[69]

Protection level	Caliber	Bullet type	Bullet mass (g)	Distance to insert (m)	Velocity (m/s)
SK L	9 mm × 19	FMJ/RN/SC, verzinnt	8.0 ± 0.1	5	360 ± 10
SK1	9 mm × 19	FMJ/RN/SC, verzinnt	8.0 ± 0,1	5	415 ± 10
		Polizei	6.0 ± 0.1	5	460 ± 10
		Polizei	6.0 ± 0.1	5	460 ± 10
SK2	357 Magnum	FMs/CB	7.1 ± 0.1	5	580 ± 10
SK3	223 Rem.	FMJ/PB/SCP	4.0 ± 0.1	10	950 ± 10
	308 Win.	FMJ/PB/SC	9.55 ± 0.1	10	830 ± 10
SK4	308 Win.	FMJ/PB/HC	9.70 ± 0.1	10	820 ± 10

Additional test procedures are given in Appendix II.

7.6.2 THE BODY ARMOR STANDARDS FOR UK POLICE 2017 (HOSDB)

Bulletproof levels classification incompliance with HOSDB Body Armor Standard 2017[63] is stated in Table 7.8, ballistic protection levels are classified into six levels. Three other protection levels are available. A bulletproof vest/ballistic plates should comply with[65]:

1. no bullet must penetrate the bulletproof insert;
2. no bullet residues can leave to the body side;
3. bullet or ceramic fragments are not allowed to leave the armor;
4. BFS cannot exceed the stated specified requirements; and
5. the bulletproof vest should be 100% waterproof.

The number of armor samples, required for the ballistic test, is given in Table 7.9, it differs from the NJI standards.

Prior to the testing, the samples should be preconditioned for 12 h at temperature 20°C ± 3°C and at relative humidity ranged between 40% and 70% rh. The number of shots depends on the size of the unformed armor: for small size, three shots to front and rear side, the incident angle 0°; for medium and large sizes, six shots at incident angle 0°, and one at incident angle 45° to the front and rear side; for formed armor of small and large sizes, four shots are required at incident angle 0°. In the case of testing the plates for protection level HO3 and HO4, three shots to front and rear side at incident angle 0° are required and for plates for protection level SG, only one shot at incident angle 0°. Shot shall be positioned on all medium and large panels submitted for testing at a position 55 ± 5 mm from the edge of the panel, minimum shot-to-shot distance is 50 mm.[65]

7.6.3 GERMAN STANDARDS

The main difference between the German SK1 standard and NIJ is the level of protection which is given in Table 7.10.[68,69] In addition, the German SK standard requires wet testing at a distance of 0 m. This applies to the protection levels SK L and SK1 in all cases the depth of the shot cannot exceed 42 mm for all protection levels.

Several other standards for the armor evaluation and testing procedures are existing, such as: Russian standards—GOST R 50744-95, China GA141-2010, VPAM 2006 and so on.[71]

KEYWORDS

- **bulletproof protection**
- **chemical resistance**
- **German Schutzklass standards**
- **NIJ standards**
- **stab resistance**
- **heat protection**

REFERENCES

1. FireHoods. https://commons.wikimedia.org/wiki/File:TFS_Nomex_Hood.JPG(accessed Feb 10, 2018).
2. KEVLAR® ARAMID fiber technical guide. http://www.dupont.com/content/dam/dupont/products-and-services/fabrics-fibers-and-nonwovens/fibers/documents/Kevlar_Technical_Guide.pdf (accessed Feb 10, 2018).
3. ASTM D 1059-01 & 2260-03, Yarn Count, Denier Count, & Filament Count of Yarns.
4. Mertová, I.; Moučková, E.; Neckář, B.; Vyšanská, M. Influence of Twist on Selected Properties of Multifilament Yarn. *AUTEX Res. J.* **2017,** 1–11.
5. ASTM D1422/D1422M - 13 Standard Test Method for Twist in Single Spun Yarns by the Untwist-Retwist Method.
6. Application Report Manufacturing of Filament Yarns Test Methods for Quality Improvement. https://www.uster.com/fileadmin/customer/Knowledge/Textile_Know_How/Yarn_testing/AR_Test_methods_for_quality_improvement.pdf (accessed Feb 10, 2018).
7. ASTM D1425/D1425M—14 Standard Test Method for Evenness of Textile Strands Using Capacitance Testing Equipment.
8. Hearle, J. Mechanical Properties of Textile Reinforcements for Composites. In *Advances in Composites Manufacturing and Process Design*; Philippe Boisse, Ed.; ISBN: 978-1-78242-307-2.
9. El Messiry, M. *Natural Fiber Reinforced Polymer Composites Engineering*; Apple Academic Press Inc.: USA, 2017.

10. ASTM D2256/D2256M—10. Standard Test Method for Tensile Properties of Yarns by the Single-Strand Method, 2015.
11. Saravanan, D. UV Protection Textile Materials. *AUTEX Res. J.* **2007,** *7* (1), 53–62.
12. Behera, B.; Militky, J.; Mishra, R.; Kremenakova, D. Modeling of Woven Fabrics Geometry and Properties. In *Woven Fabrics*; Han-Yong Jeon, Ed.; Intech, May 16, 2012.
13. El Messiry, M. Theoretical Analysis of Natural Fiber Volume Fraction of Reinforced Composites. *Am. Econ. J.* **2013,** *52* (3), 301–306.
14. Rasi, M. Permeability Properties of Paper Materials. Ph.D. Thesis, Department of Physics, University of Jyväskylä, 2013.
15. Epps, H.; Leonas, K. The Relationship Between Porosity and Air Permeability of Woven Textile Fabrics. *J. Test. Eval.* **1997,** *25* (1), 108–113.
16. Horrocks, A.; Anand, S. Handbook of Technical Textiles; Woodhead Publishing Limited in Association with the Textile Institute Abington Hall: Abington Cambridge CB1 6AH, England, 2000.
17. Zupin, Z.; Hladnik, A.; Dimitrovski, K. Prediction of One-layer Woven Fabrics Air Permeability Using Porosity Parameters. *Text. Res. J.* **2011,** *82* (2) 117–128.
18. Ogulata, R.; Mezarcioz, S. Total Porosity, Theoretical Analysis, and Prediction of the Air Permeability of Woven Fabrics. *J. Text. Inst.* **2012,** *103* (6), 654–661.
19. Ahmad, S.; Ahmad, F.; Afzal, A.; Rasheed, A.; Mohsin, M.; Ahmad, N. Effect of Weave Structure on Thermo-physiological Properties of Cotton Fabrics. *AUTEX Res. J.* *15* (1) **2015,** 30–34.
20. Umair, M.; Hussain, T.; Shaker, K.; Nawab, Y.; Maqsood, M.; Jabbar, M. Effect of Woven Fabric Structure on the Air Permeability and Moisture Management Properties. *J. Text. Inst.* **2016,** *107* (5), 596–605.
21. Zupin, Z.; Dimitrovski, K. Mechanical Properties of Fabrics Made from Cotton and Biodegradable Yarns Bamboo, SPF, PLA in Weft. In *Woven Fabric Engineering*; Polona Dobnik Dubrovski, Ed.; Sciyo, Published, 18, August 2010.
22. Geršak, J.; Gabrijelčič, H.; Černoša, E.; Dimitrovski, K. Influence of Weave and Weft Characteristics on Tensile Properties of Fabrics. *Fibres Text. East. Eur.* **2008,** *16* (2 (67)), 45–51.
23. Yüksekkaya, M.; Howard, T.; Adanur, S. Influence of the Fabric Properties on Fabric Stiffness for the Industrial Fabrics. *Text. Confect.* **2008,** *4*, 263–267.
24. Penava, Z.; Penava, D.; Nakić, M. Woven Fabrics Behavior in Pure Shear. *J. Eng. Fiber Fabr.* **2015,** *10* (4), 119–124.
25. El Messiry, M.; Sheta, A. Measurements of Fabric Shearability by Twisting Method. *Bull. Facul. Eng.* Alexandria University, 1986.
26. Peil, K.; Barbero, P.; Sosa, E. Experimental Evaluation of Shear Strength of Woven Webbings. SAMPE 2012 Conference and Exhibition, Baltimore, USA, 2012.
27. Afzal, A.; Ahmad, A. et al. Influence of Fabric Parameters on Thermal Comfort Performance of Double Layer Knitted Interlock Fabrics. *AUTEX Res. J.* **2017,** *17* (1), 20–26.
28. Cavanagh, J. Clothing Flammability and Skin Burn Injury in Normal and Micro-gravity. MSc. Thesis, Department of Mechanical Engineering, University of Saskatchewan, Saskatoon, Canada, 2004.
29. El Messiry, M. Investigation of Puncture Behavior of Flexible Silk Fabric Composites for Soft Body Armor. *Fibres Text. East. Eur.* **2014,** *22*, *5* (107), 71–76.

30. Chu, C.; Chen, Y. Ballistic-proof Effects of Various Woven Constructions. *Fibres Text. East. Eur.* **2010,** *18* (6 (83)), 63–67.
31. Braga, F.; Lima, J.; Lima, E. The Effect of Thickness on Aramid Fabric Laminates Subjected to 7.62 MM Ammunition Ballistic Impact. *Mater. Res.* 1–5. http://www.scielo.br/pdf/mr/2017nahead/1516-1439-mr-1980-5373-MR-2016-0883.pdf (accessed Feb 10, 2018).
32. Determination of Elmendorf Tearing Resistance. http://www.producttestingapparatus.com/methods/strength/tear%20meth.pdf (accessed Feb 10, 2018).
33. Horrocks, A.; Anand, S. *Handbook of Technical Textiles*; Woodhead Publishing Limited, ISBN: 978-1-85573-385-5, England, 2000.
34. Kumbasar, E. Protective Functional Textiles. MDT 'Protective textiles. http://www.2bfuntex.eu/sites/default/files/materials/PROTECTIVE%20FUNCTIONAL%20TEXTILES_%20PERRIN.pdf (accessed Feb 10, 2018).
35. Burning questions about fire. https://sewwhatinc.com/wp-content/uploads/2016/06/Standards_Watch_Spring_2008_Protocol.pdf (accessed Feb 10, 2018).
36. NEPA 2112 Standard. http://tyndaleusa.com/fr-clothing-safety-library/standards-and-test-methods-archived/flash-fire-standards/nfpa-2112/ (accessed Feb 10, 2018).
37. NEPA 2112 Standard on Flame-Resistant Clothing for Protection of Industrial Personnel Against Short-Duration Thermal Exposures from Fire. https://www.nfpa.org/codes-and-standards/all-codes-and-standards/list-of-codes-and-standards/detail?code=2112 (accessed Feb 10, 2018).
38. Stull, J; Schwope, A., Eds. Performance of Protective Clothing Six Volume; ASTM: USA, 1997.
39. Understanding the Thermal Protective Performance of Your PPE. http://www.firerescuemagazine.com/articles/print/volume-7/issue-6/firefighter-safety-and-health/understanding-the-thermal-protective-performance-of-your-ppe.html (accessed Feb 10, 2018).
40. Thermal Protective Performance (TPP). http://globeturnoutgear.com/education/standards-and-testing/thermal-protective-performance-tpp (accessed Feb 10, 2018).
41. Quantum Protective the Physics of Protection. https://www.quantumprotective.com/testing.html (accessed Feb 16, 2018).
42. Jason. What Are SFI and TPP Ratings? https://www.speedwaymotors.com/the-toolbox/what-are-sfi-and-tpp-ratings/28846 (accessed Feb 16, 2018).
43. Song, G. Performance Evaluation of Thermal Protective Clothing and Recent Development in Standards. http://www.textilescience.ca/downloads/104th_session_pres/pres_Guowen_Song.pdf (accessed Feb 16, 2018).
44. NFPA® 1971-2013 Standard on Protective Ensembles for Structural Fire Fighting and Proximity Fire Fighting. https://www.nfpa.org/codes-and-standards/all-codes-and-standards/list-of-codes-and-standards/detail?code=1971 (accessed Feb 16, 2018).
45. ASTM F 1930:2013. Standard Test Method for Evaluation of Flame Resistant Clothing for Protection Against Fire Simulations Using an Instrumented Manikin.
46. Sipe, J. Development of an Instrumented Dynamic Mannequin Test to Rate the Thermal Protection Provided by Protective Clothing. MSc. Thesis, Worcester Polytechnic Institute, Massachusetts, USA, 2004.
47. Hartman, P. Electric Arc-Flash Protective Clothing. http://www.netaworld.org/sites/default/files/public/neta-journals/NWsu04SafetyCor.pdf (accessed Feb 16, 2018).

48. NFPA 2112, Standard on Flame-Resistant Garments for Protection of Industrial Personnel Against Flash Fire. https://www.nfpa.org/codes-and-standards/all-codes-and-standards/list-of-codes-and-standards/detail?code=2112 (accessed Feb 16, 2018).
49. NFPA 2113, Standard on Selection, Care, Use, and Maintenance of Flame-Resistant Garments for Protection of Industrial Personnel Against Flash Fire (accessed Feb 16, 2018).
50. Hanover Risk Solutions. Electric Arc Flash Protective Clothing. https://www.hanover.com/linec/docs/171-1192.pdf (accessed Feb 16, 2018).
51. Matusiak, M.; Sikorski, K. Influence of the Structure of Woven Fabrics on Their Thermal Insulation Properties. *Fibres Text. East. Eur.* **2011,** *19* (5 (88)), 46–53.
52. Matusiak, M. Thermal Insulation Properties of Single and Multilayer Textiles. *Fibres Text. East. Eur.* **2006,** *4* (5(59)), 98–112.
53. Gupta, D.; Srivastava, A.; Sunil, K. Thermal Properties of Single and Double Layer Assemblies. *IJFTR* **2013,** *38* 387–394.
54. Khalil, E. A Technical Overview on Protective Clothing Against Chemical Hazards. *AASCIT J. Chem.* **2015,** *2* (3), 67–76.
55. Chemical Resistance Data. https://www.usplastic.com/catalog/files/charts/LG%20CC.pdf (accessed Feb 16, 2018).
56. Chemical Compatibility Chart. http://www2.emersonprocess.com/siteadmincenter/PM%20TopWorx%20Documents/TW_ChemicalCompatibility_chart.pdf (accessed Feb 16, 2018).
57. Personal Protective Equipment (PPE). https://chemm.nlm.nih.gov/ppe.htm (accessed Feb 16, 2018).
58. Abrasion Resistance by the Martindale Method ASTM D4966-98. https://tapp.uni.edu/pdf%20files/Abrasion%20Resistance%20-%20Martindale.pdf (accessed Feb 16, 2018).
59. U.S. Congress, Office of Technology Assessment, Police Body Armor Standards and Testing: Volume I, OTA-ISC-534; U.S. Government Printing Office: Washington, DC, August 1992.
60. Kneubuehl, B. Ballistic Protection. http://cris-technologies.com/media/c95cf8f695e58982ffff8037ffff8e64.pdf (accessed Feb 16, 2018).
61. Stab Resistance of Personal Body Armor NIJ Standard–0115.00. U.S. Department of Justice Office of Justice Programs National Institute of Justice, 2000. https://www.ncjrs.gov/pdffiles1/nij/183652.pdf (accessed Feb 16, 2018).
62. Croft, J.; Longhurst, D. HOSDB Body Armour Standards for UK Police, 2007. Part 3: Knife and Spike Resistance. http://ped-cast.homeoffice.gov.uk/standards/39-07_C_HOSDB_BodyArmourStandards_(2007)_Part3_KnifeandSpikeResistance.pdf (accessed Feb 16, 2018).
63. Payne, T.; O'Rourke, S.; Malbon, C. Body Armour Standard, 2017. http://ped-cast.homeoffice.gov.uk/standards//Home_Office_Body_Armour_Standard_[FINAL_VERSION]1.pdf (accessed Feb 16, 2018).
64. Romano, T.; Tomczyk, K.; Landseer, D.; Williams, T.; Bailey, S. Experimental-based Numerical Simulation of the Drop Test Within NIJ Standard-0101.06 for Personal Hard Armour Development. https://www.simpact.co.uk/uploads/website_text/6/files/6/41.pdf (accessed Feb 6, 2018).

65. Croft, J.; Longhurst, D. HOSDB Body Armour Standards for UK Police (2007) Part 1: General Requirements. http://ped-cast.homeoffice.gov.uk/standards//Home_Office_Body_Armour_Standard_[FINAL_VERSION]1.pdf (accessed Feb 17, 2018).
66. Ballistic Resistance of Body Armor NIJ Standard 0101.06 for flexible vest. https://www.ncjrs.gov/pdffiles1/nij/223054.pdf (accessed Feb 16, 2018).
67. V50 Ballistic Test for Armor; Department of Defense Test Method Standard. https://www.scribd.com/document/62254840/V50-BALLISTIC-TEST-FOR-ARMOR-Mil-Std-662f (accessed Feb 17, 2018).
68. Technische Richtlinie (TR)—The German Schutzklasse Standard Edition 2008. https://www.protectiongroup.dk/en/technische-richtlinie-tr-the-german-schutzklasse-standard-edition-2008-a-19 (accessed Feb 17, 2018).
69. TR "Ballistische Schutzwesten," Stand: März, 2008. https://www.tssh.com/documentation/pdf/technische_richtlinie_ballistische_schutzwesten.pdf (accessed Feb 17, 2018).
70. Committee to Review the Testing of Body Armor Materials for Use by the U.S. Army Phase II; Board on Army Science and Technology; Division on Engineering and Physical Sciences; National Research Council. file:///C:/Users/magdy/Downloads/12885%20(1).pdf (accessed Feb 17, 2018).
71. Body Armour News. https://www.bodyarmornews.com/ballistic-standards/ (accessed Feb 22, 2018).

APPENDIX I

TESTING STANDARDS FOR FIBERS, YARNS, AND PROTECTIVE FABRICS

TABLE I.1 ASTM Standards for Protective Fabric Testing.

Test	Reference test method	
Denier per filament	ASTM D 1907	Weight in grams of 9000 m of a single filament
Density of	ASTM D 1505	Fiber as weight/unit volume
Moisture	ASTM D 2654	Moisture level based on dry fiber weight
Denier (yarn)	ASTM D 1907	Weight in grams of 9000 m of yarn
Tensile properties: – Elongation – Modulus – Tenacity	ASTM D 2101 ASTM D 2101 ASTM D 2101	Tensile properties measured on yarn, filaments, or staple tow
General test methods and Specifications	ASTM F 1002 ASTM F 1506	Standard performance specifications for protective apparel
Breaking load	ASTM D 5034 Grab Test G	Load to break 1" wide section of 4" strip of fabric
Bursting strength (Mullen burst)	ASTM D 3786	Measures force to rupture knit fabric with expandable diaphragm
Fungal resistance	ASTM G 21	Measures resistance of synthetic materials to fungi
Seam slippage	ASTM D 434	Load required to separate a sewn seam as compared to unsewn fabric
Pilling (random tumble)	ASTM D 3512	Measures resistance to pilling by tumbling knit or woven fabric in a container with abrasive material
Laundering	AATCC 135	Measures shrinkage or stretching after laundering knit or woven fabrics
Dry cleaning	AATCC 158	Measures shrinkage or stretching after dry cleaning knit or woven fabrics
Antistatic performance	FTMS 191A-5931 (5 kV imposed)	Measures time to dissipate electric charge on fabric
Arc thermal performance	ASTM F 1959/F 1959M-99	Measures energy to cause second degree burn
Coefficient of liner expansion		Change in fiber length with change in temperature
Differential scanning calorimeter	ASTM E 794	Measures difference in energy inputs into a fiber and a reference material as a function of temperature. Shows melting or crystallization temperature

TABLE I.1 *(Continued)*

Test	Reference test method	
Flammability (vertical flame test)	FTMS 191A, 5903.1	Measures char length, after flame, and glow time on 3" × 12" fabric sample exposed to flame at lower edge for 12 s. Initial and after laundering or dry cleaning
Flash fire Manikin test for predicted body burn injury	ASTM F 1930	Predicts second- and third-degree body burn injury for garments on an instrumented manikin in a controlled flash-fire exposure
Arc rating of materials	ASTM F1959/ F1959M ASTM F1891	12 Standard specifications for arc and flame-resistant rainwear standard test method for determining the arc rating of materials for clothing
	ASTM F1958/ F1958M	Determining the ignitability of non-flame-resistant materials for clothing by electric arc exposure method using mannequins
	ASTM F1506- 10a	Performance specification for flame resistant and arc rated textile materials for wearing apparel for use by electrical workers exposed to momentary electric arc and related thermal hazards
	ASTM F2178	Determining the arc rating and standard specification for eye or face protective products
	ASTM F1930	Evaluation of flame resistant clothing for protection against fire simulations using an instrumented manikin
	ASTM F1939	Radiant heat resistance of flame resistant clothing materials with continuous heating
Heat resistance	NFPA 1975	Measures burning, melting, separation, and ignition on knit or woven fabrics after 5 min exposure to (260°C)
Ignitability of flammable fabrics by electric arc	ASTM F1958/F 1958M-99	Measures energy required to ignite flammable fabrics (untreated cotton, etc.)
Limiting oxygen index (LOI)	ASTM D 2863	Determines minimum oxygen content (%) in air that will sustain combustion of a material

TABLE I.1 *(Continued)*

Test	Reference test method	
Thermal conductivity	ASTM E 1530	Rate at which unit heat will flow through fiber polymer per unit temperature
Thermal protective performance (TPP)	ASTM D 4108	A fabric specimen is exposed with a combination of radiant and convective energy. The total energy required to cause second-degree burn injury to human tissue is determined based on heat transfer through the fabric specimen and the Stoll second-degree burn criteria. Single and multiple layer fabric specimens can be tested. with the heat source and exposure window modified for NFPA 1971. A spacer is used for single layer fabrics and no spacer for multilayer fabrics.
Thermal shrinkage resistance	NFPA 1975	Measures fabric shrinkage after (260°C) oven exposure for 5 min
Thermal shrinkage –in water –in dry air	ASTM D 2259	Fiber shrinkage when exposed to hot dry air or water
Thermogravimetric analysis (TGA) thermal insulating properties	ASTM F1060- 16	Measures weight loss of fiber with increasing temperature rates materials intended for use as protective clothing against exposure to hot surfaces, for their thermal insulating properties
Thermal resistance	ASTM D1518	Thermal resistance of batting systems using a hot plate
Chemical resistance	ASTM F739-12e1	Permeation of liquids and gases through protective clothing materials under conditions of continuous contact.
	ASTM F903-10	Resistance of materials used in protective clothing to penetration by liquids.
	ASTM F1001-12	Selection of chemicals to evaluate protective clothing materials.
	ASTM F1052-09	Pressure testing vapor protective ensembles.
	ASTM F1154-11	Qualitatively evaluating the comfort, fit, function, and durability of protective ensembles and ensemble components.

TABLE I.1 *(Continued)*

Test	Reference test method	
	ASTM F1186-03(2013)	Classification system for chemicals according to functional groups.
	ASTM F1194-99(2010)	Documenting the results of chemical permeation testing of materials used in protective clothing.
	ASTM F1296-08(2015)	Evaluating chemical protective clothing,
	ASTM F1359/ F1359M-13	Liquid penetration resistance of protective clothing or protective ensembles under a shower spray while on a mannequin.
	ASTM F1383-12e1	Permeation of liquids and gases through protective clothing materials under conditions of intermittent contact
	ASTM F1407-12	Resistance of chemical protective clothing materials to liquid permeation—permeation cup method.
	ASTM F2053-00(2011)	Standard guide for documenting the results of airborne particle penetration testing of protective clothing materials

APPENDIX II

ADDITIONAL TEST PROCEDURES FOR *BULLETPROOF* VESTS

TABLE II.1 Surface Areas of Armor Sizing Templates.[66]

Template	Maximum area (largest rear panel; m^2)	Minimum area (smallest front panel; m^2)
NIJ–C–1	0.0939	0.0659
NIJ–C–2	0.1354	0.1020
NIJ–C–3	0.1835	0.1443
NIJ–C–4	0.2393	0.1945
NIJ–C–5	0.3022	0.2517

TABLE II.2 P-BFS Performance Test Summary.[66]

Ballistic level	Performance requirements					Number of shots				
	Test bullet	Hits/ panel at 0° angle	Hits/panel at 30° or 45° angle	Max. BFS depth mm	Shots per panel	Panel size	Panel condition	Panel required	Shots required	Total shots required
IIA	9 mm FMJ RN	4	2	44	6	Large	New	2	24	
							conditioned	4	12	
	0.40 S&W FMJ					Small	New	2	24	
							conditioned	4	12	
II	9 mm FMJ RN	4	2	44	6	Large	New	2	24	
							conditioned	4	12	
	0.357 Magnum JSP					Small	New	2	24	
							conditioned	4	12	
IIIA	0.357 SIG FMJ FN	4	2	44	6	Large	New	2	24	144
							conditioned	4	12	
	0.44 Magnum SJHP					Small	New	2	24	
							conditioned	4	12	
III	7.62 mm NATO FMJ	4	0	44	6	All	Conditioned	4	24	24
IV	0.30 Caliber M2 AP	1–6	0	44	1–6	All	Conditioned	4–24	24	24

TABLE II.3 Baseline Ballistic Limit Determination Test Summary.[66]

Armor samples required	Test threat	Ballistic panels required	Minimum shots required	
Type IIA	Test	10	120	At least 60 stops
through IIIA approximately	Round 1			At least 30 perforations
5 armors per caliber	Test	10	120	At least 60 stops
	Round 2			At least 30 perforations
Type III	7.62 mm	4	24	6 perforations, 12 stops, 6 either, velocity range of 27 m/s
4 armors	M80 FMJ			
Type IV	0.30 caliber	2–12	12	3 perforations, 6 stops, 3 either, velocity range of 27 m/s
2 to 12 Armors	M2 AP			

TABLE II.4 Test Parameters and Requirements for Ballistic Limit Test.[66]

Parameter description	Value
Velocity of first shot	The reference velocity for the armor type and caliber
Velocity step until first reversal	−30.5 m/s if first shot was a perforation
	+ 30.5 m/s (+ 100 ft/s) if first shot was a stop
Velocity step until a second reversal	± 22.9 m/s, depending on result of previous shot
Velocity step after second reversal	± 15.2 m/s, depending on result of previous shot

Index

C

www.ingramcontent.com/pod-product-compliance
Lightning Source LLC
Chambersburg PA
CBHW060803220326
41598CB00022B/2528